A new source of funding for astronomy stemmed from the creation of the National Science Foundation (NSF) in 1950. Astronomers were quick to take advantage of the opportunities this provided to found new observatories. The science and politics of the establishment, funding, construction and operation of the Kitt Peak National Observatory (KPNO) and the Cerro Tololo Inter-American Observatory (CTIO) by the Association of Universities for Research in Astronomy (AURA), are here, seen from the unique perspective of Frank K. Edmondson, Professor Emeritus of Astronomy at Indiana University, and a former member of the AURA Board of Directors.

AURA was asked to manage the Sacramento Peak Observatory (SPO) in 1976, and in 1982 the National Solar Observatory (NSO) was formed by merging the NSO and the KPNO solar programs. KPNO, CTIO and NSO were combined in 1983 to form the National Optical Astronomy Observatories. In 1981 NASA chose AURA to establish and operate the Space Telescope Science Institute (STScI). This is a personal account of a period of major innovation in American optical astronomy.

T0184944

AURA and its US National Observatories

Kitt Peak from the air in early 1994. The dome for the new WIYN (Wisconsin, Yale, Indiana, and NOAO) 3.5 meter telescope is at the extreme left. The 150-inch (4 meter) telescope dome is at the upper right. NAO Photo Archives.

AURA and its US National Observatories

Frank K. Edmondson
Indiana University

CAMBRIDGE
UNIVERSITY PRESS

CAMBRIDGE UNIVERSITY PRESS
Cambridge, New York, Melbourne, Madrid, Cape Town, Singapore, São Paulo

Cambridge University Press
The Edinburgh Building, Cambridge CB2 2RU, UK

Published in the United States of America by Cambridge University Press, New York

www.cambridge.org
Information on this title: www.cambridge.org/9780521553452

© Frank K. Edmondson 1997

First published 1997
This digitally printed first paperback version 2005

A catalogue record for this publication is available from the British Library

ISBN-13 978-0-521-55345-2 hardback
ISBN-10 0-521-55345-8 hardback

ISBN-13 978-0-521-01918-7 paperback
ISBN-10 0-521-01918-4 paperback

Contents

Part II AURA is created   71

Part III New directions for AURA   107

Part IV Cerro Tololo's neighbors   197

Preface and acknowledgements

The earliest documented suggestion that the Association of Universities for Research in Astronomy (AURA) should have a project to preserve the history of AURA and Kitt Peak National Observatory (KPNO) was made more than 25 years ago by David L. Crawford in a letter to me, dated July 10, 1964. About 10 years later two AURA Board members, Arne Slettebak and W.A. Hiltner, spoke to me repeatedly about the need to preserve early AURA–KPNO history. The gist of their remarks was: "Frank, you've been in this from the beginning, so why don't you write down your recollections of the early history before you lose your memory." This friendly advice started my thinking about the need to do something. My first opportunity to have sufficient time to start working on this came when I reached administrative retirement in 1978 after 34 years as department chairman. I took a full year sabbatical at the usual half-pay in 1978–79. This was done partly to get out of the way of the new chairman, Hollis R. Johnson, during his first year in office, but it was primarily to start working in depth on the early history of AURA, KPNO, and CTIO (Cerro Tololo Inter-American Observatory).

The first project outline was brief. My thinking at that time was limited to archival retrieval and preservation. I planned to organize the files that I had assembled, and to add my running comments at appropriate places to help future users of the files. Also, I contemplated a half-dozen or so taped interviews with key people, an unrealistic number as it turned out; forty-six people were tape interviewed in forty-nine sessions during the sabbatical. The grand total is eighty-five persons in ninety-two interview sessions. The American Institute of Physics (AIP) Center for the History of Physics will be the archive for these tapes and transcripts. The National Science Foundation (NSF) Historian has a duplicate set.

About midway through the sabbatical my original plan to do no more than organize archival files, and to add written commentary, changed to plans to write a book, which nearly everyone I taped urged me to do. Also, it soon became clear to me that I had a moral obligation to do this. The story was much too important, interesting, and complicated to put back in the files to gather more dust.

The first archival breakthrough came eight months before the start of my sabbatical year, at the time of the symposium held at the National Academy of Sciences in Washington to celebrate the 100th anniversary of the birth of the pioneering astronomer Henry Norris Russell, my wife's father. Margaret and I checked in early at the Park Central Hotel on the afternoon of the day before the symposium, and we decided to walk across the street to see Vernice Anderson, Executive Secretary of the National Science Board. It was our good luck that she was not as busy as usual, and we had a nice visit with her. I mentioned the plans I was making for my sabbatical project, and she said: "Our NSF Historian may be able to help you." I didn't know NSF had one, and I arranged, with her help, to meet him at NSF on February 3, 1978, on my way home from a meeting at the American Institute of Physics in New York. This book would not have been possible without the help I received from the NSF Historian, Dr. J. Merton England.

My first exploratory visit to Ann Arbor to look at the Robert R. McMath Archives took place on May 17, 1978. These files had been received recently at the Bentley Historical Library and were not yet organized. Orren Mohler, a close friend and associate of McMath for many years, introduced me to Mary Jo Pugh, the Assistant Archivist, and she arranged to have these files organized ahead of schedule to help me get started. I worked there six days on two trips during the sabbatical, and five more days in 1980 and 1981.

My sabbatical leave began officially on July 1, 1978. It started with a week in Washington during which I acquired a lot of important NSF material, including the file copy, showing authorship and approvals, of the important "letter of intent" that the Director of NSF sent to Dr. McMath on October 16, 1956. I was not able to find the ribbon copy in Ann Arbor, but I did find nine thermofax copies scattered in various University of Michigan files.

I worked in the AURA corporate office in Tucson from August 10 to September 11, 1978, and again from January 6 to April 1, 1979. The AURA and KPNO archives were not well organized, and some had been destroyed. At the outset I found a file drawer containing the last part of Observatory Director Nicholas Mayall's chronological file, stored in a rented warehouse more than two miles from the KPNO headquarters. This file was marked "To be shredded," and the earlier parts could not be found. In spite of this unpromising beginning, I did find a great deal of important archival material located in the KPNO operations office, a small KPNO storage building for administrative files next to the AURA building, in the library, and in several staff offices. I got the impression that some of these files had been hidden in staff offices to save them from being shredded, and I am grateful to Lowell LeMoine and others who did this.

An informative interview on August 30, 1978 with Richard A. Harvill, President Emeritus of the University of Arizona, led me to Edward H. Spicer, Professor of Anthropology, who helped Aden Meinel get permission from the Papago Tribe to climb Kitt Peak. Dr. Spicer put me in touch with Chester Higman, who had been paid by the American Friends Service Committee to work for the Papago Tribe as Administrative Assistant to the Chairman of the Tribe and as Business Manager from February 1958 until the middle of 1961. I interviewed Mr. Higman in his home in Seattle, Washington, on July 25, 1979 during a visit with our son in Tacoma.

During the time between the two sojourns in Tucson I worked in Bloomington and made three trips for archival searches and taped interviews. The first trip (October 16–26) included five days in Washington, a day at the Ford Foundation reading their file on the $5 million grant to AURA, a day in New Haven looking through the Rupert Wildt archives, and a day in Swarthmore to see Peter van de Kamp's NSF archives. The second trip started in Ann Arbor (October 29 – November 3). I worked at the Bentley Historical Library for three full days and two half-days, looked through the equivalent of four full file drawers, and left an order for 993 photocopies. I flew to Columbus on Friday afternoon, November 3, for a weekend visit with Geoffrey Keller, my successor as Program Director for Astronomy at NSF in 1957. We recorded a one-hour tape, mostly about the Ford Foundation grant to AURA. The third trip (November 27 – December 2) started with a meeting of a committee at the AIP. I spent the morning of the 28th looking at the Ford Foundation file on the unsuccessful Carnegie Southern Observatory (CARSO) proposal for $19 million to fund the construction of a 200-inch telescope in Chile, and got 138 hard copies from the microfilm reader. I took the 1 o'clock Shuttle to Washington, and was there through Friday, December 1.

The winter sojourn in Tucson was devoted to working with the archival material that I had found during the previous summer. It also provided an opportunity for twenty taped interviews. These interviews include Paul Klopsteg, retired NSF staff member, and Gerhard Miczaika, former air force scientist, during a short trip to California in late February, and Carl Borgmann, former Ford Foundation staff member, during a stopover in Boulder on the way home in April.

Seventeen more interviews were taped before the end of the sabbatical: seven during two trips to Washington, two in Ann Arbor, one in Seattle, one in Bloomington, and six during the Montreal International Astronomy Union (IAU) meeting.

I returned to the classroom in the Fall of 1979 and taught for four more years until I became Emeritus in 1983. During this time I only had about 25 percent of my time available for the history project, so progress was much slower than it had been during the sabbatical. Thirty-four more interviews were taped during these four years.

Retirement in 1983 gave me a "permanent sabbatical" and much more time for the history project. Nine more interviews have been taped: two in late 1983, one each in 1984, 1985, 1987 and 1988, two in 1990, and one in 1991. The December 5, 1984 interview with John Denton in Berkeley provided valuable information not previously available to me. As a representative of the American Friends Service Committee, he had provided the Papago Tribe with free legal services at the time the Kitt Peak lease was signed in 1958.

I was the only member of the original AURA Board of Directors still serving when I retired. The Board graciously gave me the new title "AURA Consultant/Historian," which has been renewed annually. It has allowed me to keep in touch with what is going on, and I value it highly. In return, I have been able to serve as the corporate memory from time to time.

I wish to thank the Indiana University Department of Astronomy for allowing me the continued use of my campus office following my retirement. It would have been difficult to write this history in a different environment, and without the easy access to nineteen file drawers full of working files.

Acknowledgements and thanks are due to the many persons who made it possible for me to collect the information needed to compile and write this history:

(a) The eighty-five people, listed in Appendix I, who contributed oral histories and provided information not available elsewhere; in some cases they provided important clues about what to look for in archival files. I am deeply indebted to them.

(b) The following twenty-three institutions were archival sources:

> Arizona Historical Foundation – Goldwater papers
> Arizona State Museum – Papago Archives
> State of Arizona, Dept. of Library, Archives & Public Records, Genealogy
>   Library – Newspaper microfilm files
> University of Arizona Library – Special Collections
> Association of Universities for Research in Astronomy, Inc.
> California Institute of Technology
> Carnegie Institution of Washington
> European Southern Observatory
> Federal Records Center
> The Ford Foundation
> Indiana University – Archives
> Kitt Peak National Observatory
> Library of Congress
> Lowell Observatory
> University of Michigan – Bentley Historical Library
> Mount Wilson Observatory
> National Academy of Sciences
> National Aeronautics and Space Administration
> National Science Board
> National Science Foundation
> University of Wisconsin Library and Space Astronomy Laboratory
> Yale University Library
> Yerkes Observatory – University of Chicago

I am grateful to all of them for giving me unrestricted access to their files, including some sensitive documents, and for allowing me to have copies of everything I asked for.

(c) Twenty-six individuals, most of whom also contributed taped oral histories, were sources of archival material not contained in my voluminous personal AURA files:

Helmut A. Abt	Chester J. Higman	Leon E. Salanave
Adriaan Blaauw	Arthur A. Hoag	C. Donald Shane
Ethel Carpenter	Harold H. Lane	William Hawes Smith
John H. Denton	Lowell LeMoine	Edward H. Spicer
Elizabeth Estrada	Per Olof Lindblad	Julius A. Stratton
Muriel Fults	Nicholas U. Mayall	Joseph N. Tatarewicz
Henry L. Giclas	Paul A. McKalip	Peter van de Kamp
Leo Goldberg	Orren C. Mohler	Gordon W. Wares
Emil W. Haury	A. Keith Pierce	

Thanks are due to them for their valuable contributions to this history.

Finally, I wish to express my special thanks and gratitude to the NSF Historian, Dr. J. Merton England. His unselfish help in tracking down hard-to-find NSF documents was a crucial contribution to my work. His encouragement and support included having his very competent secretary, Justine Burton, type the transcripts of my taped interviews up to the time of her retirement in March, 1981.

Equally competent were Paula Jentgens and Brenda Records who were able to read my longhand manuscript and convert it into the final typescript. Ms. Jentgens typed the first nine chapters before she retired, and Ms. Records typed the remainder of the chapters and transcribed several of the later taped interviews. They have my gratitude for their important contribution to the production of this book.

In conclusion, I also wish to thank Dr. Robert W. Smith for taking time to read the entire manuscript with great care. He sent me two letters in late January 1995 containing four pages of detailed criticisms and suggestions. The resulting improvements in the manuscript made the copy-editor's work much easier. Dr. David DeVorkin read portions of the manuscript in the early stages, and his suggestions were also very helpful.

Foreword

"There is nothing new in this world except the history you don't know." (Harry S. Truman)

As his parting shot, the author asked me to write a chapter that connects his book to the present and future. That is a tall order for a physicist turned astronomer who lacks formal training in history, but a request from our historian, and a predecessor, should not be refused. Besides, history is fun, its lessons are useful, and its reporter may view it from a personal, if perhaps subtly biased, perspective.

First, here is a word about the Edmondsons. It is fitting that they created this book. Frank is an astronomer. For twenty-six years, Indiana University, a charter member of the consortium, reappointed him to the AURA Board. As board member, as President, and finally as historian, he led or followed its development for decades and influenced it deeply. This book reflects deep insight and personal caring, and a genuine interest in telling what really happened.

Who is Frank? For eons, he has been spouse and life's companion to Margaret, herself an astronomer's daughter. They are "The Inseparables," familiar friendly faces at countless astronomy meetings around the globe. He is also known, and feared by some, as an indomitable multilingual punster who once quipped that the polar bear "is an 'ourse of a different color'." This book is a labor of love by these two people who have dedicated their lives to each other and to astronomy. It is a tribute to Frank's talents and energy – and to Margaret's. I am deeply grateful to them both for recording the history in this book.

History reflects the competing interests of institutions, the interests and personalities of their leaders, and the environment in which they operate. The history of AURA is no exception. The book illustrates how differing and sometimes competing interests of people and their institutions shaped AURA and its activities. It shows how they responded to the US astronomy community, and how they took influence on it. Similar forces are at work today. We can learn from history as we look to the future.

It was Leo Goldberg's vision and McMath's leadership and clout in Washington that created AURA and Kitt Peak, the first national observatory. Leo called for a place to which ANY astronomer with an idea could go and observe. He put together the consortium that became AURA and enabled McMath and others to create Kitt Peak. Before then, the premier observatories had been the work of visionaries who convinced philanthropists to put up the cash to build them, and institutions to operate them so as to attract the world's best. The field advanced through extraordinary discoveries at such private centers of excellence. Thanks to the National Science Foundation, our "NSF," Kitt Peak opened up the field to researchers who had no privileged access to private observatories. Thus, a meritocracy of a different kind became possible: among individuals. As a result, competition became more vigorous: among astronomers for ideas and observations, and among their institutions for people and for the tools of the trade. These kinds of competition, usually healthful and collegial, continue to shape our science in fundamental ways. What is next?

The outlook is superb in many ways, dismal in others. It is superb because astronomy is exciting to lay audiences and to colleagues alike, and because novel technology has enabled superb facilities that will bring great progress. It is dismal because the budget for NSF's astronomy division has declined sharply within the past two decades. Kitt Peak, Cerro Tololo, and Sacramento Peak have been weakened by the loss of some 30 percent in buying power in the past decade. That decline would have been fatal without the inspired leadership and talents of Director Sidney C. Wolff. It is fortunate and gratifying that great new private and international telescopes are coming on line. Their need for modern instruments, and research funds, to take advantage of their capabilities, puts even greater pressure on NSF's shrinking astronomy budget.

If we as astronomers accept this shrinkage as given, if we compete ever more fiercely over decreasing budgets, it will be hard to reverse the trend. NSF funds for astronomy would decline further, for private and national observatories alike. If instead we take advantage, together, of astronomy's position at the frontiers of knowledge, and of unparalleled public interest and participation, we have the chance to increase astronomy's share in NSF's budget. We can do it if we unite in presenting the case for astronomy to the NSF, if we work out our disputes within the community, if we give our full support to our colleagues within NSF, and if we also refrain from undermining colleagues elsewhere and their projects as we interact with Federal policy and decision makers.

"We all must hang together, or assuredly we shall hang separately."

Benjamin Franklin

Frank captured the early history of AURA, Kitt Peak, and Cerro Tololo. We hold their future in our hands.

Goetz K. Oertel
AURA President
January 30, 1996

Part I
The beginning

1 Introduction. US astronomy in the first half of the 20th century.

American observational astronomy was dominated during the first two decades of the 20th century by the Harvard, Lick, Lowell, Yerkes, and Mount Wilson Observatories, all created by private philanthropy. Mount Wilson became the dominant observatory during the next two decades, following the completion of the 100-inch reflector at the end of World War I. Private philanthropy also led to the construction of the McDonald Observatory, which was dedicated only four months before the outbreak of World War II in 1939. Completion of the 200-inch Palomar telescope, also privately funded, was delayed by the war until 1948. Lick and Yerkes were operated by universities, one a state university and the other a private university. Lowell was a private institution supported by a modest endowment. Mount Wilson was part of the Carnegie Institution of Washington, a prestigious research organization that supported work in both the physical and the biological sciences.

The operation of the McDonald Observatory and of the Palomar Observatory was based on a new idea, cooperation in the operation of a major research facility. McDonald was owned by the University of Texas, but it was operated by the University of Chicago's astronomers, starting in 1932. The director of the Yerkes Observatory was also the director of the McDonald Observatory. The Palomar Observatory was owned by the California Institute of Technology, and it was operated in cooperation with the Mount Wilson Observatory of the Carnegie Institution of Washington, under the name Mount Wilson and Palomar Observatories, from 1948 to 1969. The first director, I.S. Bowen, was from Cal Tech; the second, Horace W. Babcock, was from Mount Wilson. The third director, Maarten Schmidt, was from Cal Tech. This arrangement, called the Hale Observatories during the years 1969–1980, was terminated in 1980. The Chicago–Texas joint operation of the McDonald Observatory ended in 1962, when the contract signed in 1932 by the two universities expired.

Otto Struve, the first director of the Yerkes and McDonald Observatories, had even more ambitious plans for cooperation. In an article "Cooperation in Astronomy", published in *Scientific Monthly*, February 1940, he described the success of the Chicago–

3

Texas operation of the McDonald Observatory. Next, he discussed the need for additional large telescopes:

> After all, the success of theoretical study depends essentially upon the supply of observational results. These now come almost entirely from a few large American Observatories . . . Is it not also rather disquieting when we contemplate the number of young astronomers who are being trained for future careers at institutions which are unable to carry on modern observational activities?

In the last part of this article he proposed:

> In order to enlarge the scope of the work of the McDonald Observatory it would be reasonable to invite the collaboration of other institutions. Let us suppose that a plan of collaboration could be worked out which would be satisfactory to all participating institutions. We should then be able to organize jointly an observing station in the Texas mountains, where the McDonald Observatory is located, which would be much more powerful than the present McDonald Observatory alone . . . It is my strong conviction that astronomers and University administrators should seriously consider such a project.

A letter from Struve to Edmondson, dated March 19, 1940,[1] contains a specific proposal:

> Assuming that we can get together five or six collaborating universities, we shall all make a joint application to the Carnegie Foundation or the Rockefeller Foundation for a grant of about a quarter of a million dollars to build a 72-inch Schmidt telescope to be used in addition to the 82-inch reflector. Assuming further that one of the foundations will make the grant, we propose to organize a cooperative observatory in conjunction with the McDonald Observatory. The maintenance of the two large telescopes is to be borne by the collaborating universities. The cost of such maintenance appears to be about $75 per night per telescope. This includes general maintenance, photographic supplies, repairs, small accessory equipment such as spectrographs, photometers, etc., as well as the salaries of six observing assistants who are to secure the plates for the collaborating institutions. The reason for this latter arrangement is that most university departments cannot send astronomers to Texas except during vacation. I am reasonably certain that at least two-thirds of the nights will be useable – and possibly more. Hence, participation in the plan of the extent of 30 nights with one telescope would cost about $2,250.
>
> The management of the new observatory is to be vested in an observatory council, consisting of members of the participating universities, and the council is also to decide on such matters as acceptance of new members, etc. If you are interested, I shall be glad to hear from you so that I can provide you with more detailed information.

The President of Indiana University, Herman B Wells, wrote to Struve on July 17,

1940[2] expressing the interest of Indiana University in this proposal. Struve acknowledged the letter on July 24. The plan had been presented to the University of Chicago President, Robert M. Hutchins, on March 16, 1940[3] in a memorandum: "Plan for Astronomical Collaboration in Connection with the McDonald Observatory." Hutchins forwarded it to Homer P. Rainey, President of the University of Texas.[4] Hutchins and Rainey were not opposed to the idea, but the time was not ripe for executing it. World War II was in progress, and it was clear that the United States would have to become involved before long. Struve wrote again to Wells, on December 27, 1940:[5]

> Because of the heavy rearmament program, it now seems out of the
> question that a new large telescope could be built until normal conditions
> have been reestablished. Hence, even if it should be possible to obtain a
> grant from one of the foundations for the construction of such an
> instrument, it would, I believe, be unwise to make any definite plans and
> after discussing the matter with President Hutchins and Dean Compton, it
> seems best to postpone this phase of the project.

Struve was far ahead of his time when he made this proposal. His idea was the spiritual ancestor of AURA and the Kitt Peak National Observatory, but with some differences of detail. Private funding was replaced by federal funding through an agency, the National Science Foundation, which was created ten years after Struve's paper was published in *Scientific Monthly*. A small group of universities cooperating in a joint enterprise for their mutual benefit was replaced by a consortium of universities, AURA, Inc., organized to provide management for a national observatory that would be available to the entire astronomical community, not just the members of the consortium.

2   Planning for the growing needs of US astronomy in the post–war period. The important role of the National Science Foundation.

The bill that created the National Science Foundation was passed by the House of Representatives on April 27, 1950, and by the Senate the next day. President Truman signed it into law on May 10. This was the successful conclusion of a vigorous debate that started in 1945. The interesting and complicated story has been told in J. Merton England's book, *A Patron for Pure Science*, covering the period 1945–1957;[1] an earlier informal history covering the first 25 years of the National Science Foundation, *A Minor Miracle*, was written by Milton Lomask.[2] These books are the source for some of the following discussion. The National Science Foundation's first formal activity was on December 12, 1950 when the National Science Board (NSB), having been chosen by the President and confirmed by the Senate, held its first meeting. The law that created NSF required the President to consider the National Science Board's suggestions for a director when he filled that post. A special committee initially prepared a list of 40 names, and cut it down to 11 to be discussed by the Board at its second meeting on January 3, 1951. The name of Alan T. Waterman, Deputy Director and Chief Scientist of the Office of Naval Research (ONR), was on the final list of three, and he received the first formal offer of the NSF directorship. He accepted, and the appointment was made by President Truman on March 9, 1951. Dr. Waterman was reappointed by President Eisenhower in 1957, and President Kennedy allowed him to finish his second term by waiving the compulsory retirement rule when he turned 70 in 1962.

Waterman's first task was to assemble a staff, including a deputy director, and several assistant directors who would be in charge of the operating divisions of the Foundation. Program directors, advisory panels, and consultants also had to be appointed. His second task was to persuade Congress to appropriate funds sufficient to allow the Foundation to initiate programs to support research and education in the sciences. The start-up budget for FY1951 was $225,000, and the NSF Law had set a ceiling of $15 million on later NSF appropriations.

The Bureau of the Budget approved a $14 million budget request for FY1952, the fiscal year starting July 1, 1951, and this was in the budget sent to Congress. On August

17 the House Appropriations Committee recommended $300,000, a 98 percent cut, and that is what the House voted to include in the FY1952 budget. Waterman immediately appealed to the Senate Appropriations Committee to restore most of the cut, and asked for a hearing. In September the Senate voted approval of $6.3 million, and the compromise finally approved on October 20 was $3.5 million. The late date, one-third of the way through FY1952, and the small size of the appropriation served to delay the start of the grants and the fellowship programs. Twenty-eight research grants in biological and medical sciences finally received Board approval on February 1, 1952. There was one grant in astronomy, $8,000 for two years to Willem J. Luyten of the University of Minnesota to support "Astronomical Research: Motions of the Stars." The first awards in a scaled-down fellowship program were announced in April 1952.

The potential of the National Science Foundation as a source of funds to support astronomical research, including facilities, was quickly recognized by American astronomers. The American Astronomical Society (AAS) appointed a special committee in 1950, chaired by C.D. Shane, to study relations between astronomers and NSF. A draft report, written by Jesse Greenstein, was circulated to the Committee on July 6, and Shane sent a revised version to R.C. Gibbs at the National Research Council on July 20. After further revisions the report of the committee was published in late 1951.[3] It proposed a National Science Foundation budget for astronomy of $380,000 in four main categories for the fiscal year ending June 30, 1952: A. Fellowships, $35,000; B. Publications, $35,000; C. Short-term Research Projects, $160,000; D. Capital Equipment Grants, $150,000. The report did not include funds for large new facilities in the first year, but said these would be needed later. The committee's estimate of the "requirements in longer-term grants for urgently needed expansion of capital equipment" was a modest additional $200,000 per year.

The report also contained some administrative suggestions that NSF quite properly chose to ignore. The Office of Naval Research had started a program of small grants in astronomy in 1948 to provide interim support until the political problems of creating the National Science Foundation could be resolved.[4] The AAS report proposed that NSF should use the ONR Committee on Astronomy in an advisory capacity to evaluate the scientific merit of research proposals, and to give advice on other astronomical matters. An upper limit of $10,000 per project was recommended, twice the size of the largest ONR grants. The members of the ONR Committee had been named by the Council of the American Astronomical Society, and NSF preferred to choose its own advisors. Dr. Paul Klopsteg, the NSF Assistant Director for Mathematical, Physical, and Engineering Sciences (MPE), attended the meeting of the ONR Committee on January 19, 1952 to observe its mode of operation. In private conversation he rejected the AAS proposal in strong terms: "Who the hell do they think is running the NSF?"[5]

The first "Ad Hoc Meeting of Astronomical Consultants" took place on August 1, 1952;[6] the name was changed to "Advisory Panel for Astronomy" at the second meeting on February 5 and 6, 1953. The members were: Lawrence H. Aller, Dirk Brouwer, Jesse L. Greenstein, Gerald E. Kron, Gerard P. Kuiper, Martin Schwarzschild, and Fred L. Whipple. Joel Stebbins was an invited guest and Leo Goldberg represented the MPE Divisional Committee. The NSF staff members in attendance were: Alan T. Waterman (Director), Paul E. Klopsteg (Assistant Director for MPE), Harry C. Kelly

(Assistant Director for Scientific Personnel and Education – SPE), Ralph A. Morgen (Program Director for Engineering and Metallurgy – MPE), Raymond J. Seeger (Program Director for Physics and Astronomy – MPE), and Robert R. Stoll (MPE).

The meeting began with a general statement by Dr. Waterman about the purpose of the Foundation. In response to a question about NSF's possible action in relation to ONR's termination of its support of astronomy, he said that NSF would try to avoid having the support of any worthwhile project abruptly terminated. Dr. Kelly described the various fellowship programs, and Dr. Klopsteg summarized the grants program for fiscal year 1952. He also said the $4.75 million appropriation for fiscal year 1953 would permit research expenditures in MPE of $800,000. Allocation to the different disciplines would depend on the spread and pattern of the proposals that were received.

Dr. Seeger reviewed his activities as Program Director for Physics and Astronomy. The nine astronomy proposals that had been received were reviewed by 22 reviewers. These proposals were 2.4 percent of the 369 received by MPE, and the two astronomy grants were the same percentage of the 85 grants that had been made. The morning session ended with a report by Dr. Goldberg on the recent meeting of the MPE Divisional Committee.

The afternoon session was devoted to a discussion of the first four of the five questions listed on the agenda. It was agreed that the consultants should meet twice a year, and that they should evaluate all proposals. Other reviewers should be used only in exceptional cases. The third question was: "In what way can NSF support the building of observatories? Their overall programs?" Following some general discussion the group voted in favor of a grant of $10,000 – $12,500 to Vanderbilt University, half of the requested $25,000. A proposal from the University of Arizona, Ohio State University, and Indiana University was discussed in detail. The vote "favored rejection of this specific proposal, although most were favorable to the idea of a photoelectric observatory and to cooperation of several institutions for major installations." (This proposal, its background, and subsequent actions by NSF will be taken up in Chapter 4.) A radio astronomy proposal from Bok for $32,000 was discussed, and culminated in a tie vote (with Whipple absent); however, it had already been recommended favorably to the National Science Board by the Director. An Advisory Panel for Radio Astronomy was established in early 1954, and it reviewed later proposals in radio astronomy until the two panels were merged in 1957.

3   The AUI site survey for a National Radio Astronomy Observatory. The controversy over the selection of AUI to build and operate NRAO.

Radio astronomy was already two decades old by the time the first NSF research grants were made. However, the 21-centimeter line from neutral hydrogen in interstellar space was detected on March 25, 1951,[1] only 16 days after the appointment of Alan T. Waterman as the first Director of the National Science Foundation. Thus, the Foundation and 21-centimeter radio astronomy, an exciting new development in a relatively new branch of astronomy, began life together. The first proposal to NSF for a radio astronomy project was submitted by Bart J. Bok in June, 1951 and a revised version was received on July 3[2] in which $32,000 was requested to cover part of the cost of constructing a 24-foot equatorially mounted radio telescope designed for 21-centimeter observations; some private funding had already been obtained. Upon receipt of the revised proposal, the Director recommended favorable action to the National Science Board. Nevertheless, support for this proposal received only a tie vote after discussion by the astronomical consultants on August 1, 1951. Raymond J. Seeger, at that time Program Director for Physics and Astronomy, recalls that later "the Panel jumped on me"[3] about making this grant, the beginning of support for radio astronomy by NSF. Approval by the National Science Board came in September at its 16th meeting, and the grant was made in October. A later and larger grant, in March 1955, provided funds for a 60-foot equatorially mounted radio telescope at Harvard's Agassiz Station.

The concept of a National Radio Astronomy Observatory grew out of two conferences. The first of these was held in Washington, DC on January 4–7, 1954.[4] It was an interdisciplinary conference on radio astronomy, sponsored by NSF, the Carnegie Institution of Washington and Cal Tech, that included discussion of the need for large radio telescopes in the United States. Further discussion took place in a small meeting that was requested by NSF after the conference.[5] One immediate result from this meeting was the appointment by the NSF Director on May 3, 1954 of an Advisory Panel for Radio Astronomy, consisting of Merle A. Tuve (chairman), Bart J. Bok, Jesse L. Greenstein, John P. Hagen, John D. Kraus, Rudolph L.B. Minkowski, and E.M. Purcell. Tuve was a major figure in the reemergence of radio astronomy in the United States. Bok had started the radio astronomy program at Harvard, and Greenstein was one of the first to appreciate the astronomical importance of Grote Reber's pioneer observations. Hagen was deeply involved in radio astronomy at the Naval Research Laboratory (NRL), Kraus had designed and was building a large antenna at Ohio State, and Purcell was co-discoverer of the 21-centimeter line of neutral hydrogen. Minkowski was playing a major role in the optical identification of radio sources, using the Palomar 200-inch telescope.

In the meantime, Donald H. Menzel and several other scientists from the Massachusetts Institute of Technology (MIT), Harvard, and NRL, following a suggestion by Julius Stratton, requested Associated Universities, Inc. (AUI) to explore with NSF the feasibility of a national radio astronomy facility. AUI called a conference of 37 scientists with an interest in radio astronomy to discuss this proposal on May 20, 1954 in New York. NSF personnel in attendance were Paul Klopsteg, Associate Director; Raymond J. Seeger, Acting Assistant Director for MPE; and Peter van de Kamp, recently appointed as the first Program Director for Astronomy. The President of AUI, Lloyd V. Berkner, was chairman of the conference. A recommendation was approved that AUI should request funds from NSF to support "basic preliminary planning and preliminary feasibility studies – looking toward the creation of a large facility." A 12-member ad hoc committee was appointed by AUI to guide the project, and it met on July 26, 1954 in New York. A draft proposal for $105,000 to be submitted to NSF by AUI was approved.

The AUI proposal was discussed by the Radio Astronomy Panel at its first meeting on November 18–19, 1954. Those in attendance were Tuve (chairman), Bok, Hagen, Kraus, Minkowski, and van de Kamp. The Panel reviewed the AUI proposal and recommended a grant of $85,000, which was made in January 1955. The Panel also stipulated that the site should be within 300 miles of Washington, in order to provide an eastern astronomical facility to complement the western optical observatories.

The first AUI report, including recommendation of a 140-foot dish as the principal instrument for the observatory, was submitted to NSF in early May 1955, to be available for the NSB meeting on May 19–20, its 34th meeting. An important and far-reaching recommendation from the MPE Committee of the Board was adopted by the NSB at this meeting:

> 1. The NSF should recommend as a national policy the desirability of government support of large-scale basic scientific facilities when the need is clear and it is in the national interest, when the merit is endorsed by panels of experts, and when funds are not readily available from other sources.
> 2. A national astronomical observatory, a major radio astronomy facility, and university research installations of computers, accelerators and reactors are examples of such desirable activities for NSF.
> 3. Funds for such large-scale projects should be handled under special budgets.

The "special budgets" provision was meant to prevent any encroachment "upon the regular established programs of the Foundation."

AUI next submitted a request for $234,000, in July 1955, to support continuation of the studies for establishment of the radio astronomy facility. In September the Radio Astronomy Panel cut this to $140,500, and a grant of this amount was made on October

11, 1955. Two months later Green Bank, West Virginia was designated by the AUI Steering Committee as the best site, and this was approved by the Radio Astronomy Panel in mid-January 1956.

Tuve's personal hostility to Berkner, and to AUI as manager of the radio astronomy facility, began to surface at the first meeting of the Radio Astronomy Panel. The full story has been told by Allan A. Needell.[6] The National Science Board's MPE Committee decided that a conference should be called to discuss the AUI feasibility study "and this whole question of a pattern for a truly national organization to further radio astronomy research."[7]

The "Conference on Radio Astronomy Facility" was held in Washington on July 11, 1956, and it was a baptism of fire for the new Program Director for Astronomy, the author of this book, who was responsible for arranging the logistics of the meeting. The morning session was quiet and straightforward, with no warning of what would take place in the afternoon session. It began with an official welcome by Detlev W. Bronk, Chairman of the National Science Board. Waterman discussed the "Purpose of the Conference," and described the present state of the radio astronomy facility in the Foundation. A separate item for facilities had been included in the NSF budget request for fiscal year 1957, and this facility would be the first major effort to be supported by NSF. Tuve discussed "U.S. Needs in Radio Astronomy," and Bok surveyed "Radio Astronomy Research in the United States." Berkner made some introductory remarks about the "NSF Feasibility Study by AUI," and Richard M. Emberson, Berkner's assistant, gave a detailed account of the site survey and design studies for the 140-foot radio telescope. Minkowski spoke on the "Liaison of the Radio Astronomy Facility with Astronomers," and Goldberg discussed the "Liaison of the Radio Astronomy Facility with Universities." The final topic, discussed by Hagen, was "Expediting the U.S. Program."

The afternoon session was concerned with the "Possible Organization and Management to Insure National Character of the Radio Astronomy Facility." Two proposals were presented. AUI, which had been encouraged by NSF grants to carry on the site survey and radio telescope design studies, proposed that it would construct the facility for NSF and operate it for NSF under a management contract. A plan proposed by William G. Pollard, Executive Director of the Oak Ridge Institute of Nuclear Studies (ORINS) proposed that universities with an interest in radio astronomy should form a new group and incorporate under the title "Association of Universities for Radio Astronomy." AURA would construct the facility under a management contract with NSF, and then operate the facility under conditional title, with NSF support through grants. The official minutes[8] and a 70-page transcript[9] give a sanitized version of the points made by speakers favoring the AUI plan and by speakers favoring the ORINS plan. The discussion was personal and bitter. For example, Tuve faced the AUI representatives and called them "self approved and self approving people." Finally, Edward Reynolds, Harvard's Vice-President for Administration and an AUI trustee, could stand it no longer and asked for the floor. A handsome man, he resembled Michelangelo's bust of Jove, and when he stood up, extended his arm at full length, and pointed at Tuve, he was for the moment Jove hurling thunderbolts from Olympus. At the start he said: "The atmosphere here has disturbed me very, very much indeed." He concluded by saying: "We don't want this contract under this atmosphere. You are going to have to want us enthusiastically if we are to take it. Then we can do a good job for you." This restored the meeting to civility, and it ended with an announcement that the Radio Astronomy Panel and the National Astronomical Observatory Panel would be asked to have a joint meeting as soon as arrangements could be made.

The joint meeting, which was attended by five members of the Advisory Panel for Astronomy, was held in Ann Arbor on July 23, 1956. NSF staff members in attendance were: Waterman, Klopsteg, Seeger, and Edmondson.[10] Waterman began the meeting by pointing out that "The general purpose of the present meeting was to consider and advise the NSF as to the most effective means of ensuring the soundness of the programs at the contemplated astronomy and radio astronomy facilities in the best interests of astronomical research and of the astronomers throughout the country." He discussed the functions of the three panels, and made it clear that they were purely advisory to the National Science Foundation. Close liaison, or alternatively partial or complete merger of the panels, would be needed to preserve unity in astronomy. It was the consensus of the meeting that unification of this kind was desirable, if not essential, to achieve optimum progress. The Advisory Panel for Astronomy should be kept fully informed of the activities of the special panels for the astronomical observatory and radio astronomy.

The joint meeting of the three panels helped clear the atmosphere, but did not restrain Tuve's efforts to stop AUI from getting the contract to build and operate the radio astronomy facility. On July 16, 1956, William G. Pollard[11] wrote to nine astronomers and three university presidents inviting them to serve as incorporators for a "new corporate body of universities under the name of 'Association of Universities for Research in Astronomy' (or such other name as the twelve of you might choose prior to incorporation) which would be incorporated under the laws of the State of West Virginia." A draft "Agreement for Incorporation" and draft "By-Laws of the Association of Universities for Research in Astronomy" were attached. Also attached was a draft of a proposed letter of invitation for membership in AURA that would be sent to 24 universities and two non-academic organizations by Dr. Irvin Stewart, President of the University of West Virginia. It was proposed that the incorporators would meet at 8:00 a.m. on August 21 to elect officers, adopt the by-laws, and elect the universities which had accepted the invitation to join. This would have made it possible to present a proposal from AURA at the August 24 meeting of the National Science Board.

Pollard's letter also said: "The proposed name for the association has been changed [from his suggestion at the July 11 meeting] so as to cover the whole of astronomy." Article III of the draft "Agreement for Incorporation" read:

> This corporation is formed for the following specific purposes:
> 1. To acquire, construct, and operate a national facility for astronomical research to be located in the Green Bank Valley area of Pocahontas County, West Virginia, and such other facilities for astronomical and scientific research at such other sites as it may from time to time determine to undertake;
> 2. To assist its member universities and other institutions and individuals in their own programs of optical and radio astronomy and related activities;

3. To foster and encourage advancement of knowledge concerning astronomy, radioastronomy, and related fields, together with the education and training of personnel in these fields;

All for the advancement of learning and scientific knowledge.

The implied competition with the McMath Panel did not materialize because this version of AURA didn't come into being. Only the name and acronym survived. Goldberg had received a copy of Pollard's letter, and remembered the proposed name: "I saw at once that Association of Universities for Research in Astronomy spelled AURA, and I think I proposed that name at one of the meetings of the organizing committee, but the idea came unwittingly from Pollard at Oak Ridge, Tennessee."[12]

Pollard sent a copy of this material to Waterman on August 2, and he sent copies to members of the Board's MPE Committee on August 16. Pollard's cover letter said that Tuve had told him that NSF wanted them "to hold off on taking any final steps for the incorporation of the proposed Association of Universities for Research in Astronomy until the NSF had an opportunity to reach a final decision on the management of the radio astronomy facility."

The NSB's discussion considered three choices: (1) Management by a single university (West Virginia or Virginia), (2) Management by a new organization (the Pollard–Tuve consortium), and (3) Management by AUI. The agreement by AUI that three "trustees-at-large" would be added to the AUI Board was a factor in the decision of the NSB at its 41st meeting on August 24, 1956 to award a five-year contract to AUI. The Board made it clear that serious consideration would be given to "the possibility of establishing at the end of that time a common management for the Radio Astronomy Observatory and the Optical Astronomy Observatory."

The site survey for the optical observatory was started in early 1955, at nearly the same time as the first grant to AUI. The search for the optical site was still in progress when the contract with AUI was signed. AURA was not incorporated until 14 months later, and Kitt Peak was chosen 4 months after that.

4 John B. Irwin's paper on "Optimum Location of a Photoelectric Observatory," and the proposal to the National Science Foundation from the University of Arizona, Ohio State University and Indiana University. Reaction of the NSF Advisory Panel for Astronomy. The Flagstaff "Astronomical Photoelectric Conference," and consequent actions by NSF.

The first step in the series of events leading to the establishment of AURA and the Kitt Peak National Observatory was the publication of a paper by John B. Irwin, Professor of Astronomy at Indiana University, entitled "Optimum Location of a Photoelectric Observatory".[1] He had worked up this material for a paper which he gave at a meeting of the Neighborhood Astronomers at Ohio University, Athens, Ohio on October 13, 1951. Geoffrey Keller, Professor of Astronomy at Ohio State University and Director of the Perkins Observatory (Ohio Wesleyan University), felt strongly that this was an important paper and should be published. Irwin responded to his urging and sent it to *SCIENCE*, which seemed to be the journal that would reach the widest spectrum of scientists. Irwin says it is his only paper whose publication date he remembers (February 29, 1952).

The second step originated in the continuing efforts of Edwin F. Carpenter, Director of the Steward Observatory of the University of Arizona, to overcome the interference by supermarket and other commercial searchlights with his photographic work on faint galaxies. In his 1953–54 annual report[2] he went so far as to suggest that "the Observatory be allotted a special operating item, say about $200 per year, or maybe $300, by which we could rent the sky from station KTKT on specific nights when the searchlight would be run. .... At about $10 per night, we could rescue 20 or 30 nights per year." Irwin's paper attracted his attention, because he had also proposed to the University of Arizona that the Steward Observatory telescope, a 36-inch reflector, be moved to a new location away from all city lights. Carpenter had paid a visit to Bloomington in the summer of 1951, when Irwin was collecting material for his paper. While there he initiated some informal discussions of ways in which Indiana University, and at most one or two other universities, could share the costs of a site survey to be followed by moving the Steward Observatory telescope to a new location. Telescope time and operating costs would be shared by the participating institutions.

On April 23, less than two months after Irwin's paper was published, representatives of the University of Arizona, Ohio State University, and Indiana University visited the National Science Foundation, at its second home on California Street, to discuss

14

the possibility of a proposal from the three universities for funds to conduct Carpenter's hoped-for site survey. We met with Paul Klopsteg, Assistant Director for Mathematical, Physical and Engineering Sciences (MPE) and Raymond J. Seeger, Program Director for Physics and Astronomy.[3] Seeger encouraged us to submit a proposal, saying that the only way the Foundation could know how much money astronomers needed to support their research would be from their proposals asking for NSF support. If astronomers were timid, the money would go to other disciplines that were more aggressive.

A proposal was quickly prepared, mostly by Keller,[4] with the title: "A Search for a Suitable Location for a New Astronomical Observatory to be Used Jointly by the University of Arizona, Indiana University, and Ohio State University."[5] It was submitted to NSF on May 15, 1952 by the University of Arizona with Carpenter as principal investigator. Irwin and Keller were listed as representing Indiana University and Ohio State University. The proposal stated:

> It is the desire of the three Universities mentioned in the title to establish a cooperative observatory in the most favorable location available in the United States. Due consideration would have to be given to practical matters such as cost and accessibility. It is hoped that, if the venture were successful, other institutions would join in the operation of the observatory and that the facilities of the observatory would gradually be expanded. Further details concerning a possible future proposal for the observatory buildings and equipment will be found in Appendix A. The present proposal is intended to cover only those investigations which must be made before a suitable site can be chosen.

The one-year budget proposed for the site survey was $21,200, and overhead was explicitly not requested. Salaries for two observers, a part-time machinist, and an electronics technician totalled $10,000. Permanent equipment, including a $6,000 truck, was $9,000; expendable equipment, supplies, and travel for Irwin and Keller to Arizona amounted to $1600. A sum of $500 was included for repairs and contingencies, and $200 was listed for consultation fees (legal and engineering). [Note: These numbers add up to $21,300.] Appendix A was entitled: "Outline of a Possible Proposal for the Establishment of a Southwestern Observatory to be operated jointly by the University of Arizona, Indiana University and Ohio State University." The opening section reads:

> *Name and Address of Institution:*
> It is suggested that the construction of the observatory buildings and grounds and their subsequent operation be placed under the control of a non-profit corporation. The members of the board of this corporation would be selected by the three participating institutions. Equipment brought to the observatory by one institution alone would remain the property of that institution. Property contributed by the NSF would become the property of the corporation. The director of the observatory would be appointed by the board of the corporation.
> The name, location and details of the articles of corporation would have to be worked out between the three cooperating universities. Presumably

> this corporation would be in existence before the proposal was formally submitted.
>
> It is assumed that after the completion of the observatory and its first year of full operation, the maintenance costs of the observatory will be borne entirely by the cooperating institutions. Other institutions would be encouraged to join the corporation on a maintenance cost sharing basis, provided the facilities were adequate.

The suggested five year budget included $70,000 to move the Steward Observatory telescope and to build a dome to house it, $62,500 for a new "light collector"-type 36-inch telescope and dome, $52,500 for other buildings, $2000 for landscaping, and $18,500 for utilities. Initial operating expenses were $6,000 for travel for 3 years, $1,000 for telephone and telegraph for 5 years, $9000 for the salary of a caretaker for 3 years, and $20,000 for repairs and contingencies. The total five-year budget was $241,500.

This proposal repeated many of Struve's earlier ideas about a cooperative observatory, including the concept that operating and maintenance costs would be paid by the cooperating universities after an outside agency had provided the capital funding. It was NSF's first proposal for a research facility, arriving more than a month before Bok's proposal for funds for the Harvard 25-foot radio telescope. Both were discussed at the August 1, 1952 meeting of the Advisory Panel for Astronomy. The minutes[6] report: "The Ohio–Indiana–Arizona proposal was discussed in detail and a vote favored the rejection of this specific proposal, although most were favorable for the idea of a photoelectric observatory and to cooperation of several institutions for major installations." (As mentioned earlier, there was a tie vote on the Bok proposal.) Some of the letters sent to Seeger before the meeting make interesting reading. For example Stebbins wrote on July 15: "The proposal of the Universities of Ohio, Indiana, and Arizona is fundamentally a good one, but even the initial budget proposed is so extravagant that I fear they have weakened their case."[7] The proposal was formally declined in a letter from Waterman to Carpenter, dated September 11, 1952.[8]

Seeger and others immediately realized that the Arizona–Ohio State–Indiana proposal had raised some important issues, and that NSF should look into the question of the needs for new optical astronomical facilities. Klopsteg muddied the waters by writing to Roger Lowell Putnam, sole trustee of the Lowell Observatory, "to inquire about the possibility of a plan by which the facilities of the Lowell Observatory might be used by visiting astronomers."[9] Klopsteg and Putnam probably became acquainted in 1951–52 during the period of Putnam's service in the Economic Stabilization Administration, the final year of the Truman administration.

Seeger lost no time in appointing an ad hoc Panel on Astronomical Facilities, consisting of Robert R. McMath (Chairman), I. S. Bowen, Otto Struve, and Albert E. Whitford. The Panel, minus Bowen, met in Washington on February 6, 1953. Mr. Putnam had been invited by Klopsteg to meet with them,"[10] and this annoyed McMath."[11] The Panel members and Mr. Putnam met with the Advisory Panel for Astronomy in the afternoon, and McMath gave a report of their discussions. The ad hoc Panel "recommended that a conference on photo-electric methods be held to see whether a new facility for such investigations should be established. Dr. Whitford had agreed to be chairman of the organizing committee for this conference. It is proposed that NSF sponsor the conference to be held (at Boulder or possibly at Lowell Observatory) after the AAS meeting in Colorado in August 1953."[12]

McMath had begun a two-year term as President of the American Astronomical Society in June 1952. A short time before the February 6 meeting at NSF, he had appointed an ad hoc AAS committee consisting of Struve and Kuiper to obtain information about research projects and facilities that were under consideration and that would require support from United States Government sources. The committee sent a circular letter to most of the active observatories in the United States on March 9, 1953.[13] Struve reported to McMath on May 21[14] that a considerable amount of material had been received in response to this letter, and that he would like to turn it over to McMath for presentation to the AAS Council at the August meeting in Boulder. With McMath's approval, Struve sent the original letters from 24 institutions plus a résumé on June 11.[15] McMath immediately offered to send a copy of the résumé to Klopsteg.[16] Seeger acknowledged receipt of this document on August 5,[17] and said it would serve NSF as a budgetary guide for the needs in astronomical facilities.

The conference that had been recommended by the McMath Panel at the meeting on February 6 was held at the Lowell Observatory on August 31 – September 1, 1953. The Organizing Committee consisted of Albert E. Whitford (Chairman), John B. Irwin (Secretary and Editor of the Proceedings), Gerald E. Kron, and Raymond J. Seeger. The NSF staff had decided that "the Foundation should arrange the conference itself, rather than ask an institution to do so,"[18] and Waterman authorized Seeger to "obligate an amount not to exceed $3,600 to defray the expenses of approximately twenty-five persons to attend the conference."[19] This was increased later to $5,000.[20] The invitations were sent by NSF over Waterman's signature, and the expenses of the 34 participants were reimbursed directly by NSF. Whitford's claim for reimbursement said he was allowed $220.00 and spent $220.07, and he could certify that "all of the funds granted were used to defray travel expenses."[21]

The Organizing Committee started work after Seeger wrote to Whitford[22] confirming the $3,600 appropriation by NSF to support the conference, the dates and place of the conference, and the membership of the organizing committee. Seeger asked Whitford to consult the Committee to produce a list of twenty-five astronomers to be invited to the conference. In a later letter[23] he suggested seven names to be put on the list. He suggested that the chairman for the topic "Discussion of possible cooperative telescope" should be McMath, Stebbins, or Struve. The last paragraph of the letter said:

> With respect to announcements to those for whom travel funds cannot be offered, I urge you to make this list entirely invitational and small. It is my belief that the worth-whileness of such a conference depends upon the freedom and opportunity for discussion which, I believe, is inversely proportional to the number of participants. Please let me have your recommendations as soon as possible.

During the rest of the month of May, there was a flurry of correspondence[24–31] related to some concerns about why the Lowell Observatory had been chosen as the site of the conference. One source for concern was an indiscreet luncheon table remark in Pasadena by A. G. Wilson, incoming Director of the Lowell Observatory, that NSF "is going to" build a photoelectric telescope at Lowell.[25] Another source was McMath's festering unhappiness with Putnam's participation in the February 6 meeting, as written in a letter to Whitford: "As you already know, I did not quite like Klopsteg's bringing Mr. Putnam into our ad hoc Committee meeting, nor did I quite like Putnam's attending that afternoon session of the regular panel. I have told Klopsteg of my feeling in this matter."[27] Whitford's reply included: "At the Midwest meeting in Evanston last Saturday, I found a strong rumor that NSF "wanted to do something" for the Lowell Observatory, and a little resentment that Dr. Klopsteg was inclined to launch out on a venture of this sort without proper sounding-out of astronomical opinion."[28] McMath wrote that he had heard the same rumor, and "since seeing you I have again seen Klopsteg and have told him that I shall be entirely governed by whatever your conference brings forth."[30]

Putnam had also heard these rumors, and this is the complete text of his letter to Whitford:[24]

> I have just come back from a trip to the Pacific Coast, and I heard a few things that disturbed me a little. I wasn't primarily associating with the astronomers, but I got a little repercussion that there was some resentment on picking Flagstaff as the site for your September 1st meeting. There seemed just a little feeling that this was a cut and dried program that the Lowell Observatory was going to slip over on the astronomical profession. I hate to have that feeling abroad, because you and I both know that the whole idea of this photo-electric instrumentation started elsewhere. The first any of us at Lowell heard about it was when Paul Klopsteg got hold of me in Washington last year, to ascertain if we would be willing to work in on such a program. The next that happened was the February meeting, in which the National Science Foundation invited me to participate. You and I both realize that the Lowell Observatory is ready and willing and anxious to cooperate with the whole astronomical profession and with the National Science Foundation, to work out what is going to be best for the advancement of knowledge. I believe we have facilities that could be very useful to such a project, and we are actively interested and working on research in those directions; but the basic thought of a cooperative venture comes from such people as Dr. Klopsteg and Dr. McMath, as well as from you.
>
> What is disturbing to me now is, how do we cure the misapprehension. I wonder if the people you are inviting to the conference would feel better if it were held at Boulder, or if they were polled to see what site they would prefer for the conference. I am still just as anxious as ever, and so are the people at Lowell, to be hosts to a meeting, and will contribute whatever we can, but I don't want any possible resentment to upset the progress of science, so I will leave the problem in your lap, to do with as you think wisest.

Whitford's reply[25] stated that he saw no problem with holding the meeting in Flagstaff. He also wrote:

> At the meeting in Washington, I sensed some resentment on the part of

the members of the regular advisory panel, who wondered why an Ad Hoc
Committee had to be appointed to administer something that possibly could
not be gotten through the regular channels. I took pains to mention what
was in the wind when writing to several leading astronomers about other
matters, and to give them a chance to express dissent if they thought that
a photoelectric telescope was not the best first project for instrumental aid
from NSF. There has been no reaction. I wrote to Dr. Greenstein at Cal
Tech, the chairman of the regular panel, and asked him to ascertain if
there was any feeling on the panel against a project along the lines
proposed. He gave an opportunity for such expression by means of a
circular letter and to my knowledge there has been no response. His letter
did reflect the feeling that there might have been informal commitments
by NSF to the Lowell Observatory.

Putnam's immediate answer[26] included:

> I sensed, as you did, a little feeling on the part of the regular Advisory
> Panel, and perhaps especially by its Chairman, that we were interlopers,
> and I had a little fear that some of them might wonder whether my
> position in Washington last year had not been used to pressure N.S.F.
> Actually it was all the other way around, as I am sure you know.
> I have sometimes had to watch in my business connections that my
> willingness to undertake controversial government jobs didn't cause them
> harm, and I hope very much the same thing will not be true of the Lowell
> Observatory. Astronomers, I believe, are just as human as everybody else –
> they want to protect their own institutions and particularly their own
> reputations – and it is perfectly easy for a feeling of jealousy to grow up.
> Naturally I am anxious to build the prestige of Lowell, but even more, to
> increase the opportunities of the good young men we are getting there
> now – but I don't want that natural desire of mine to be in any way a
> stumbling block to a basically sound advancement of science.
> I am sure you understand our feeling, and perhaps all we have to get
> across to most of the people is the fact that this was started by N.S.F., and
> not by any one at Lowell – though, of course, it is true that when the ball
> was tossed our way, we were thoroughly ready to grab it and run with it,
> but we didn't call the signals.

McMath had received a copy of Whitford's letter to Putnam, and liked it.[27,31] His letter
to Whitford continued:

> I spent February in the Pasadena area on a mixture of astronomy and
> guided missiles. I was lucky enough to see Bowen a good many times, and
> we did have some conversation about a possible photoelectric telescope. I
> think we both felt that it might be unfortunate to have the NSF make a
> grant of this size to an institution such as Lowell, which is represented
> regularly by just one man. On the other hand, one wonders what the
> situation would be should a new photoelectric telescope be given to the
> University of Wisconsin. The funds will be Federal in their origin, and

I am still puzzled as to how such an instrument should be operated. There has been some thought that perhaps the American Astronomical Society should receive gifts of this type and allow, say, the council to operate the instrument. I am definitely against this, as I think it would complicate matters to a point where the usefulness of the instrument might be impaired.

This is a rather rambling letter; I suppose I am really thinking out loud. In any event I hope you can hold the proposed conference and that under your leadership whatever conclusions may be reached will be "fearless" and based solely upon the best end result to astronomy. If needs must, I am quite prepared to tell the Klopsteg–Seeger combination that Mr. Putnam had better do his own work insofar as the rehabilitation of Lowell is concerned.

I did not sense any feeling at 813 Santa Barbara Street that they were being robbed by Lowell. My own feeling is that a photoelectric telescope is a good first project for big instrumental aid from NSF. I now have no feeling about its location except that it should, if possible, be attached to some operating observatory.

Whitford's reply[28] was direct and to the point:

I am not clear about the meaning of your statements as to your attitude following the talks with Dr. Bowen. If you and Dr. Bowen now take a rather negative attitude toward recommending that NSF build a cooperative telescope at Lowell Observatory, would it not be well to have some kind of a round robin discussion among the astronomical members of the ad hoc committee and inform Dr. Seeger and Dr. Klopsteg of any revised consensus that may come out of it? I must say that I can see objections to putting the telescope at any other places that might be considered as alternatives to Flagstaff. It should be somewhere with a fairly high percentage of photometric nights, which means the West or Southwest. It was Dr. Struve's view that it should be adjacent to an observatory with an administrative staff accustomed to, and equipped for instrumental maintenance. I strongly second this view. If you call the roll, we would have, first, McDonald, already scheduled for a similar telescope from the University of Texas funds. Second would be Sacramento Peak, an excellent site but probably too involved in armed forces security measures and red tape to be useful. There would be the University of Arizona, but they would have to go quite a ways to find a good station and I for one doubt whether the present staff is anywhere near as well equipped to handle the engineering and maintenance aspects as the Lowell group. There should preferably be someone around who understands electronics. It might be Palomar but here the acceptance of visitors would be tied in with the deliberations of the Observatory Council and there would be complications involving Monastery board bills. Mount Wilson is too close to Los Angeles and its smog. Lick Observatory will soon have the greatest variety of photoelectric installations, including the 36-inch Crossley and a 22-inch reflector. They have expressed willingness to accept properly qualified visitors. The climate is good but it has poor distribution between one long cloudy season and one long clear season.

Perhaps you have thought this thing through and come out with quite a different answer. I had always thought that Wisconsin could never qualify for a semi-national cooperative telescope of the type envisaged for the first large contribution from NSF to the instruments of astronomy. And, it was clearly stated in the meeting that day that NSF was dragging its feet on proposals to build standard pieces of equipment that would bring this or that institution up to the level of its neighbors. I am therefore at a loss to understand your reference to Wisconsin, unless you thought of us as an administrator of a telescope elsewhere. This is the sort of action at a distance that I question whether we should undertake without a great deal more backing than is now in sight. I hope you will pardon this lengthy out-pouring but the strong suggestion of a reversal in the opinion of the committee meeting of February 6 leaves me a little perplexed. Plans for a conference at Flagstaff are well advanced and it will certainly do no harm to go through with it whatever happens.

Whitford also wrote to Struve[29] telling him that he was the unanimous choice of the Organizing Committee to chair the session on "Discussion of a possible cooperative telescope." Sub-topics might be: (1) potential demand; (2) size and type of telescope; (3) site and plan of operation. Struve replied that he had an observing run scheduled at Mount Wilson during the time of the conference, and would not be able to attend.[32]

Whitford's letter to Struve also summarized the questions that should be answered by the conference:

The questions to be answered are, it seems to me: (1) Do astronomers feel that the time is ripe for NSF to put fairly large sums of money into instruments, in addition to those for annual projects? (2) Is an additional telescope at a first-rate site, primarily for photoelectric observing, as good a first project as can be found? (3) If the photoelectric telescope is the first choice, where should it be placed? I find that some astronomers are unhappy because only a few of the projects recommended by the regular panel in February have as yet been instituted due to lack of funds. This may be only a temporary situation. Dr. McMath says that the budgetary prospects for NSF are better for the coming year than was first thought and there may be as much as ten million dollars. I believe he has been instrumental in improving this outlook. On the second question, I agree with NSF in their feeling that they want to make the first instrumental projects cooperative or semi-national in character in order to avoid the pressures and difficult decisions that would go to helping this or that institution to bring their general instrumental facilities up to the level of others. On the third question, my own feeling is that Flagstaff is about as good a suggestion as has been made when one compares the climate and the local maintenance facilities with that at alternative sites.

McMath's deteriorating health caused his physician to warn him not to risk 7000 feet above sea level, the altitude of Flagstaff, for more than a few hours. For this reason he wrote to Waterman on June 23[33] declining the invitation to attend the conference, and he also informed Struve[34] and Whitford[35] of his decision. Whitford wrote at once to Seeger[36] pointing out that McMath's withdrawal left the University of Michigan without representation. He suggested that Goldberg be invited as a replacement for McMath, with the privilege of naming an alternate if he could not attend.

NSF sent an invitation to Goldberg, and Whitford wrote to him on August 12, 1953[37] to express his pleasure. He told Goldberg that Struve, Stromgren, and Bowen were unable to attend, and that the meeting would be "short on astronomers with a wide general background, and particularly those with experience in building and managing telescopes." He also remarked that Goldberg would "probably have a better understanding of the National Science Foundation than any other astronomer who will be present." Goldberg was a member of the MPE Divisional Committee at the time. He was the first astronomer on the Committee (1952–54), joined by Struve (1954–56) and followed by McMath (1955–57). Whitford asked Goldberg to comment at the meeting on the probable usefulness of the cooperative photoelectric telescope to places like Michigan and other centers of graduate instruction. Stebbins, the chairman of the session on "A Possible Cooperative Telescope", also wrote to him[38] asking him to contribute "anything you like in one or both hemispheres." Goldberg's reply[39] included: "I am afraid, however, that my thinking is much more ambitious than the proponents of the photoelectric telescope, and if you think it proper I should like to express my views at Flagstaff." He also said that the size of the present NSF budget should not be a matter of concern when planning for the future. The NSF budget limitation of $15 million maximum had been removed in a bill that was on President Eisenhower's desk for signature, and the FY 1954 appropriation was $8 million, an increase of $3.3 million from the previous year.

Thirty-five astronomers attended the conference, representing all of the larger US observatories and many of the smaller ones. In addition, the Dominion Astrophysical Observatory, Canada, the Commonwealth Observatory, Australia, and the Royal Observatory, Cape of Good Hope were represented. The program was divided into four sessions: (1) Present and potential fields in photoelectric photometry (five papers); (2) Instrumental and observational techniques (six papers); (3) Atmospheric and climatic effects (seven papers); and, (4) A possible cooperative telescope (three papers). Goldberg's discussion in the closing session[40] turned the conference around when he said: "I think what this country needs is a truly National Observatory to which every astronomer with ability and a first class problem can come on leave from his university." His ideas about telescope size were: "I don't think we should confine our thinking to a small photoelectric telescope. Personally I would like to see another instrument with about a 100-inch aperture established at the National Observatory, and of course, one naturally thinks of the Lick telescope which is now in the process of completion. The very natural thing would be to work toward a duplicate of the 120-inch Lick telescope. At the same time a photoelectric telescope with perhaps a 36-inch aperture would provide a very useful auxiliary instrument. I am sure that the photoelectric observers would not be averse occasionally to a telescope with a much larger aperture." The location he proposed for the observatory was: "In the southwest partly because the weather cycle is out of phase with California and it is interesting to note that all of the great work on external galaxies that has been done has been done in California

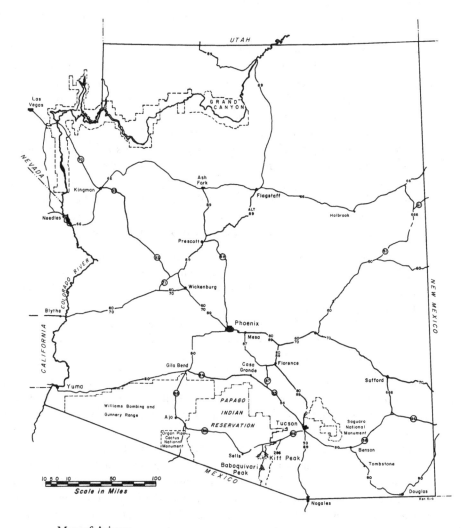

Map of Arizona.

in the worse possible observing seasons. Greater advances could be made if similar observations were carried on under more favorable observing conditions." This suggestion was based on discussions he had with Walter Baade in Ann Arbor during a special 1953 summer school just before the Flagstaff Conference.[41] Baade was one of the foremost observational astronomers of the middle of the 20th century. His work with the 200-inch Palomar telescope doubled the cosmic distance scale in 1952, and he was the first to suggest the creation of the European Southern Observatory (ESO), as described in Chapter 18.

Dr. Cecilia Payne-Gaposhkin spoke during the Final Discussion[42] and stated that the Harvard Observatory Council "would be happy to explore the possibilities of making available one or more of its instruments – primarily its mirrors and telescopes – for the purposes of the Observatory which we may be envisioning." Whitford's remarks in the Final Discussion[43] compared the problems of organization and management for a large and for a small facility (i.e. a National Observatory and a photoelectric telescope). Giclas recalls[44] Irwin's comments after the meeting: "Well, they've just taken it away from me", and "They've gotten too big." Irwin felt that the possible transfer of the Harvard 60-inch reflector and Goldberg's proposal for a National Observatory were "against the whole sense of this meeting, which was to discuss the establishment of a photoelectric observatory."[45] He felt that the Kitt Peak National Observatory, although it may have been an outgrowth from his 1952 paper, was "not the photoelectric observatory that I and Geoff Keller, and I guess Ed Carpenter were thinking about at that time. But we have no bad feelings about it whatsoever. At least I don't, certainly."

The Report of the Flagstaff Conference was presented to the National Science Foundation on November 6, 1953 in the form of answers to five questions that had been raised by the NSF Staff.[46] The answer to the final two questions was:

> Answer: These two questions can be answered together. The Conference feels that further study of the technical requirements for a new telescope and of the choice of a site can best be done by a committee. Therefore, the present Organizing Committee, on behalf of the Conference, respectfully recommends that the National Science Foundation, in consultation with the regular Advisory Panel for Astronomy, appoint such a committee to study these questions.

A brief summary of the conference was also published in the January 1955 issue of the *Astronomical Journal*[47] together with summaries of all the papers given in the first three sessions and two of the papers presented in the fourth session. The Proceedings[48] were published in October 1955.

Whitford wrote to Goldberg[49] almost two months after the Conference to say: "I find myself more and more drawn to your proposal for a major observing center in the southwest, presumably in the Arizona plateau region." The letter includes mention of "a committee to carry forward the recommendation of the Conference," and several management possibilities including a "cooperative association of universities." Goldberg got a favorable reaction at the November 20, 1953 meeting of the MPE Divisional Committee when he discussed the possibility of NSF assisting in the planning of a "National Observatory."[50]

The Advisory Panel for Astronomy endorsed the plans for an "inter-university Astronomical Observatory" during its January 7–8, 1954 meeting.[51] A list of names for membership on the planning committee was agreed to. Bart Bok and Bengt Strömgren were to be added to the earlier ad hoc committee, and McMath was to be asked to serve as chairman. Seeger phoned McMath on January 13[52] to discuss the proposed membership of the committee, to be called the NSF Advisory Panel for a National Astronomical Observatory, and "he was enthusiastic about everyone." Seeger told McMath the charter of the Panel "could be as broad as he would like to have it." NSF had in mind "not merely the consideration of a photoelectric telescope in the southwest but [also] a comprehensive plan for the future development of astronomy." McMath agreed to serve as chairman of the Panel, and confirmed this in a letter written the next day,[53] but he had second thoughts about Bok being a member of the Panel:

"Supplementing our telephone conversation of January 13, it would seem unwise to me to appoint Dr. Bok a member of this committee at this time. It is possible that the new director of the Harvard College Observatory could be reasonably expert in the instrumental field and that he is the man we should appoint to this committee if Harvard representation is desired by the NSF. As we all know, there is no real instrumental tradition at Harvard." Seeger's letter acknowledging McMath's acceptance[54] included a paragraph with the charge to the panel.

The NSF Director promptly appointed a five-member "Advisory Panel for National Astronomical Observatory" consisting of McMath, chairman, Bowen, Stromgren, Struve, and Whitford. Seeger wrote to McMath on April 19[55] to inform him that "all members have accepted." He also informed McMath that Peter van de Kamp would join the MPE staff at the end of May for a period of fifteen months as the first Program Director for Astronomy. Thus the stage was set for a major increase in support of astronomy by the National Science Foundation.

5 The role of Robert R. McMath as a trusted advisor to the Director of the National Science Foundation, and as an influential advocate for increased funding for NSF. The NAO Panel and the site survey for the location of a National Astronomical Observatory.

The selection of McMath to chair the 1953 ad hoc "Panel on Astronomical Facilities" and the new "Advisory Panel for National Astronomical Observatory" was not surprising to anyone familiar with his background in engineering, business, government service, and astronomy.[1] Waterman and Klopsteg[2] were longtime friends, and they respected his judgement and ability to get things done. He was the astronomer they knew best.

Born in Detroit in 1891, McMath graduated from the University of Michigan with a bachelor's degree in civil engineering in 1913. He worked as an engineer for two bridge companies after graduation, and was associated with Motors Metal Manufacturing Company for thirty-two years, starting as Assistant Manager in 1922 and ending as Chairman of the Board in 1954. McMath and his father, Francis C. McMath took 16 millimeter motion pictures of sunrise on the moon in August 1928 using a 4-inch refractor located at his summer home at Deer Lake near Clarkston, Michigan. These pictures attracted the attention of Ralph H. Curtiss, Director of the University of Michigan Observatories, who encouraged them to photograph the rotation of planets and solar phenomena. They and their amateur astronomer friend, Judge Henry S. Hulburt, installed a $10\frac{1}{2}$-inch reflector at McMath's new summer home on the north shore of Lake Angelus, and made their first observations on the evening of July 1, 1930. This marks the beginning of the McMath–Hulbert Observatory.

Heber D. Curtis became Director following the untimely death of Ralph Curtiss on December 25, 1929. He was quick to recognize McMath's talents as a developer of new instruments, and urged him to continue his work on stop motion photography of celestial phenomena. In the course of a few years the observatory evolved from its original amateur status to a leading professional solar observatory. This took place after the observatory was given to the University of Michigan in 1931. Substantial contributions from industry, private foundations, and the university's research funds supported the construction of new equipment for motion picture photography of the sun. The instrument, called a spectroheliokinematograph, was completed in July 1932. Two years later the first completely successful motion picture photograph of a solar prominence was achieved. This led to the construction of a 50-foot tower telescope designed primarily for motion picture photography of the sun. A larger and more sophisticated facility, the 70-foot McGregor Tower and Building, was completed in 1940, but little work could be done during World War II.

McMath recognized quality when he saw it, and after the war he assembled a staff at Lake Angelus that included Leo Goldberg, Orren Mohler, Helen Dodson, and Keith Pierce. He was appointed Director of the Observatory by the University of Michigan at the time it was given to the university in 1931, and was appointed Professor of Astronomy in 1951. He became Director and Professor Emeritus in 1961.

McMath received many honors for his contributions to astronomy, including the John Price Wetherill Medal of the Franklin Institute (1933), the Society of Motion Picture Engineers Journal Award (1940), and membership in the American Philosophical Society and the National Academy of Sciences. He was President of the American Astronomical Society (1952–54) at the time he was asked to chair the NAO Panel.

His wartime government service* and his connections in the business world gave him access to influential people, and he was able to enlist their support for science and for astronomy. For example, he was a friend of Joseph M. Dodge, Director of the Bureau of the Budget in the first Eisenhower administration, and he made an appointment to see Dodge in early May 1953, shortly after he took office.[3] After an exchange of pleasantries, Dodge said: "Well, Bob, what are you really here for?" McMath replied: "I'm here in behalf of the National Science Foundation." "What's that?", said Dodge, and McMath explained. Dodge had his secretary bring in the NSF folder, and then told McMath: "They have requested a big increase in their budget. I've cut that out and kept them where they were." McMath was able to persuade him to approve the requested increase, from $4.75 million up to $12.25 million. The House cut this to $5.724 million, and the Senate approved $10.00 million. The final compromise was $8.0 million, an increase of 70 percent from the previous year.

McMath had enlisted his close friend Walker Cisler, President of Detroit Edison, to help lobby for the NSF budget increase. Their efforts produced an additional success when Congress removed the $15 million ceiling on NSF appropriations.[4] Klopsteg wrote to McMath[5] on August 5, 1953: "Considering all of the obstacles and hurdles this is a record of substantial success, in the achievement of which my friend Bob deserves a lot of credit and thanks." McMath's reply[6] said: "I don't know how much credit I deserve. One thing I can say wholeheartedly, that I was working for the NSF and my good friends. It was probably of great importance to let Mr. Dodge know that the NSF and its top personnel are very well thought of by people in whom he had confidence."

The first meeting of the "NSF Advisory Panel for National Astronomical Observatory" took place in Ann Arbor on November 4–5, 1954.[7] During the nine months before this meeting there was a considerable amount of informal communication among the prospective members of the Panel. For example, McMath wrote to Struve[8] two weeks after he had agreed to chair the Panel, and the first sentence of Struve's reply[9] says: "For the purpose of discussion I believe it would be desirable to broaden the

---

* He received the President's Medal for Merit in 1948, in recognition of his work with the Office of Scientific Research and Development during World War II.

scope of our planning in regard to the national observatory and to investigate whether this observatory shall be built in the United States or in the southern hemisphere. . . . I am not certain in my own mind whether I would advocate such a plan but I believe we should consider it, and we should do so in connection with the European plan for a cooperative observatory in South Africa." Oort had written to Struve about the European plan on February 11, just two and a half weeks after the first formal meeting to discuss a European observatory in the southern hemisphere.

McMath spent the month of February in California, and discussed plans for the Panel with Struve in Berkeley and Bowen in Pasadena. Waterman was also in California during the second week of February, and arranged to meet with McMath and Bowen in Pasadena.[10] They discussed whether NSF could provide a building and telescope, retain title, and have those who operate it be outside the government. Waterman said this was a matter for counsel to determine, and said he would look into it. He wrote to McMath[11] on March 8, 1954:

> With regard to the legal point which arose in connection with the planning of your committee, Counsel informs me that it is legally possible for the National Science Foundation to build a laboratory and retain title on the part of the Government, while contracting for its operation outside the Government. Plans along this line should be worked out with some care, and it is Counsel's opinion that the most straightforward example might be patterned after Associated Universities Incorporated in their operation of Brookhaven. It would be well to discuss this further at some point.

AUI was the first organization to benefit from this legal opinion when the contract to build and operate NRAO was executed in November 1956, almost a year before the incorporation of AURA.

Federal law required security clearances for members of all NSF Panels, and this produced some irritating moments for two members of the NAO Panel before their clearances were finally approved. McMath was still not cleared eight months after he had agreed to chair the Panel,[12] and Bowen was having a problem as late as October.[13] Clearances for Struve[14] and Whitford were taken care of by their previous NSF appointments as members of the MPE Divisional Committee and the Advisory Panel for Astronomy. Stromgren seems to have been cleared without any reported difficulty.

As the time of the first meeting of the NAO Panel approached, a new problem surfaced. The Panel had been appointed by the Director of the National Science Foundation, and everyone had assumed that reimbursement of NAO Panel members for travel and subsistence would be paid directly by NSF, in the same way the members of the Advisory Panel for Astronomy were reimbursed. Unfortunately, the NSF funds budgeted for panel and staff travel were so severely limited by Congress that direct reimbursement of the NAO Panel would have caused a fiscal hardship for these NSF activities. The solution was to change the NAO Panel to a contract panel funded by a grant to the University of Michigan.[15] A proposal for a grant of $5,750.00 to cover the cost of three meetings of the Panel was written by Goldberg[16] and transmitted to van de Kamp by E. A. Cummiskey, the University of Michigan attorney.[17] NSF added panel salaries at $40 per day, and rounded the new total to $9,700 for eighteen months. The grant letter[18] was mailed on November 1, three days before the Panel meeting.

The Panel meeting took place in Ann Arbor on November 4–5, 1954. The five members were joined by van de Kamp, representing NSF, and Goldberg (part-time). The discussion covered a wide range of topics:[19] (1) the need for a national observatory, and astronomical manpower; (2) site investigation; (3) proposed instruments; (4) radio astronomy; (5) administration of NAO; (6) Smithsonian Astrophysical Observatory; (7) solar astronomy and seeing; and (8) new business. The conclusions were:

(1) The need for a national observatory was answered in the affirmative by all members of the Panel, who felt that the national observatory would stimulate interest in astronomy and would be helpful in the training of young astronomers.

(2) There was a consensus that the site investigation should be the broadest possible, and that seeing conditions for both day and night work should be studied. There was no recommendation at this time with regard to the southern hemisphere. The most important action was to recommend "the appointment of an executive secretary to act under the instruction and orders of the Panel for the purpose of making a broad study of possible sites and seeing."[20] Dr. Aden B. Meinel, then at the Yerkes Observatory, was the first choice of the Panel from a list of five names. The Panel recommended that NSF include $200,000 in its fiscal 1956 budget to fund the site and seeing investigations.

(3) The Panel recommended that a 30-inch telescope be installed as soon as possible, and that an 80-inch telescope be the second instrument. A decision concerning a very large telescope should come later. It was also agreed that Theodore Dunham's proposed 70-inch coude telescope (possibly in the southern hemisphere) was not of interest to the National Observatory.

(4) There was a consensus that the NSF plans for radio astronomy should be supported.

(5) The Panel recommended that the NAO "be operated by a board appointed by the National Science Foundation on a rotating basis. . . . The Board is to review proposed research programs, allocate telescope time, and establish policies as needed. The Board should be fully autonomous, and the Panel would like to avoid any semblance of government control." A minimum annual budget of $75,000, $51,000 for salaries and $24,000 for operating expenses, was proposed. Two astronomers, one engineer, and six support staff were planned.

(6) McMath reported that Dr. Leonard P. Carmichael, the Secretary of the Smithsonian Institution, had talked to him about the future of the Smithsonian Astrophysical Observatory,[21] and would be grateful for any comments from the NAO Panel. The consensus was that it "should stay where it is at the moment, but later be moved to the National Observatory site."

(7) McMath's discussion of his dream solar telescope led to a general discussion of daytime seeing. Separate sites might have to be considered for solar and stellar observatories.

(8) Electronic computing machines for astronomy were briefly discussed under "New Business."

When the Panel meeting ended, it was understood that van de Kamp would go back to Washington and explore NSF reaction to the various recommendations, and especially the funding of the position of Executive Secretary of the Panel. However, Stromgren did not wait for this to happen before informing Meinel about the Panel recommendation to hire him as Executive Secretary. Meinel soon complained to McMath[22] that "... up to the time of writing [this letter] I have received no information directly from NSF." This showed a complete misunderstanding of the relationship of the Panel to NSF. However, there were valid personal reasons why he wanted quick action. He was deeply involved in the NSF auroral program for the International Geophysical Year (IGY), and would have to arrange to transfer these responsibilities to someone else at an early date. He would also have to suspend his research activities in order to concentrate on completing the modernization work he had started at the Yerkes and McDonald observatories. McMath's reply[23] summarized the recommendation of the Panel. He also said he had not had any word from van de Kamp, and there was no point for them to discuss the NAO plan until NSF made a decision about funding.

McMath's subsequent letter to van de Kamp[24] described the situation: "Apparently Bengt Stromgren went straight to Dr. Meinel when he got home from our meeting, and Meinel is now writing to know when the appointment will come through." He reminded van de Kamp that he was supposed to go back to Washington and "scout this situation for the Panel." The letter ended: "Please let me hear as soon as convenient, as we have stirred up Dr. Meinel to quite an extent." van de Kamp's reply[25] said: "It seems to me that Bengt Stromgren acted a bit hurriedly." He said he would take up the Executive Secretary recommendation "as soon as possible," but this was already a month after the Panel meeting.

Meinel and Bowen had an unexpected meeting at Grand Central Station in New York in mid-December. Meinel's follow-up letter to Bowen[26] said: "In view of our discussion I am greatly encouraged to consider favorably the nomination the Panel has extended." He also commented on "the ominous aspects of government ownership of pure research facilities," and his feeling that the observatory "must not be relegated to the status of an observing station." He pointed out that he had as much interest in "administrative policy formation" as he had in "instrumentation," and may not be "the most ideal selection if you want someone merely to carry out the decisions of the Panel." He concluded with these remarks: "I will follow the policies of the Panel but where the need arises for a modification I would expect to be free to advise the Panel of the need and give suggestions of the mechanics of the modification. I believe in a driving leadership rather than one that is driven by casual circumstances of a laissez-faire situation. If this is not the type that the Panel wishes then it should reconsider its nomination immediately."

McMath's reaction to his carbon copy of Meinel's letter was to tell Bowen:[27] "His statement of the job proposed for him coincides quite exactly with my own, and I am today writing to him to that effect." His letter to Meinel[28] reads in full:

> Thank you for the carbon copy of your letter of December 15 to
> Dr. Bowen. I have read your letter with great interest.
> In the concluding paragraphs of your letter you raise certain points
> concerning a type of leadership, etc., that our NAO panel would expect.

I am in entire agreement with your statement and might add that it was just this sort of thing that led the panel to choose you.

I would most certainly expect, as Chairman, to consult with you whenever it is necessary and would hope that you would ask advice when you are sure it is needed. And, of course, from the NSF's standpoint your procedures and reports would naturally have to be approved by the Government panel. Here I am sure you would have no difficulty, as the panel has no intention of ever giving you anything which might remotely resemble "close supervision and direction."

Bowen's reply to Meinel[29] is mostly concerned with the question of government support. His view was that the government could give funds for capital expenditures, but could not be relied on for long-term operating support. He did not respond directly to Meinel's ideas about the role of the Executive Secretary, but made a very general statement:

The exact relationship of the Executive Secretary to the Panel as a whole has not been precisely defined, and will also doubtless change as the program shifts from initial fundamental planning, to construction and finally to operation. Certainly in the latter stages I would envision the relationship as somewhat similar to that of a college president to his board of trustees, namely most of the detailed planning and the execution would be in the hands of the executive officer with major policy matters subject to review by the board. In all of the above it should of course be understood that I am giving my own thoughts and am speaking not for the Panel. In any case, I very much hope you can see your way clear to help the committee on this problem. I am sure your experience with astronomical and optical problems would be of great assistance in getting the program under way.

van de Kamp wrote to McMath[30] on December 23, 1955, nearly two months after the Panel meeting: "I agree with you that we should act promptly, and we here at the NSF are now discussing the ways and means to proceed." He also said a rough estimate of the cost of the NAO was needed to assist NSF planning. McMath's reply[31] was blunt: "The entire panel joins me in hoping that you at NSF can do something and do it promptly." He estimated the cost of the present plans for NAO to be $3.5 million, and said substantial sums would be needed later for a large (120-inch) telescope and his "dream solar telescope." The letter ended: "There has been some correspondence between Dr. Meinel and Dr. Bowen and the writer. I do hope that you could find some way to get this set up soon." This brought a telephone response from van de Kamp, and it was followed by a quickening of the glacial pace of NSF.

At the request of NSF Goldberg sent a proposal to van de Kamp[32] on December 30. $11,724.98 was requested to cover the salaries of the Executive Secretary and an office secretary, plus funds for travel, telephone, postage, and drafting assistance for a five-month period starting February 1, 1955. The proposal stated that the Executive Secretary would "act under the instruction and orders of the Panel for the purpose of making a broad study of possible sites and seeing." van de Kamp informed Goldberg by telephone[33] on January 25, 1955 that the grant had been approved by the NSF staff, and that the grant letter from Dr. Waterman would be forthcoming in a week or ten days. The grant was rounded to $11,700, and the grant letter, dated February 1, 1955, was acknowledged by the University of Michigan on February 7.[34] McMath was the official Principal Investigator on the grant, but the day-to-day administrative details were to be handled in the Department of Astronomy office by Goldberg and his secretary, Kathryn Weddell.[35] Their contact in the administration was the University of Michigan Controller, Gilbert L. Lee, Jr., who was later a member of the AURA Board (1957–1972) and the first full-time salaried President of AURA (1972–1977).

The members of the Panel were in agreement that Meinel should visit each of them at an early date. Whitford wrote:[36] "I would certainly favor having the Secretary come to Ann Arbor as soon as he begins work for a session with you as panel chairman and with Dr. Goldberg as the administrator of the project." Meinel was in California in late January, but missed seeing Struve who was in Washington.[37] However, Struve hoped to see him when he visited the Yerkes Observatory on March 22–23. Struve advised McMath:

> It seems to me that he should make an extended visit to Michigan and go over the whole problem of the NAO with you. Since he is inclined to be somewhat impetuous, I believe you will wish to instruct him fully. He should understand that
> 1) He is working under the committee and is expected to carry out general directives as will be made known to him from time to time by you, and
> 2) He is not to initiate any new steps or to make significant departures from the directives of the committee without submitting them in advance to you.
> 3) He should at once organize the tests for seeing so that they can start as soon as the funds are available.

Some serious misunderstandings would have been avoided if McMath had been able to meet with Meinel for an extended briefing session, as recommended by Whitford and Struve. Instead, Meinel was off and running even before the NSF funds for his position had been approved. He sent McMath a ten-page report[38] describing his activities during a trip to California and Arizona in the last half of January 1955. His discussions with Shane, Bowen and others in California included such topics as engineering staff, telescope design, towers for testing seeing, and a "Southwestern Astrophysical Institute." In Flagstaff he talked to the Lowell Observatory Director, Albert G. Wilson, and its Trustee, Roger Lowell Putnam. He also had a brief discussion in Flagstaff with John G. Babbitt, a member of the Board of Regents of the University of Arizona.[39] The cover letter with the report mentions contacts with Senator Barry Goldwater, and with the President of the University of Arizona, Richard A. Harvill. The letter ends:

> I am very happy to report that a number of other problems facing us now appear to be nearing solution. The possibility of the creation of a new group to absorb the basic support responsibilities rather than asking hard-pressed existing groups looks bright. The idea that the state of Arizona establish a new campus on the order of the Lick arrangement

seems to be making progress. President Harvill just telephoned me that
Governor McFarland had asked a meeting of the Board of Regents to
consider whether they can make a proposal to your Panel to ensure the
basic support of the National Observatory. Undoubtedly we still have a
long way to go to iron out the details but it is a start in the right direction.

The background for this paragraph was finally given in a letter written more than two
weeks later.[40]

McMath acknowledged receipt of Meinel's report, but did not comment on its con-
tent.[41] He did urge Meinel to come and see him "as soon as is practicable." Meinel's
reply[42] said their sixth child was due "within 10 days," and he would try to come
shortly after that. McMath passed this information along in a round robin memo to
the "NAO–NSF Panel,"[43] and mentioned Meinel's report on the assumption that all
members of the Panel had received copies. A separate letter from McMath to Bowen[44]
began with the plaintive sentence: "I have been trying to get Meinel to come and see
me, which so far he has not been able to do." He also told Bowen that he was pleased
by Meinel's statement that he had "leaned heavily" on Bowen for advice "each time
an opportunity has arisen." Bowen replied:[45] "I find it a little hard to understand his
statement that he has leaned heavily on my advice. I have seen him twice since the
meeting of the Panel." The first meeting was by accident at Grand Central Terminal
in New York, and the second was "three or four weeks ago when Meinel dropped into
my office quite unannounced. . . . He left my office after fifteen minutes to a half hour
conversation." Bowen also felt that "every effort should be made to encourage Meinel
to get instructions from the Committee through you rather than from the individual
members of the Committees," and "In picking Meinel I think we all realized that we
might have some problems of this type. However, I suspect it will be easier for the
Committee to hold him down a bit when he attempts to go too fast than it would be
to provide the push for an individual of quite the opposite temperament." McMath's
reply[46] expressed agreement with this point of view. He also wrote:

> At the moment, he appears to have picked up a football and run all over
> the field with it, with no tackles available. As far as I can see, he has the
> Governor of the State of Arizona and the United States Senators from
> Arizona all agreeing to back Arizona as the site for the NAO. Additionally,
> he seems to have tried to arrange with the President of the University of
> Arizona to set up a separate campus which will be the site of the NAO. He
> admits that things may be a little difficult with the Tucson astronomers but
> thinks this can all be smoothed out.

Meinel's unauthorized activities began to bear fruit when the President of the University of Arizona, Richard A. Harvill, wrote to Seeger at NSF[47] and to McMath.[48]
Harvill told McMath that the proposed observatory had been discussed at a recent
meeting of the Board of Regents of the University and State Colleges of Arizona. "Eight
members of the Board including the Governor of Arizona attended the meeting." The
press was not present, and there were no publicity releases. "It was agreed that an
arrangement could be effected whereby the University of Arizona could take ownership
and responsibility for the management of the Observatory, on the basis of conditions

to be agreed upon between the Board of Regents of the University and State Colleges of Arizona and the National Science Foundation. It was pointed out that a separate division of the University similar to that which obtains with regard to other observatories located in universities could be arranged here at the University of Arizona." Harvill also said: "there would be no effort on the part of any individual or sectional group to secure an advantage with regard to the exact location of the Observatory. This, it was agreed, should be determined by the Panel on the basis of relevant scientific considerations." He asked McMath for his reactions with respect to the most appropriate procedure from this point, and expressed the hope that they might meet for a personal discussion.

McMath reacted quickly to Harvill's letter,[49] telling the Panel: "I have today received a letter from President Harvill of the University of Arizona, which disturbs me greatly. I very much regret that Dr. Meinel stepped out of bounds in January before his appointment was confirmed, indeed before any funds were available. I would of course much appreciate any comments you care to make after reading President Harvill's letter." His letter to Harvill[50] was characteristically blunt; this is the complete text:

> Dear Dr. Harvill:
> I have your letter of February 24 concerning the National Astronomical Observatory and its possible location in Arizona.
>
> Dr. Meinel was acting without any consultation with me as Chairman of this Panel and, so far as I can determine, he discussed this matter with no other member of the Panel. Briefly, the Panel's plan is to try and find the best site for both a stellar observatory and a solar observatory located in the continental United States. For obvious reasons this rules out, because of climatic conditions, any location east of the Mississippi River. We do wish to explore thoroughly Arizona, New Mexico, northwestern Texas, and of course a location on the West Coast has definitely not been ruled out from consideration.
>
> The Panel last met formally early in November 1954. It requested funds from the NSF for the purpose of employing an executive secretary who would devote full time to this site and concomitant seeing problem. Dr. Meinel was not appointed until February 1 of this year, when the funds became available through a grant to the University of Michigan. Dr. Meinel has reported to me and the members of the panel about his conference with you late in January of this year.
>
> The Panel has also requested from the National Science Foundation a rather large sum of money in order to make it possible to properly carry out this site and seeing investigation. Obviously this investigation cannot start until these funds are available.
>
> My secretary has copied your letter of the 24th, and it is being sent today to the other members of the Panel. Referring to your last paragraph, in which you cordially volunteer additional information if such is needed, or to try to answer questions which might arise, I think you will agree with me that in view of the above there is nothing needed at this moment.
>
> I am delighted with your statement regarding the exact location of the

observatory, to the effect that this location should be determined by the Panel on the basis of relevant scientific considerations. I hope very much that this project will not be discussed and that you will do all in your power to keep political considerations from interfering with the work of the Panel.

Harvill's reply[51] was conciliatory, and he also phoned Waterman[52] "to apologize for any unfortunate interpretations of his action with respect to the plans of the Panel headed by Dr. McMath". He recalled this twenty-three years later during a taped oral history interview:[53] "Dr. Meinel appeared in my office one day, and we had an interesting conversation." At Meinel's suggestion he wrote to McMath expressing the interest of the University of Arizona in the NAO project, and: "I got a letter back from Dr. McMath, and it said the best thing to do is do nothing." He met McMath later, and "we became good friends."

McMath's letter to the Panel also said that van de Kamp had suggested holding a meeting in March or early April, prior to the May meeting of the National Science Board. The main purpose of the meeting would be to prepare the justification for the NAO Fiscal 1956 budget request. McMath would make the presentation to the NSB on May 20. Finding a date presented serious scheduling problems for the busy members of the Panel, but March 30 and 31 proved to be possible. McMath sent these dates to van de Kamp,[54] and in a separate letter[55] said the Panel had voted to make Goldberg an official consultant because he was handling the administrative details of the NAO grant to the University of Michigan. Approval of the appointment by NSF was requested. van de Kamp got this information when he phoned McMath before receiving the letters.[56]

The problem of getting Meinel to visit McMath was still unresolved when McMath received a letter from Bowen[57] reporting a request he had received from Meinel to borrow two 4-inch survey telescopes, and asking "whether this is authorized and whether we should go ahead with the shipment as requested." McMath's reply[58] said: "This is another example of Meinel's unilateral actions," and suggested not shipping the telescopes before the March 30 meeting. He enclosed a copy of a letter[59] he had sent to Meinel and commented: "I have given up any expectation of his coming to see me. . . . It still seems to me that the most natural thing for a new employee to do would be to go and see the Chairman of the Panel which is employing him. I never thought it was my duty to chase him; rather, the exact opposite."

The letter to Meinel acknowledged a short note from Meinel, and mentioned Bowen's inquiry. It continued: "Inasmuch as you have not talked to your Chairman nor have you requested instructions from the Chairman by letter or telephone since your appointment of last February 1, it now seems to the writer that he must perforce spell out for you his understanding of the decisions reached by the NAO Panel at its meeting last November." This was done in detail, and the letter ended: "In conclusion, I repeat what I have said several times before, that you should come to see me as soon as possible." Meinel replied immediately by letter[60] and telephone. He agreed to visit McMath on March 17.[61]

Meinel's visit turned out to be shorter than expected, because he chose to travel by train instead of flying from Milwaukee in the morning. The train was not due in Ann

Arbor until 1:45 p.m.,[62] and his meeting with McMath began about 3:30 p.m. McMath took Goldberg and Meinel home to dinner, hoping to get more done. A joint letter to Bowen and Struve[63] described the visit: "I am very much puzzled. He arrived with a briefcase containing all of the Panel's correspondence, which he had of course received from Bengt. His second question to me was, 'When do I get elected to the Panel so that I can vote?' In retrospect, this question just about portrays his entire attitude." During the meeting Meinel lectured McMath: "The Panel had done no forward think-ing and no forward planning. We should at once proceed to establish an engineering office on the West Coast – presumably with Baustian on a substantial part-time basis. We would have to set our proposed observatory up in connection with some existing State university. And so it went. He told me that he would have ready for March 30 next all of the necessary financial and budgetary data in detail, with which to go before the NSF in connection with our request for a large grant." The letter concludes: "There is no point in my attempting to go further. I wanted to give each of you two good friends some idea as to Meinel's attitude so that you could, if you wish, watch for this sort of thing and say the right thing at the right moment during our forth-coming Panel meetings. It may easily be that he will adopt a different attitude when faced with the entire Panel. On the constructive side, he has agreed to visit me at least once a month. I told him emphatically that he was working for the Panel and that all decisions would be Panel decisions. I also said that in the interests of good organization the Panel had delegated me the job of working with him directly between Panel meet-ings." Bowen's reply[64] included: "No one should go ahead making fundamental policy decisions, and in particular taking action on them without the approval of the whole Committee. It seems to me that this can and should be very thoroughly impressed on Meinel by all of us."

The NAO–NSF Panel met in Ann Arbor on March 29–31, 1955. The minutes of the meeting were written by McMath[65] using Meinel's notes taken at the meeting. The first evening was devoted to discussing policy and inter-disciplinary relations, including the relationship of the NAO to the NSF Large-Scale Facilities Program (LSF). The pending appearance of the Chairman and Panel representatives at the meeting of the MPE Sub-committee of the National Science Board on May 19, and possibly at the NSB meeting on May 20 was discussed.

Discussion during the next two days centered on: (1) research justification and edu-cational justification[66] for the NAO; (2) the objectives of the NAO; (3) the site problem including the possibility of a southern hemisphere site; (4) a seeing survey and other technical questions relating to the FY 56–57 budgets; and (5) anticipated budget requirements for FY 58, 59, and 60.

The discussion of items (1) and (2) showed that there had been an important change in the Panel's thinking since the first meeting. "It was realized that the original 'observing station' concept of the NAO was too narrow from the NSF standpoint. The development of a stable scientific center should be the objective, although it should necessarily grow from the basic observing facilities." The NAO would be "a new astro-nomical scientific center."

It was agreed that the site should be "out of phase" weatherwise with the west coast observatories. A continental United States site was deemed to be the only solution under the terms of the Panel's charter from NSF. "The possibility of assisting US astronomers in the use of a cooperative southern hemisphere observatory should not be excluded by precedent through any decision by this Panel regarding the NAO."

The Panel recommended that the site survey should be carried out using automatic instruments mounted on towers. Phase I would be the investigation of 30 sites, using towers 10 feet high. Phase II would be the investigation of six sites, using towers 60 feet high. A height of 100 feet had been selected at the first Panel meeting. Phase III would be the final evaluation of three sites.

Meinel suggested that the proposed 30-inch telescope be increased to 36 inches. An unused 36-inch mirror blank is available at the Yerkes Observatory, and a 36-inch mounting had already been engineered and was under construction for the McDonald Observatory. The Panel accepted this suggestion.

The accelerated schedule for the 36-inch telescope caused the Panel to recommend that $10,000 be allocated in FY 56 for a preliminary study of the 80-inch telescope. The minutes go on to state: "The Panel unanimously believes that a much larger telescope will ultimately be necessary. The telescope is not included in any of the budgets." The proposed budgets were $300,478 for FY 56 and $578,160 for FY 57.

This meeting brought Meinel and the Panel together in a satisfactory working relationship, and a week later Meinel wrote to McMath[67] asking approval to employ Helmut Abt to assist him in expediting preliminary work for the site survey. Funds for Abt's part-time employment and associated travel would be available in the grant from NSF. McMath's letter[68] giving approval came by return mail. A follow-up letter from Goldberg[69] formally authorized the expenditure of Abt's salary and travel expenses from the NSF grant, and said "Submission of the usual travel vouchers" would be necessary. Meinel told McMath that he planned to meet Abt at the McDonald Observatory on April 25, and they would examine possible sites during the bright of the moon in May, and perhaps also in June.

Meinel's upper-atmosphere work gave him access to some of the first rocket photographs taken at White Sands,[70] "with some excellent pictures during perfectly clear weather of the entire region from mountains of Baja California, California, Arizona, New Mexico on the photographs." They "noticed for the first time the range west of Tucson with Baboquivari showing rather plainly on it and some mountains in that vicinity." Aeronautical maps, which had 1,000-feet contours were also used.[71] Arrangements for a reconnaissance flight were made with John O. Casparis, a pilot from Alpine, Texas to fly in a Cessna 140, a two-passenger, single-engine plane.[72] He charged $.10 per mile plus meals and lodging; this included his services, the plane, and the gasoline. Casparis and Abt covered about 2,000 miles in three days, starting at 7:15 a.m. on May 7, 1955.

They covered the mountains in the western part of Texas and the southern part of New Mexico on the first day. Abt's detailed notes[73] report the first look at Baboquivari and Kitt Peak. Baboquivari did not look promising, and "Kitt Peak from NW. Looks better – more trees. Investigate." The cost of the overnight stop in Tucson for two persons was: Supper $1.70, Motel $7.14, Movie $2.00, Breakfast $0.75. The second day was spent exploring mountains in southern Arizona. The cost of the overnight stop in Prescott for two was: Dinner $2.28, Hotel $7.14, Breakfast $1.00. The taxi to the airport cost $3.00. The third day started in northern and eastern Arizona, and then continued back to Texas through southern New Mexico. Engine trouble forced them to land in a deserted area about 100 miles west of Alpine. They managed to find a telephone, and called Mrs. Casparis to come and get them.

Abt's first jeep trip started on May 12, using his own jeep. He camped out the first night, and stayed in a motel the second. He took many photographs along the way of mountains that had been seen during the Cessna flight, and arrived in Tucson the evening of May 14, where the motel cost $2.55. He took pictures of Kitt Peak on May 15 from roads on the north and west sides, and also looked at other mountains in the Tucson area. He ended the day at the Signal Peak Fire Lookout Site in the Pinal Mountains south of Globe, and made notes about visible city lights that night. He made a two-day trip to Pasadena from Globe and return, and was back at the McDonald Observatory on May 20.

Abt was joined by Meinel for another jeep trip, May 24–29. They were on Signal Peak after dark on the first night. The next day they noticed Chevalon Butte on the way to Winslow, and thought it looked promising. They explored the Flagstaff area on the third day, and spent the night in Prescott. They explored the area between Prescott and Show Low on the fourth day, and were back at the McDonald Observatory on the 29th.

During April and May, important discussions relating to the plans and funding for the NAO were taking place in Washington. Goldberg represented McMath and the Panel at a meeting of the Executive Committee of the MPE Divisional Committee on April 24.[74] He had expected to answer questions, but was called on to make a complete presentation of the NAO project. The radio astronomy group and the computational center group made similar presentations, and the meeting adjourned without further discussion. After the meeting, van de Kamp sent a draft of a staff paper[75] entitled "Plan for a Cooperative Astronomical Observatory" to all members of the MPE Divisional Committee, which included Goldberg and Struve. The historical section of the paper began with the Arizona, Indiana, Ohio State proposal in 1952, and ended with a summary of the second meeting of the NAO Panel. The section entitled "Recommendation for NSF Action" listed six steps that "should be taken to implement the recommendations of the NSF Advisory Panel for the National Astronomical Observatory." Three of them dealt with a proposal that NSF should encourage interested educational institutions to form an incorporated group to own and operate a cooperative astronomical observatory established initially from government funds. Two financial recommendations mentioned the proposed $300,000 grant from FY 56 funds to the University of Michigan to continue the work of the NAO Panel, and recommended that NSF request $3,600,000 in the FY 57 budget to fund the construction of the cooperative astronomical observatory and its operation through June 30, 1960. The final recommendation proposed that the Panel should continue to study the need and plans for the construction of a 120-inch telescope and a solar telescope at the site of the cooperative astronomical observatory, or elsewhere, beginning in FY 61.

The apparent change in the ground rules, namely ownership and name of the observatory, was a matter of immediate concern to McMath and some members of the Panel.[76] At the May 19 meeting with the MPE Subcommittee of the National Science Board, McMath made a short statement[77] concerning "the parts of the staff paper to which he and Dr. Bowen jointly objected." NSF Counsel replied that "the problem was one of semantics." Most of the thirty-minute time slot allotted to McMath and Bowen was devoted to a question and answer period. They joined all of the NSB members in the evening for cocktails and dinner, but it was not necessary for them to meet with the full Board the next day. McMath was told privately that the NAO project "appeared to be on the rails and moving" insofar as the discussion in the executive session was concerned.

The staff paper also elicited a critical comment[78] from the Program Director for Engineering Sciences, G.H. Hickox: "As representatives of the Engineering Sciences Program, we do not feel qualified to comment in detail on the proposal except to note that it appears to have had its origin with astronomers, to have been considered by a panel of astronomers, and finally recommended by them. There is some question as to whether this represents an unbiased considered judgement."

A week after the NSB meeting Seeger called McMath[79] to confirm that he would visit McMath and Goldberg on June 2 for further discussion of the NAO project. McMath told him that "the Panel is unanimously agreed, if no funds are available in 1956, to resign." He also pointed out that he would need to know within two weeks whether funds would be available to continue Dr. Meinel as Executive Secretary beginning July 1, 1956. Seeger also asked McMath if he would be willing to serve on the MPE Divisional Committee, replacing Goldberg whose term would end on June 30. He accepted, after asking Bowen, Struve, and Goldberg for their advice.

Seeger brought good news and bad news to the meeting with McMath and Goldberg at Lake Angelus.[80] The bad news was that Congress had not yet appropriated the FY 56 funds for NSF. Seeger suggested that the Panel should accept a sum of money sufficient to bridge the gap between July 1 and September 1. Meinel thought $50,000 would enable him to get started, if it was backed up by the remaining $250,000 on September 1. The good news was that the NSF Staff had recommended that the rest of the money for FY 57, 58, 59, and 60 be requested from Congress in one lump sum of $3,600,000. It was suggested that Goldberg, acting for the University of Michigan, should go ahead and submit a proposal promptly for the $300,000 needed in FY 56.

The Panel had proposed, at its first meeting, that the observatory should be "operated by a Board appointed by the National Science Foundation on a rotating basis." Seeger told McMath and Goldberg that NSF "can legally neither operate nor appear to operate" a facility such as the Panel had proposed. NSF could "appropriate funds only to a properly organized institution." This appears to be exactly what Waterman had already told McMath in a letter dated March 8, 1954.[81] Nevertheless, Seeger's message evoked complaints from Whitford,[82] Bowen,[83] and McMath[84] that the Panel had not been informed earlier of this legal limitation on what NSF could do.

There was also some discussion in these letters of the name of the observatory. Some NSF staff members had objected to the use of "national" in the name, and suggested that "American" be substituted. McMath accepted this, but somewhat reluctantly. Whitford didn't like the change, but felt that "American" was better than "Inter-university," another title used by NSF.[85]

McMath summarized the new state of affairs in a long memorandum to the Panel[86] after he heard from Whitford and Bowen. The three most important points were: (1) NSF would retain title to the observatory; (2) a group of universities would be put together during FY 56 to provide a proper recipient for NSF funds to operate the observatory; and (3) the NSF, at its May meeting, had accepted the principle of "permanent support" of large-scale facilities. McMath told Seeger that a minimum of $75,000 per year would be needed for operating the observatory. The other points dealt with financial details for FY 56 and FY 57 that had been agreed to by McMath and Seeger.

Struve wrote to McMath[87] after he returned from an observing run at Mount Wilson. He did not like "American Astronomical Observatory" as the name ("all other observatories in the U.S.A. are 'American'."), and suggested naming it for the site that was chosen. "There is a certain advantage in specifying the location of an observatory in its name." This, of course, is exactly what was done three years later.

Stromgren's reply to McMath,[88] characteristically, was the last to be received. He felt that the operating organization for the new observatory was something that should be thoroughly discussed at the next Panel meeting. He did not like the name "American" for the observatory and thought Struve's suggestion was better.

Goldberg prepared the recommended proposal for $300,000, but the University of Michigan administration was not willing to act on such a large item without referring it to the Regents of the University, and their next meeting was not until July 21.[89] As an interim measure, Goldberg submitted a proposal for $29,000[90] to cover expenses for the period July 1 – September 30, 1955. This would allow Meinel to test a prototype 30-foot seeing tower and associated equipment at the Yerkes Observatory during the summer. It would also allow Abt and Meinel to make a further exploration by jeep for the purpose of choosing 30 sites for further study from the 66 sites for which they had obtained detailed information. The grant letter was dated June 30, and was acknowledged by the University of Michigan on July 6.[91]

Seeger's two-page form letter[92] welcoming McMath as a new member of the MPE Divisional Committee was acknowledged by McMath,[93] and in a second letter[94] to Seeger he addressed the question of NSF allocation of funds to astronomy and the MPE Division. MPE had been increased by less than 10 percent in FY 56, while the Foundation's appropriation had increased by one-third. van de Kamp had asked Goldberg how much less than the Panel's request for FY 56 would be acceptable. McMath told Seeger: "I would be willing to accept a fall grant in the amount of $250,000, which with the $29,000 already granted, would amount to a total of $279,000. Anything less than this sum means that the NAO–NSF Advisory Panel will have to string out its work in time, and this I am not willing to do. . . . Either the NSF wants the NAO–NSF Advisory Panel to go ahead with the job and can and will support it, or else there isn't any great point in the Panel's proceeding at all." McMath also wrote to Waterman[95] and protested vigorously about the inadequacy of the NSF allocation of funds to support research grants in astronomy in FY 56.

The final version of the proposal, increased to $314,738, was entitled "Studies Leading to the Establishment of an American Astronomical Observatory." It was sent to NSF by the University of Michigan on July 27, following approval by the Regents.[96] The grant was reduced by NSF to $250,000 for nine months. The grant letter was sent to the University of Michigan on August 26, and Waterman sent cables to McMath[97] and Goldberg the same day in care of the International Astronomical Union in Dublin. The university acknowledged the grant on September 1;[98] McMath wrote a letter thanking Waterman[99] as soon as he returned from the IAU meeting.

The Panel, with Goldberg in attendance as consultant, met in Dublin on August

31, 1955, following the receipt of the good news about the $250,000 grant.[100] The Panel instructed the Chairman to write to the Director of the National Science Foundation and inform him that it was "the Panel's unanimous decision that it would hesitate to proceed with the NAO program were the regular grants-in-aid program to be in any way affected." McMath's letter[101] brought a prompt response from Waterman,[102] who pointed out that the NAO would be a separate budget item in FY 1957, but there had been no such provision in the FY 1956 budget. The Foundation felt that the urgency of getting on with the plans for NAO was sufficient to justify a temporary inconvenience in FY 1956. Waterman wrote again[103] more optimistically: "We expect, before the year is out, that the ordinary grants-in-aid program [for FY 56] will not be diminished." At Seeger's suggestion, McMath also had written to the Chairman of the MPE Divisional Committee[104] asking that the agenda for the October 27–28 meeting include a discussion of the FY 56 allocation for grants-in-aid for astronomy, reported to be cut 25 percent from the FY 55 allocation. Goldberg also wrote a long letter to the Chairman of the MPE Divisional Committee,[105] addressing the scientific issues as well as the fiscal issues.

Goldberg recalled in later years[106] that the members of the Divisional Committee "had no special desire to support Astronomy", but they did think that the precedents for large facilities being created by the astronomers would help them later on. One member called it the "ratchet effect."

Abt and Meinel began a final reconnaisance trip on August 3, leaving the Yerkes Observatory in a new jeep purchased using NAO project funds. They left the McDonald Observatory on August 7 and made stops at Chevalon Butte, Flagstaff, Hualapai Peak near Kingman, Pasadena, Palomar, and the Tucson area. Meinel wrote:[107] "Kitt Peak appears rather hard for a good road. It has numerous rock outcroppings, indicating a hard core." They learned that Carpenter had been looking at sites on the slopes of the Sierrita Mountains for a new location for the Steward Observatory 36-inch telescope. When he was told about the possible NAO interest in the Tucson area, he decided to postpone his decision about a site in the hope that some kind of arrangement could be worked out with the NAO. Meinel returned to Yerkes by train on August 18, and Abt drove the jeep to McDonald.

Meinel had promptly reported the results of the trips he and Abt had made in May to McMath and the Panel.[108] Before starting the August trip with Abt he suggested to McMath[109] that the time had come to establish a field office in Phoenix to handle "all operations pertaining to site evaluation and engineering." As soon as the $250,000 grant was approved Meinel asked for approval of a Phoenix office with a staff of seven, including himself, Abt, and Harold Thompson.[110] Thompson was a design engineer who had been working for Meinel at the Yerkes Observatory. The remaining staff were draftsman, secretary, accounts secretary, and purchasing agent. McMath told the Panel[111] that he approved establishing the Phoenix office, but thought a staff of four people would be adequate at the start. He said Meinel would be in Ann Arbor on September 29 to work out the details with the University of Michigan administration.

Meinel arrived in Ann Arbor on September 29 on the early morning train from Chicago, and spent the day with McMath and Goldberg. They had a productive two-hour conference with representatives of the administration of the university: G.L. Lee, Jr., University Controller, and E.A. Cummiskey, University Attorney.[112] Lee and

Cummiskey agreed that the establishment of an operating office in Phoenix by about October 1 was both proper and practical. However, the by-laws of the University required that a full-time representative of the University business office must be in residence in Phoenix to sign vouchers, issue requests for quotations, etc. Lee and Cummiskey agreed that the university would provide such a person. In addition to the representative of the university business office, Meinel would start with a staff of four: Meinel, Abt, Thompson, and an office secretary. Staff could be added later on the basis of demonstrated need.

Back in Washington an important staff change had taken place at NSF. van de Kamp had been given a leave of absence from Swarthmore College for the academic year 1954–55, and he continued to serve as NSF Program Director for Astronomy through the summer of 1955. His successor, Dr. Helen Sawyer Hogg, took up her duties in mid-September[113] following her return from the IAU meeting in Dublin. She followed the practice, started by van de Kamp, of sending an invitation to submit research proposals to all members of the American Astronomical Society.[114] The opening paragraph read: "It is my privilege to write you on behalf of the National Science Foundation which I am serving this year on leave of absence from the University of Toronto, replacing Dr. Peter van de Kamp who has returned to his work at Swarthmore College." The names of the nine members of the Advisory Panel for Astronomy were given, and also the deadline for proposals to be considered at the January 1956 meeting of the Panel. The final paragraph read: "During my sojourn with the National Science Foundation I shall be glad to help you in any astronomical fashion that I can."

A copy of Hogg's memorandum was sent to the Director of the National Science Foundation with an unsigned message typed at the end: "Why is a Canadian (or any other foreigner) "Program Director for Astronomy" of the National Science Foundation? Can no American be found to serve in this office?" Waterman passed this on to Clyde C. Hall, NSF Public Information Officer, "for your information and any action you regard as advisable." Hall's reply ("SUBJECT: Scurrilous note re Dr. Hogg") included: "My advice in this matter is that we ignore it. We stand firmly in the knowledge that Dr. Hogg is a talented and honorable woman, and, indeed, a citizen of the United States. . . . If there are still persons in the United States who are genuinely concerned about this matter, we are well forewarned. We might, therefore, consider what kind of response we would make to a member of Congress who might be prodded into asking you such a question when next you are called upon to defend our budget.[115] There is nothing in the record to show that Hogg was ever told about this, and the matter was laid to rest.

The new Program Director for Astronomy had more than enough to keep her fully occupied. The grants program had received increased funding in FY 1955, and this would mean more paperwork to process proposals. The AUI search for a site for the radio astronomy observatory was actively under way (see Chapter 3), and the controversy over the eventual management was warming up. Last but not least, steps were under way to establish a Phoenix office as the headquarters for the NAO site survey.

Meinel, Abt, and Thompson left the Yerkes Observatory in Abt's Chevy convertible on October 9. They were in Phoenix October 13–17, and were back at Yerkes on the 20th. Arrangements were made to rent office space in Phoenix at 221 East Camelback Road.[116] Meinel and Thompson bought houses and moved their families to Phoenix in November, while Abt found a conveniently located apartment.

The proposal for $314,738,[117] written by Goldberg on June 13 and sent to NSF on July 27, included the statement: "Dr. Meinel informs us that all of the sites selected for the seeing survey are on federal or state-owned land, which is under control of the U.S. Forest Service." By the time Meinel and Abt made their August trip they had become aware that Kitt Peak was on the Papago Reservation and was not under control of the Forest Service. They visited the Office of the Bureau of Indian Affairs in Phoenix[118] to obtain information about how to get access to Kitt Peak. Meinel summarized the site survey plans in a letter to F.M. Haverland, the BIA Area Director in Phoenix,[119] and asked Waterman for assistance from NSF in securing approval from the Department of the Interior of any arrangements negotiated with the Area Director and the Papago Tribal Council.[120] Waterman responded by writing to Glenn L. Emmons, Commissioner, Bureau of Indian Affairs, asking him to be of assistance to Meinel.[121] Haverland forwarded Meinel's letter to Albert M. Hawley, Superintendent of the Papago Agency. Hawley replied:[122]

> I discussed Mr. Meinel's request with Mr. Mark Manuel, who brought the proposition up before the Papago Council on September 2nd. The Papago Council referred the request to the Schuk Toak District Council for their consideration. Kitt Peak in the Quinlan Mountains is in this district and the district council has final say in assignment of land for beneficial use and occupancy. Tribal Council reaction was favorable and we feel that the District Council will be in accord with the granting of his request. It will take from three to four weeks to get an answer from Schuk Toak District Council.

Meinel's letter to Haverland had included a request for permission to place a primitive road up Kitt Peak and to erect a tower containing instruments for testing the quality of the site. The Schuk Toak District Council's vote on this request was negative.[123] At this point, President Harvill suggested to Carpenter that Professor Edward H. Spicer and other anthropologists from the University of Arizona should be asked to help in overcoming this serious problem.[124] During a meeting in the Spicer home Carpenter talked about the problem with Emil Haury, Head of the Anthropology Department, and the Spicers.[125] Their discussion led to the conclusion that Meinel's problem was one of communication. The suggestion emerged that a visit of Tribal Officials to the Steward Observatory might be the best way to solve this problem.

Mrs. Spicer is also an anthropologist, and she had worked for nearly a year in the 1940s in the Papago Village of Topawa, the home of Mark Manuel who was now the Chairman of the Tribe. The personal relationship of the Spicers and Mark Manuel was more than a little helpful in bringing about the visit of members of the Schuk Toak District Council and the Papago Tribal Council to the Steward Observatory. Their informal approach to Mark Manuel was followed by a formal invitation from Carpenter, and the visit took place on October 28, 1955.

Meinel and Abt were back at Yerkes packing for their move to Phoenix, and none of the anthropologists were able to be at the Steward Observatory for the big occasion. Meinel recalls Carpenter's description of what happened:[126]

> As Dr. Carpenter told me (I was not present at that time), he said there was a bit of a problem finding which one wanted first to look through the

telescope. They were very impressed and awed by this big, dark structure in the middle of the dome, and the platform, and the whirring of the weight-driven clock that was down below on it which they could see twirling around as they came up the stairway up to this telescope and to look into it. And they were particularly struck by the fact that when Dr. Carpenter would look into it, his eye would gleam as he would look at the moon, as though something strange was happening to his eye. And it was with great reluctance that one of them first went up and looked through it. It was not the chairman or anybody. It was just a verbal indication of disbelief at what he was seeing as he looked in, and the others standing around could see his eye light up – the light of the moon went into it – and the exclamations that he made. And then, as Ed said, he recoiled back from it and then went back up and looked through it again, to look at it again and began talking in Papago to his colleagues that were there. And then one by one they went up to the telescope, backed off to the background of it and began to laugh then as the next ones went up to it, because they had been initiated now into what they would see. Very impressive. And finally the entire group had looked at the telescope and were very conversant in Papago and in their English as to what they saw. And it was very clear that it was a very dramatic occasion for all of them to look out the dome and see the moon there and to point out to it and then tell what they had seen through the telescope.

The visit of Papago officials to the Steward Observatory produced the hoped-for result. Harry Gilmore, who had replaced Hawley as Superintendent of the Papago Agency in Sells, wrote to Haverland[127] on December 13:

> The Chairman of the Papago Council informs me that he was notified this morning by the Chairman of the Schuk Toak District Council that that body has agreed to make Kitt Peak available to the National Science Foundation for observatory purposes, and will so recommend to the Papago Council at its next meeting, which will be on January 6, 1956. Since section III of Ordinance No. 15 provides, in effect, that the Papago Council shall be governed by the recommendation of the District Council in the matter of leases, this means that the Foundation may now be assured of its lease. It goes without saying that Mr. Manuel and I are very much elated over this almost unexpected turn of events. I think a great deal of credit is due the Council Officers and the two Council delegates from Schuk Toak District, who persisted in reopening the question after the District Council had twice turned it down. I believe that the final decision was due in large part to the kindness of Dr. Edwin F. Carpenter, Director of the Steward Observatory at the University of Arizona, who gave an evening to explaining to the District Council the purposes of astronomy and the modus operandi of an observatory.

Haverland immediately sent a copy to Abt at the new Phoenix office.[128] Meinel expressed his pleasure when he acknowledged the letter,[129] and McMath thanked

Carpenter for his important contribution:[130] "I am sure that we never would have obtained permission had it not been for your efforts."

The testing of Kitt Peak as a site for the NAO was approved by the Papago Council at its regular meeting on January 6, 1956,[131] when it accepted the recommendation of the Schuk Toak District Council. Meinel and Abt had already attempted to climb Kitt Peak on December 20, but had to turn back from a point 600 feet below the summit.[132] At the suggestion of Gilmore, Meinel and Abt met with Mark Manuel and the Schuk Toak District Council on January 22. Access to Kitt Peak was discussed, and arrangements were made for pack horses and Papago guides to assist in the next attempt to reach the top. Eight horses and five persons (Meinel, Thompson, Ramon Lopez, Al Martinez, and a Tucson newspaper photographer – Cliff Abbott) started the successful ascent on March 14. During the descent on the following day, Meinel's horse slipped and fell. He escaped serious injury, but his left hand and fractured wrist were in a cast while broken bones mended. He liked what he saw on the top of the mountain, but wrote to Hogg:[133] "The difficulty of ascent is so great that it is utterly impracticable to carry the necessary equipment to the top with horses." A caterpiller tractor trail solved this problem,[134] and the site survey was under way.

6   The site survey. Planning for a large solar telescope.

The use of tall towers ("perhaps six 100-foot steel towers, probably modifications of already standardized oil-drilling towers") to carry automatic seeing recording equipment was initially proposed by McMath in a letter to Bowen written two and a half months before the first meeting of the NAO Panel.[1] This letter also acknowledged receipt of a long report giving Bowen's ideas about equipment (including solar instruments), location, and financial and administrative arrangements for the proposed observatory.[2] Bowen's report was written in response to a memo McMath had sent to the members of the Panel on May 12, 1954, shortly after their provisional appointment by NSF.[3]

McMath sent copies of Bowen's report to Stromgren, Struve, and Whitford, and his cover memo[4] concluded: "The writer supposes that it will be necessary to investigate seeing at many locations from towers of, say, 100 feet high. This in itself is quite an undertaking." The memo also mentioned the long-time interest of McMath and Bowen in finding a site with "a total of 30 hours of one-second of arc solar seeing per year," because this would justify construction of a 60-inch aperture F/30 solar telescope. He had also mentioned this more than a year earlier in a letter to Struve.[5]

McMath wrote a follow-up letter to Bowen a few days later,[6] and sent a copy of this letter to Struve. He restated his suggestion made in the earlier letter that NSF should contract with a big organization, such as Northrup Aviation, to investigate seeing at several selected sites. The contractor would provide the towers, design and construct the test equipment, and possibly reduce the observations for inclusion in the final report. "The contract would, of course, be supervised by our commmittee, or perhaps even a narrower interpretation to the effect that it would be supervised by you and me."[7] This idea was also the subject of a letter to Struve,[8] written in reply to Struve's comments[9] about McMath's memorandum of August 18 to the NAO Committee and his letter of August 21 to Bowen.

McMath presented his proposed way to conduct the site survey at the first meeting of the NAO Panel.[10] It was the consensus of the Panel that McMath's plan should be followed. The Panel also recommended the appointment of Aden B. Meinel as

Executive Secretary "to act under the instructions and orders of the Panel for the purpose of making a broad study of possible sites and seeing."

The plan to employ a west coast aerospace firm to provide the towers and conduct the site survey seems to have been abandoned quietly between the first and second Panel meetings, and it was agreed to reduce the height of the tall towers to 60 feet.[11] Meinel had designed modular units 5 feet high that could be assembled into 10-foot or 60-foot towers, and he expressed a desire to conduct his duties and make the preliminary engineering studies and tests from headquarters at the Yerkes Observatory during the summer of 1955. This arrangement was handled administratively through a University of Michigan subcontract with the University of Chicago. NSF supported the summer project by providing a $29,000 supplement to the Michigan grant in order to fund the construction and testing of a 30-foot experimental tower and associated instrumentation.[12] Contracts for the construction of the towers, and the telescopes and electronics to be installed in the towers, were let in late 1955, following the transfer of the project headquarters to Phoenix.[13]

Abt, who was on leave of absence from the Yerkes Observatory, had been voted a position there as assistant professor in June 1955.[14] The site reconnaissance was nearing its end, and he felt that his academic advancement might be jeopardized if he stayed on to participate in the seeing tests after the move to Phoenix. Meinel and McMath accepted Abt's decision,[15] and he returned to his position at Yerkes at the end of January 1956. He came back to the project for two months in April and May, and had no further connection with the NAO until he joined the Kitt Peak National Observatory staff in September 1, 1959.[16]

The aerial reconnaissance had narrowed the potential sites to 150. Examination from the ground had eliminated all but a dozen of these, and five were selected in mid-1956 for initial testing.[17] Four of the sites were in Arizona: Kitt Peak in southern Arizona, the Sierra Ancha region in central Arizona, Chevelon Butte in northern Arizona, and the Hualapai region in northwestern Arizona. The fifth site, Junipero Serra Peak, in California was thought of as a potential site for the large solar telescope.[18]

Abt made a trip to California in mid-January, shortly before he left the project, to look at Junipero Serra Peak.[19] His guide was Leon Salanave, who had visited this mountain in May 1955 in preparation for observing Mars during the opposition in the Summer of 1956. Salanave's reports of solar day-time seeing observations made during the Mars opposition[20] led to his employment by Meinel later that year.

Abt also made two trips to the Sierra Ancha mountains in late April and early May, and returned on May 17 to put in a north–south line at a site on McFadden Peak. A concrete foundation for a 60-foot tower was installed at this site by a local contractor, and Abt checked its orientation on May 29, a few days before his return to the Yerkes Observatory.[21] Unfortunately, it turned out that the foundation had been put on private land without permission of the owners, and a lawsuit was threatened. It was settled out of court when the owners agreed to accept payment of $200 for a two-year lease of the site.[22]

In his August 14 letter Meinel requested McMath's permission "to settle this matter and then abandon this site." McMath replied: "You not only have my permission to abandon the Sierra Ancha site, but I will go further and direct you to abandon this."[23] McMath sent the legal papers to Goldberg for processing, and Goldberg's secretary

E.F. Carpenter, F.K. Edmondson, J.C. Duncan, and A.B. Meinel on Kitt
Peak, 14 June 1957. Automatic Polaris seeing monitors were in the 10-ft and
60-ft towers, and meteorological equipment was mounted on the mast. The
caterpillar tractor was the only vehicle that could negotiate the steep primitive trail up the mountain. NOAO Photo Archives.

attached a short memo to clarify what she perceived to be McMath's confusion about
whom the lawyer was representing.[24]

The short-lived Sierra Ancha site was replaced by Summit Mountain and it was not
even mentioned as a possible site in Meinel's final report.[25] In July 60-foot towers were
erected on Chevalon Butte and in August on Kitt Peak, followed by the Hualapai site
and Summit Mountain, south of Williams, Arizona. The Kitt Peak and Hualapai 60-
foot towers were the only ones equipped with automatic Polaris monitors, and 10-foot
towers were erected only at these two sites. Summit Mountain and Chevalon Butte
were eliminated after a short period of testing. Mormon Mountain, south of Flagstaff,

was selected to replace them, but it was never tested, owing to heavy snow that made winter access impossible. Some testing was done on Slate Mountain, north of Flagstaff, but it was never seriously considered owing to its small useable area at the top.

Interest in testing Junipero Serra Peak developed in early 1956, following Abt's reconnaissance in January. Meinel's report to the NAO Panel in February[26] was soon followed by the discovery that the Air Force was interested in using Junipero Serra Peak as a site for a radar station. McMath wrote to Dr. C.C. Furnas, Assistant Secretary of Defense for Research and Development, asking for reconsideration because of the NAO interest. The reply from the Air Force said this was being studied and that Dr. Furnas would be notified of the result.[27]

Meinel began to have reservations about testing Junipero Serra Peak after he investigated the problem of access during the August 1956 meeting of the American Astronomical Society in Berkeley.[28] A fifth tower would be needed to test this site, because the first four had been erected at the Arizona sites. He favored "a transfer of this site to the solar project at an early date."

Leon Salanave seemed to be the logical choice to conduct the testing on Junipero Serra Peak, and he was hired as a research associate on the solar project in late October.[29] His duties were to include testing seeing at sites in Hawaii and other Pacific islands in addition to Junipero Serra Peak. He began by spending a week in January 1957 at the Phoenix headquarters[30] working with Meinel, learning about the site testing apparatus and helping in the preparations to set up the first Polaris monitor on Kitt Peak.

McMath invited Salanave to describe his 1956 seeing observations on Junipero Serra Peak at the NAO Panel meeting in Phoenix on February 25–26, 1957. Salanave and Meinel also described the plans and the proposed portable test equipment to be used for the preliminary survey in Hawaii.[31] Salanave visited sites on Mount Haleakala, Mauna Loa, and Canton Island during the period March 29 – April 23.[32] He ruled out Mauna Kea as "practically inaccessible," and said that "from a general topographical and meteorological point of view [it] would not seem significantly different from Mauna Loa." His preference was for Haleakala.

W.W. (Bill) Baustian had come from the Lick Observtory to join the staff in the Phoenix office as Chief Engineer on September 1, 1956.[33] He went up the west slope of Junipero Serra Peak with Salanave on January 3–4, 1957 to inspect possible routes for the all-year access road.[34] Keith Pierce, Thompson and Salanave got to the top of Junipero Serra Peak on foot in March,[35] and this was followed by another visit by Thompson in July.[36]

Meinel and Pierce visited Junipero Serra Peak at the end of August.[37] They also visited other Arizona sites on the way, and Meinel recalled in later years standing on the Hualapai Mountain site "looking west to a perfectly clear sky, starting about 30 miles west of us, stretching as far as we could see at a time when all the Arizona sites were clouded in by our summer weather that we have here."[38] He also remembered the visit to Junipero Serra Peak as "a wild jeep ride up over a mountain." During the night they looked at Mars, and Meinel recalled: "I had never seen such a magnificant sight in my life." It was potentially a good site, but with some severe logistics problems that had to be solved before testing could begin. The first hint that there was a problem with the towers was in Meinel's report on site operational problems at the fifth meeting

Tractor and U-Haul trailer. F.K. Edmondson and A.B. Meinel on tractor, J.C. Duncan, E.F. Carpenter and Mrs. Carpenter on trailer, 14 June, 1957. Edmondson photo.

of the NAO Panel in Phoenix on February 25, 1957.[39] The serious nature of the problem was described to McMath a month later: "The wind tremor problem, mentioned at the panel meeting, is more serious than we had expected. A wider base to the Kitt tower has only reduced the amplitude by 50 percent. The remaining effect appears to be due to "air coupling" between the instrument tower frame-work and the outer covered towers."[40] McMath immediately wrote to the panel: "I hate to think that the 60-foot seeing towers have turned out badly."[41] McMath was also worried about the effect of wind on seeing, and suggested that the meteorological records at Palomar, Mount Wilson, and Lick could provide information about this. He sent copies of Meinel's letter and his memo to C. D. Shane, Director of the Lick Observatory and a consultant to the Panel. In a separate letter McMath suggested to Bowen that it might be useful for Meinel to come to see him.[42] Meinel phoned Bowen on May 14 to discuss the availability of wind velocity records at Mount Wilson and Palomar, and he told Bowen that he would look up the information for Lick and McDonald.[43] They arranged to meet for further discussion at the meeting of the Astronomical Society of the Pacific in Flagstaff on June 17–19. Whitford[44] and Struve[45] soon expressed their concerns to McMath, and there was active correspondence during the next few weeks about what should be done to cope with the problem.

Meinel decided the solution was to add mass to the instrument tower to reduce the amplitude of the sway.[46] He described the structure of the tower: It "has three, not two concentric structures. The intermediate tower is covered and open at both ends, but free-standing from the outer tower 'sleeve' which is heavily guyed. The instrument tower is free-standing inside the intermediate tower. There are no physical connections between any of the towers and each has a separate isolated foundation. I must confess, I felt that such precautions would meet the task, but deflections in seconds of arc are far beyond the normal range of expected tolerances in tower structures." He proposed "to add mass by concrete slabs attached to the sides to raise the total mass of the top of the instrument tower by 4,000 lbs. The desired reduction in sway should meet our requirements." This would double the cost of the tower if the work was done by project personnel; it would be a factor of four if the work were done by a contractor. A few days later he told Pierce[47] that 2 × 6 steel bars would be used instead of concrete slabs. Other causes and cures were suggested,[48] and it was decided to make the first experimental changes on the Hualapai tower. Meinel wrote to McMath:[49]

> The modifications to the Hualapai tower will be completed next week so that we will know if the wind problem is solvable. The mass and damping addition was completed this week and looks and feels very satisfying, but the telescope cannot be added until all the work is done. We feel that the decision to experiment on Hualapai rather than Kitt has been justified. The field crews (other than us observers) absolutely abhor Kitt because of access difficulties over the present trail. This reluctance is not timidity, but simply due to the realization that a single error of judgment with heavy equipment on 78 percent grades will mean a fatality.

The initial failure of the towers increased the importance of using other methods to evaluate the quality of the seeing. Meinel wrote to McMath: "The equatorial telescopes (one 6 inch and two 16 inchers) that were originally to supplement the tower telescopes now appear to rank as the prime survey instruments."[50] Several students were hired by Meinel for three months of summer work in 1957, $1.00 per hour for undergraduates and $2.00 per hour for graduate students.[51] Two students from Tucson assisted Claude Knuckles on Kitt Peak,[52] and J.C. Golson was assisted by students at the Hualapai site.[53] Edward C. Olson, a graduate student from Indiana University, participated in observing at the Kitt Peak and Hualapai sites, and also made short visits to the Chevelon Butte and Summit Mountain sites to test seeing with a 6-inch catadioptric refractor. He also worked on data reduction methods at the Phoenix headquarters.[54]

Meinel reviewed the site evaluation program in a letter to McMath when the first of the 16-inch telescopes was ready for mountain use.[55] He said the tower problem seemed to have been solved, at least for winds up to 15 miles per hour. He wrote: "The chief result of our survey to date is that Kitt Peak is far superior to our other Arizona sites. . . . Chevelon is the poorest, with Summit a close second. Hualapai is occasionally good, but is the windiest site. Both Hualapai and Summit are within 15 miles of the nearest town. This fact coupled with the inferior seeing makes me question whether we could accept either as the final site." He also pointed out that Salanave's reports were the only information to date about Junipero Serra Peak. He asked for permission to abandon the Chevelon and Summit sites, and said he was inclined to recommend dropping the Hualapai site but would "carry it on a lower priority than Kitt Peak and Junipero Serra Peak as a comparison." This letter brought strong protests from Whitford[56] and Bowen.[57] Whitford wrote: "I am a little hesitant about cutting back the testing program so severely after only a few months and scarcely anything

Night Assistant J.C. Golson observing with one of the 16-inch portable site survey telescopes. NOAO Photo Archives.

but subjective tests made at or very near ground level. We did start with the proposition that this site survey was going to be a thorough one, and that the final choice was going to be based on measurements and not opinions." Bowen wrote: "I feel that the site selection problem may well involve the most important single decision to be made by the Panel. . . . However abandoning practically all of the sites except Kitt Peak is essentially selecting Kitt Peak as the site of the Observatory." Stromgren alone voted to abandon the sites.[58] McMath wrote to the Panel[59] after talking to Meinel on August 6: "With reference to his desire practically to select Kitt Peak now, I told him I was entirely unwilling to consider [this]. . . . Our objectives were clearly stated two or three years ago and I do not think that Meinel has in any way reached them. I did ask Meinel to prepare a technical report covering all of the sites. I asked him to include

Tractor and trailer on steep part of road. Edmondson photo.

every piece of information he had on any site. I also asked him to distribute this technical report to each member of the Panel."

Harold Johnson, a staff member of the Lowell Observatory, was very critical about the way the site survey was being conducted:[60]

> I recall that, at the beginning of the NAO project, there was a lot of ballyhoo to the effect that for the first time a really thorough site survey would be made. I, for one, do not regard a survey in which seeing observations are made only on one site, except for very occasional spot checks on others, as fulfilling this high promise. ... I have grave doubts as to the scientific objectivity of the site survey that is now being conducted. It seems obvious to me (and to Hoag and others) that Kitt Peak has already been chosen as the site for the NAO even before the site survey has gotten well under way.

Meinel mailed the requested technical report[61] to McMath and the members of the panel on August 14, 1957. The cover letter[62] expressed pleasure with the performance of the 16-inch telescopes, and stated: "I personally have more confidence in the visual and photographic records being taken with the 16-inchers than in the automatic Polaris telescopes." The letter included the two recommendations made in the report and added a third: (1) principal operations will be done on Kitt Peak and Junipero Serra Peak; (2) secondary operations will be done at the Hualapai site and on Cone Peak (west of Junipero Serra Peak); (3) random checks of seeing will be done on Summit Mountain, Chevelon Butte, and at the Lowell Observatory as time permits. The report

Tractor trail near top of Kitt Peak. Edmondson photo.

raised once again the question of whether the Observatory should be in the California or the Arizona weather cycle.

The report was anecdotal without any hard figures, and it drew immediate unfavorable responses from McMath, Bowen and Whitford.[63] Meinel's reply[64] included comments about the five sites, and said: "The 'number' report requested by you and Dr. Bowen will be assembled, but in some aspects it is laborious to prepare in as great a detail as can be gained from direct inspection of such things as the microthermal records." He felt it would be a waste of $8000 to bring the Summit Mountain and Chevelon Butte towers to full operation. He also said it would take three weeks to get bids and eight weeks more for road construction to Junipero Serra Peak. The process should be started at once to insure completion before the rainy season begins.

McMath perceived that Meinel "might like to recommend abandoning the towers as an 'unsuccessful experiment'," and wrote that he was sure the Panel would "insist on a good series of visual observations made by competent observers" to take the place of the tower data.[65] Meinel's reply[66] included: "The Kitt tower is now performing reasonably well, and the Hualapai tower has started operation. The chief complaint that I have is that maintenance of automatic equipment is almost as costly as using observers." The letter continues: "In your absence I discussed the question of development of Junipero Serra Peak with Dr. Bowen today. While he said that he could not pretend to give an opinion for anyone other than himself, he said he personally would be unwilling to recommend extensive development on JSP at present. As a consequence we will not attempt to take the 16-inch with us next week since we could not get it safely down the existing trail due to the high center of gravity of the trailer."

Whitford summarized his view of the state of the site survey in a three-page letter to McMath.[67] His comments include: "We have adequate astronomical data for something less than half a year on only one site, Kitt Peak," a somewhat negative comment about Junipero Serra Peak, a favorable comment about Kitt Peak seeing results to date, and "The towers, which represent a considerable investment in money and man hours, may have been a mistake. At any rate, their height was made as great as it was with the solar site in mind. Night time data could have been gotten with a shorter and much more stable structure."

A very large solar telescope had been McMath's "dream telescope" for many years, and he had discussed it frequently with Walter S. Adams and Bowen.[68] The possibility of including a solar telescope in the plans for the National Astronomical Observatory was certainly a factor in McMath's acceptance of the chairmanship of the NAO Panel.[69] Indeed, McMath's first formal communication to the Panel[70] included "a very large and modern solar telescope" as part of the equipment for the NAO. He said he had included it "at the suggestion of Dr. Struve," and added: "The necessary quartz for the big mirrors is available."

Three experimental quartz mirror blanks (a 68-inch, a 65-inch, and a 63-inch) had been made by General Electric (GE) as part of the 200-inch telescope project. They were stored at the GE factory in East Lynn, Massachusetts until early 1950. GE needed the storage space, and Bowen telephoned McMath and asked if he could use them. McMath replied: "Not now, but sometime in the future." Bowen arranged with GE to have the three blanks given to the University of Michigan, and McMath sent a truck to East Lynn to pick them up.[71] The University of Michigan Regents transferred the title to these quartz blanks to AURA in 1960 for $575, the cost of the truck trip.

The solar telescope was the major item on the agenda of the third meeting of the NAO Panel[72] held during the Dublin meeting of the International Astronomical Union. Struve opened the discussion and said: "A big solar telescope should now have a very important priority in our planning." The Panel concurred. McMath was instructed to write to Meinel that an investigation of daytime solar seeing was of equal importance with night-time stellar seeing. A change of plans to make it possible for the big solar telescope to come along early in the program was also discussed. Meinel misunderstood this and thought the 80-inch telescope would be side-tracked for the solar telescope.[73] Struve calmed his fears by explaining what the Panel had decided about starting early planning for the solar telescope.

McMath sent three preliminary drawings of "the great solar telescope" to Hogg[74] following discussion of the project at the MPE Divisional Committee meeting on March 9, 1956. This letter also mentions sending Pierce to the Pic du Midi for a month to evaluate solar seeing,[75] and the desirability of hiring a recent PhD to investigate island and atoll seeing. This could all be done within the $30,000 that had been suggested by Seeger and Hogg in their report to the MPE Committee. Hogg suggested[76] that the $30,000 be added to the planned proposal for $600,000 for the NAO project. "The solar telescope project will be considered separately here, but it should simplify administrative procedure at Michigan to have these proposals come in together." The proposal was modified to include this suggestion,[77] but the National Science Board was forced to reduce the amount of the grant to $515,000 to meet a restriction imposed in the House version of the budget bill.[78]

Hogg sent the proposal to the Advisory Panel for Astronomy for review of the solar telescope part. The ratings were: 1 excellent, 3 very good, 1 good, and 4 not rated.[79] The $30,000 for the solar telescope study was reinstated and added to the $515,000 previously approved by the NSB.[80] The NSF press release[81] made specific mention that the grant "also includes $30,000 for a study leading to the erection of a solar telescope." The $30,000 was used to employ Salanave to work on solar seeing testing[82] and paid for Pierce's visit to several European observatories to check on solar seeing.[83]

The joint meeting of the three NSF Astronomy Panels in Ann Arbor on July 23, 1956[84] gave members of the NAO Panel an opportunity to discuss the next step in solar telescope funding with Seeger. McMath informed the Panel that he had followed this with a request to NSF for $180,000 for FY 1958.[85] This sum was "divided into three parts, viz. $60,000 each for preliminary engineering, metal-based mirror development, and site and seeing surveys." Nothing was heard from Seeger prior to the October 25–26 meeting of the MPE Divisional Committee, when the question was again raised by McMath. Edmondson telephoned Goldberg shortly after the meeting and suggested that the Panel should file a proposal to be considered by the Advisory Panel for Astronomy. The proposal was prepared by Goldberg and McMath,[86] but was held until McMath could poll the NAO Panel. McMath believed this request from NSF meant the money would be taken from the grants program, so he asked the Panel to vote on whether to send the proposal ("Yes") or drop the matter for the time being ("No").[87] Struve voted "No," Stromgren voted "Yes", Bowen voted "Yes, if impractical to obtain as observatory budget item", and Whitford voted "It all depends."[88] McMath interpreted the vote to mean send the proposal, but don't accept money from the grants program.[89]

The Executive Committee of the MPE Divisional Committee met in McMath's Detroit office on December 1.[90] There was considerable discussion of the big solar telescope and its timing, and McMath "stated flatly that the Panel would not accept money for engineering from the regular grants program."[91] He wrote to Seeger after the meeting[92] "to confirm my understanding of our discussion of the big solar telescope and its place in your time schedule." He said the proposal for $60,000 would be mailed in a few days. A copy of this letter was sent with McMath's memorandum to the Panel.[93]

The $60,000 grant[94] had a new grant number, G-3443, but retained the title of the previous grant, "Studies Leading to the Establishment of an American Astronomical Observatory." A personal copy of the grant letter was sent to McMath, and he responded by pointing out that a different title had been used in the proposal.[95] He added: "So I assume that the Foundation is merely backing up the whole project and that my understanding, i.e. that the most recent grant is to be devoted to the solar aspects, is correct."

The NAO Panel had requested $180,000 for the solar telescope project following the July 23, 1956 meeting in Ann Arbor.[96] Of this, $30,000 had been added to G-2622[97] and $60,000 more had been provided in G-3443.[98] McMath wanted the rest of the money without further delay, and he wrote to Seeger on July 5, 1957:[99] "May I take this occasion to remind you that there is still $90,000 due on the proposal given to you last year for the solar project." He added: "I suppose this letter should have been addressed to Dr. Eckhardt." There was a new chain of command at this critical time. A.E. Eckhardt, age 69 and a retired geophysicist who had been Director of Research for the Gulf Research and Development Company, was appointed Assistant Director for MPE for a one-year term starting June 1, 1957.[100] Seeger reverted to Deputy Assistant Director after having served as Acting Assistant Director for four and a half years following Klopsteg's elevation to Associate Director of NSF on December 1, 1952.

The seventh meeting of the Advisory Panel for Astronomy was held at the University of Illinois on August 17–18, 1957, in connection with a meeting of the American Astronomical Society. A discussion of eight research facilities, of which five were being considered for support by NSF, led to two resolutions assigning priorities in the following order: the radio astronomy observatory and the optical observatory (tied), the automatic measuring machine for the Lick Observatory proper motion program, and the solar observatory.[101] The consensus of the Panel was that "the site survey for the Solar Observatory is urgent and should be supported out of FY 58 facility funds if at all possible."

By the time of this meeting Goldberg had been warned in a letter from Eckhardt[102] that "Congress has instructed the Foundation not to start a facility unless enough money has been appropriated to finish it. . . . In effect Congress has said that when we first ask for money for a facility, we must ask for enough to complete it. This precludes going to Congress year by year with supplementary estimates in the case of facilities." Eckhardt dropped the other shoe when he wrote to McMath three weeks later.[103] He referred to what he had said in the letter to Goldberg and wrote: "This means that, without prejudice, we have decided to postpone further consideration of the solar observatory. We have not included any sum for it in our FY 59 estimates, and we have no funds for it in the current fiscal year. . . . We assume that with the suspension of immediate plans for the solar observatory the site testing can also be postponed, and the project can be frozen at its present state without detriment to its completion, if and when it is reactivated." McMath's reply[104] said: "I am interpreting this to mean that I am at liberty to spend the remainder of the two solar grants in order to get the most information from those committed funds. With luck we should be able to do a fair amount of testing on the sites which are now under active investigation for the optical observatory. If this interpretation is not correct, please advise me promptly." Eckhardt's reply to McMath[105] suggested alternate uses, but said that if McMath wanted to use the money for further solar site testing "I would not object to your doing so." McMath thanked him[106] "for an entirely adequate answer to my letter of September 12."

Salanave learned the bad news from Pierce.[107] He was in the process of selling his home in San Francisco and moving to King City, the town nearest Junipero Serra Peak.[108] A month later Pierce wrote that McMath had decided to hold the remaining solar funds in reserve and vigorously pursue the survey for the stellar site selection.[109]

This would put Salanave 100 percent under Meinel's administration, and it was planned for him to have the responsibility for setting up the Mormon Mountain Station. Pierce said: "I would suggest that you get in touch with Aden as quickly as possible and I would very much like to have you bring the jeep, the trailer and your family to Flagstaff and set-up housekeeping in the trailer, while you find an apartment to stay in Flagstaff." Three weeks after this Salanave wrote to Meinel and Pierce[110] that the problem with his San Francisco house was nearly resolved and "myself and family stand ready to shift to Flagstaff whenever you say."

Salanave moved to Flagstaff after AURA was incorporated and had taken over from the NAO Panel.[111] He did some testing on Slate Mountain[112] using a borrowed 16-inch reflector and a 6.5-inch refractor, but his status became uncertain[113] after Kitt Peak was chosen by the AURA Scientific Committee on March 1, 1958. He moved to Tucson and continued with AURA until September 1958, when he obtained a position at the University of Arizona Applied Research Laboratory.[114]

The solar telescope project was down but not out. The first Soviet Sputnik satellite was launched on October 4, 1957, three and a half weeks before the incorporation of AURA. The National Science Foundation was given a supplementary appropriation of close to $10 million dollars in the panic that followed, and the prospect for funding the solar telescope looked better.

7 Discussion of the purpose and goals of the National Astronomical Observatory by the Panel. Communication with the astronomical community and the public. The evolution of plans for the management and operation of the NAO.

The charge to the NAO Panel[1] was that it should "advise the NSF of the general astronomical needs that can be met by a specific plan for a national astronomical observatory." The Panel discussed the need for a national observatory at its first meeting, November 4–5, 1954,[2] and discussed the research justification and educational justification of the observatory at the second meeting, March 29–31, 1955.[3]

Between these meetings Struve presented a thoughtful memorandum about the needs of astronomy at the January 15, 1955 meeting of the NSF Advisory Panel for Astronomy.[4] This was revised to incorporate suggestions from members of the Advisory Panel, and it was then published in the Publications of the Astronomical Society of the Pacific.[5] The memorandum made a strong case that there is a need for planning in astronomy, citing as an example Henry Norris Russell's 1919 article[6] about the aims of astronomy and specific problems that required solution. Struve concluded with ten questions addressed to the Panel, and this led to the adoption of a resolution: "The NSF Advisory Panel for Astronomy favors a study of the need for planning in Astronomy, and we advocate that such a study be made." An ad hoc committee on "The Needs of Astronomy" was established with Panel member Fred Whipple as chairman. Thirty-nine papers were solicited and published under the title "New Horizons in Astronomy."[7] Reading these papers thirty-five years later, one is struck by the absence of any serious discussion about the needs of astronomy. They are excellent reviews about the state of astronomy in 1956, but it was not until 1964 that the general needs of astronomy were addressed in the Whitford Report.[8]

After the Flagstaff Conference there was a surprising lack of communication with the astronomical community by the NSF and the NAO Panel. For example, Irwin wrote to McMath four months after the NAO Panel had been appointed:[9] "I have just learned somewhat incidentally that you are Chairman of an NSF Committee in connection with a cooperative observatory." He offered to send McMath copies of letters he had received from astronomers before and after the Flagstaff Conference, and to be of any other assistance McMath might wish to have. McMath's reply[10] said: "I would be delighted to have any information which you can send me concerning the Flagstaff

Observatory." He thanked Irwin for his offer of assistance, and very briefly described the charge to the NAO Panel. Irwin sent the letters and copies of the Flagstaff Conference papers to McMath,[11] and said NSF had provided funds to publish summaries of the papers in the *Astronomical Journal*.[12]

Communication with the astronomical community finally took place on November 10, 1955 at a meeting of the American Astronomical Society in Troy, New York.[13] Whitford, Chairman of the NSF Advisory Panel for Astronomy and also a member of the NAO Panel, gave an authoritative after-dinner talk that summarized the events leading to the establishment of the NAO Panel, the purpose and goals of the NAO, and the present activities of the Panel. Seeger deserves credit for taking the initiative that led to this talk being given.[14] It was well received by those who heard it, and McMath urged that it be published in order to reach a larger audience.[15]

Not all was sweetness and light behind the scenes at the Troy meeting.[16] During a working dinner on Friday evening, the members of the Advisory Panel for Astronomy concluded that the Needs of Astronomy Committee had served its purpose with the publication of *New Horizons* and that an actual work project in the needs field should be set up. The Panel felt that Struve was the best person to head this, and that he should be asked to submit a proposal for a grant to the University of California to pay the costs of such a study.

The dinner meeting of the Panel was scheduled to merge into a joint meeting with the Needs of Astronomy Committee at 8 o'clock, with Whipple (Chairman of the Needs Committee) and Menzel (representing the AAS) present. No other members of the Needs Committee were in Troy. To the astonishment of Hogg and the Astronomy Panel, Whipple and Menzel had invited the entire Council of the AAS to attend this meeting.[17] Furthermore, the members of the Council were asked to vote along with members of the Panel. Whipple resisted the proposal by Whitford and Clemence that he step down from his duties as Chairman of the Needs Committee, and said he wanted to have another meeting of the Committee. Hogg doubted that NSF funds for direct support of Panel travel could be made available, so Whipple asked the entire group (eight AAS Councilors and three Astronomy Panel members) to vote whether Smithsonian should ask NSF for a grant to cover travel and publication funds. Whipple declared the motion had passed unanimously, and the joint meeting ended. Menzel and the AAS Council continued their business after Hogg and the NSF Panel members left. Whitford mentioned this episode in a letter to McMath two months later.[18] Whitford also told McMath: "Dr. Seeger, of NSF, is unhappy about the lack of focus in the present year's effort to get on with formulating an all-inclusive appraisal of the needs of astronomy in the next decade."

The NSF Public Information Office became interested in the news value of the NAO project in early 1956 after learning about the plans to erect the 60-foot tower on Kitt Peak.[19] Clyde C. Hall, NSF Public Information Officer, felt that the NAO story would "convince the folks in Congress that we are spending public monies wisely and well; something they can see and touch; something that will help the Director measureably when he appears on the Hill." He wanted the National Science Foundation to "receive as much credit as possible for what is essentially its own undertaking." He asked Lee Anna Embrey, Deputy Public Information Officer, "to move into the story immediately." Ten days later, she sent Meinel a draft of a proposed press release and said: "Mr. Hall would like to have me on the spot in Arizona for about two weeks [when the tower is erected] to take care of press inquiries on behalf of the Foundation."[20] Meinel's reply[21] said there would be "a reconnaisance to lay exact plans about March 15, and the date should be known by then." He felt the last week in May would be realistic.

Embrey soon became concerned that several news stories had already appeared, including one in the *Durham Morning Herald*. She felt the NSF press release should be issued in the near future, and sent a revised version to Hall.[22] She also wrote to Meinel[23] that they were concerned at NSF about "how much impact a national release is going to have when the major elements of the story are trickling out via the Tucson and Phoenix papers." She also said: "We have observed that if one or two reporters or journals obtain what amounts to an 'exclusive' on an important story, it may create some resentment among the other media, or in any case discourage their interest in playing it up."[24] She told him that some members of the House Subcommittee on Appropriations "are apparently displeased by widespread publicity attaching to projects for which they have not yet appropriated funds." Meinel's reply[25] was defensive, and concluded: "I really do not know how to cope with your problems of orderly information release. The sloppy publicity that has appeared around here has not been too serious locally. . . . Thus far I did not give much worry since I thought the distribution was strictly localized."

Hogg had visited the Phoenix NAO office on March 1, and she wrote to Meinel on March 12 to resolve her confusion about the date when the seeing tower would be carried up Kitt Peak.[26] Meinel replied[27] that their first ascent of Kitt Peak on March 14–15, 1956 showed that "it is utterly impracticable to carry the necessary equipment to the top with horses." Erection of the tower would have to wait until the hazardous stretches of the trail could be improved. He asked Hogg to "give Miss Embrey the sad news."

Hall felt it was necessary to go to the top at this point, and he sent McMath a copy of Embrey's proposed press release with a cover letter[28] beginning with a comment about the publicity that "is finding its way out of Arizona and into the press of the Nation on a kind of hit-or-miss basis." He told McMath he would be sending Embrey to Arizona "about the time Dr. Meinel plans to move the tower to the mountain top in order that there may be on hand a well qualified representative of the Foundation to help press and magazine representatives." McMath told the NAO Panel:[29] "Dr. Seeger has insisted that this [press release] be cleared by me." His reply to Hall[30] suggested adding Struve's title at the University of California plus a description of the University of Michigan's role in supporting the NAO project.

A few days later Hall informed Waterman[31] about the delay in installing the Kitt Peak tower and said: "There seems little point in dispatching anyone to the site at this juncture." The Kitt Peak tower was finally erected in mid-August,[32] but no one from NSF was there.[33] "The experience was most harrowing" was Meinel's description of moving the tower parts up the mountain. He said two photographers for *LIFE* magazine "(arranged with Miss Embrey) quickly lost interest and returned to Tucson."

Bowen sent McMath a story from the April 11, 1956 *Los Angeles Times*[34] with the comment: "Apparently this started on the basis of stories from Arizona. I did my best to 'scotch' it with Barton [the reporter] but as you see I did not succeed very well. At least the story does not say that the NSF is about to start a 220-inch telescope which was in the Arizona dispatches." Stories like this continued to be of serious concern to Hall and Embrey, largely because of the potential adverse effect on members of the House Appropriations Committee. Embrey wrote a long letter to McMath[35] on August 1, 1956 stating her complaints in detail. She said that with the new grant of $545,000 being imminent "it would be a good time to take stock of the situation public relation-wise and place the whole matter on a firmer basis of understanding." McMath's reply[36] was sympathetic, and he suggested that Embrey should "arrange with Dr. Meinel a trip to Phoenix so that you may become fully acquainted with our operations there." Meinel wrote to McMath:[37] "I have your letter of August 8 and I am always glad to cooperate with Miss Embrey. In view of the heavy load that our small staff must cope with, I do not personally have time to devote to the initiation of publicity."

Embrey followed her August 1 letter to McMath by sending him a draft of a press release about the $545,000 grant to the University of Michigan.[38] She pointed out that "this will be the first official release covering this important project" and asked McMath to check the details. For example, she was not clear "whether four seeing towers are now in place or only one." McMath was on vacation and his reply[39] said: "Dr. Mohler telephoned me and read me your news release. There were a few very obvious changes which I understand Dr. Mohler gave to you on the 13th. Otherwise I think the release is very good indeed." The letter concluded: "Should you have any more 'problems' with our Phoenix office, please let me know. Instructions have been made very clear."

Embrey sent the final version of the news release to McMath on August 21.[40] It was scheduled for release on Sunday, August 26, 1956, and listed five sites to be tested, including "the Sierra Ancha Region (7,500 feet) in central Arizona." This site was abandoned on August 18.[41] McMath's reply[42] ended: "I hope that you will be able to visit the Phoenix office in the not-too-distant future."

Embrey made the trip to Arizona the week of September 24, 1956.[43] Tuesday, September 25 was spent in the Phoenix office, and Meinel took her to the Hualapai and Summit Mountain sites on the 26th. They continued on to Flagstaff, where she got to see Mars through the 24-inch telescope at the Lowell Observatory. The following day included visits to the Lowell Observatory, the U.S. Naval Observatory, the Arizona State College Observatory, and the Museum of Northern Arizona. She flew to Tucson on the 28th, where she visited Dr. Edwin Carpenter at the Steward Observatory, and talked about NSF programs to people in other departments at the University of Arizona. Meinel's "boundless and tireless energy" and the "risky nature of much of the work" impressed her, and she felt that part of the public relations difficulty "stems from the fact that Dr. Meinel is too busy and too rushed to take the necessary time to apprise us of newsworthy developments as they occur." Meinel told her that McMath had advised him that NSF had imposed a complete news "blackout." She was at a total loss to understand this, and explained that NSF wanted "to give out such news as was available," stressing the need for him to keep NSF informed about newsworthy developments, but with the limitation that NSF could not mention "budget items before the budget has been approved by the President and submitted to the Congress." Overall, she summarized her meeting with Meinel as having "the most satisfactory results." Her letter to McMath[44] describing her trip said: "I am sure that as a result of our frank discussion of our mutual problems and responsibilities, there should be no further problems relating to the public relations of the project."

Meinel was to be in New York in the middle of October for a meeting of the AUI Advisory Panel, and Embrey arranged for him to stop over in Washington and give a special colloquium at the Foundation and show his motion pictures of placing the towers.[45] This was scheduled at 10:30 a.m. on Thursday, October 18.[46] Edmondson reported to McMath:[47] "I had a good chance to talk with Meinel in New York at the AUI meeting on Tuesday and Wednesday. He was here today and gave a colloquium this morning. It was very well received and performed a valuable educational function for several staff members. There was even a man from the Bureau of the Budget present."

Seeger followed the February 25–26, 1957 meeting of the NAO Panel in Phoenix with a visit to the University of Arizona on March 1, 1957. He made an interesting suggestion to Carpenter during a visit to Kitt Peak the next day.[48] Before Seeger arrived Meinel had written to Embrey:[49] "If he still wants to go up when we get there he will have a ride to remember." Seeger felt that Kitt Peak "appeared to be an excellent site" but worried about "the height to which smog from the East arises." He did not say anything about the ride.

The Optical Society of America invited Meinel to give an after dinner address about the National Astronomical Observatory at its winter meeting on Friday, March 8, 1957. The press release[50] prepared by Eugene H. Kone, Director of Public Relations of the American Institute of Physics, said that Meinel was "Director of the National Astronomical Observatory" and that "the new national observatory will be located in Flagstaff, Arizona." Embrey learned of these serious errors in time to have a correction to the press release issued before Meinel's talk was given.[51] She told Kone: "I must confess that I was nonplused to learn that so important an organization as the American Institute of Physics was completely unaware of our association with the Observatory project." She said she would add him to the NSF mailing list, after finding that the *American Journal of Physics* and *Physics Today* were on the mailing list but that the Institute, as such, was not.

Discussion of the purpose and goals of the NAO soon became intertwined with the discussion of the instrumentation and the management and operation of the observatory. Struve's April 15, 1955 memorandum[52] on "The Educational Value of the National Astronomical Observatory" included the statement: "One of the principal purposes of the National Observatory is to provide additional means for carrying on research by capable astronomers who are connected with eastern and midwestern institutions and who do not now have access to adequate instruments located in a good climate." He also wrote: "The field of solar physics is in some respects even more important than stellar astronomy with large reflectors." Whitford's talk at the November 10, 1955 AAS meeting[53] included: "Present thinking of the Panel leans towards a cooperative association of universities. These universities would be representative of those in the eastern and midwestern sections of the country which would be the most frequent users of the telescopes." This statement was quoted in the Associated Press release about the site survey. He also said: "Use of the facilities should be granted on the basis of merit rather than on the institutional connection of the applicant."

84-inch (2.1 meter) telescope. NOAO Photo Archives.

Soon after the AAS meeting Irwin communicated his concerns to Whitford:[54] (1) "Although the Observatory is for the 'have nots', they are represented on the Committee only by Wisconsin and Michigan." (2) The plans "do not seem adequate as to numbers of telescopes." He said the observatory should be able to serve the needs of at least a dozen universities plus the small resident staff, and he suggested there should be two 36-inch telescopes at the start, plus a 24-inch designed especially for photoelectric work. Nevertheless, he was "highly delighted with the prospect of something as large as an 80-inch and the possibility of something even larger later." He was also delighted that "the telescopic needs of us 'easterners' have been brought to the attention of the NSF."

The NAO Panel at its first meeting[55] proposed a 30-inch and an 80-inch telescope as the first instruments of the observatory, and delayed a recommendatiion for a very large telescope. The possibility of a large solar telescope was also discussed. At its second meeting the NAO Panel accepted Meinel's recommendation "that a 36-inch telescope be considered, instead of a 30-inch, since the engineering had already been

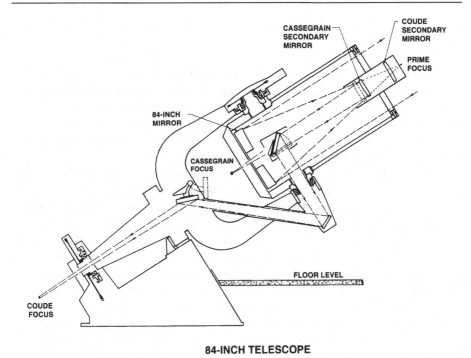

84-INCH TELESCOPE

84-inch (2.1 meter) telescope diagram.

undertaken and is nearing completion for the 36-inch McDonald telescope.[56] Further-more, an unused 36-inch mirror blank is available at Yerkes Observatory and all the tooling for the optical work is at hand." This would save time, and the telescope could be used at a temporary site if it were ready before the final site was chosen. The Panel recommended that funds for construction be included in the FY56 budget, and funds for temporary erection in the FY57 budget. The Panel also recommended that $10,000 be allocated in FY56 for a preliminary design study of the 80-inch telescope. The third meeting of the NAO Panel[57] took place in Dublin, Ireland on August 31, 1955, during the General Assembly of the International Astronomical Union. The Panel concurred with Struve that "a big solar telescope should now have a very important priority in our planning."

The fourth meeting of the NAO Panel[58] took place in Pasadena on February 25–26, 1956, after the Phoenix office had been set up. The Panel "agreed that the order of development of the observatory be: first the 36-inch, then the 80-inch, followed by the big solar telescope." The minutes also report: "It was the opinion of the Panel that the NAO is to be regarded as an observing station in contradistinction to the usual establishment which is called an observatory." This action reversed a decision made at the second meeting,[59] and Meinel objected.[60] McMath's semantic reply[61] defined an "observing station" as any fully equipped and staffed observatory which is not connec-ted with a teaching university or college. . . . No one on the Panel has even considered setting up another McDonald Observatory". Struve agreed[62] and added "I do not believe this observatory should resemble in structure the existing large observatories, but should carry out the original purpose of providing facilities for astronomers in the east and middle west."

The plan to use the engineering design of the McDonald Observatory 36-inch telescope mounting did not work out, owing to problems in coming to satisfactory terms with Joseph Nunn, the engineer who designed it.[63] The arrangement with Nunn was terminated and an engineering design contract was signed with Charles Jones.[64] The project engineer, who was hired by Meinel arrived to join the staff in Phoenix on September 1, 1956, was W.W. (Bill) Baustian, previously at the Lick Observatory.[65]

Meinel changed his mind about the 12-inch photoelectric site testing telescopes that were originally budgeted by the Panel. He felt they were too small and carried through the design of a 16-inch mounting to an advanced stage without consulting the Panel in advance.[66] When he asked McMath to authorize two of these telescopes, McMath was agreeable,[67] but Bowen had reservations.[68] This led McMath to do an "about face" and he wrote to Meinel:[69] "I am entirely unwilling to go contrary to Dr. Bowen's wishes in this matter. So I believe you had better consider this, as well as an earlier letter, instructions to proceed with 12-inch optics." Meinel was very upset and phoned Bowen about this. Having made his point about the need for prior consultation with the Panel, Bowen withdrew his objection to the 16-inch telescopes.[70]

Meinel and Harold Johnson reacted to Bowen's criticism of the reasons for the choice of the size of the 36-inch by suggesting that it be replaced by two 20-inch telescopes.[71] McMath reported all of this to the Panel.[72] Bowen clarified his point of view in a long letter to McMath.[73] "I have been very dubious about the value of these small telescopes as part of the permanent equipment of the observatory. This is based on our experience with the Palomar 20-inch and the Mount Wilson 60-inch. Because of this I have been somewhat concerned about the growth in size of the 36-inch and in particular about the introduction of two 16-inch telescopes as part of the permanent equipment." A second clarifying letter[74] said that "the present plans to construct a 36-inch should be continued" and that he had "no strong feelings as to the exact size (12- or 16-inch) of the photoelectric telescopes."

McMath discussed all of this with Goldberg and then wrote to Bowen:[75] "We were entirely in agreement about the 36-inch. This instrument is in our first plans and has been defended twice by the Foundation before the Bureau of the Budget and Congress. . . . And, both Leo and I are certain that there are many eastern astronomers who would be delighted to get their hands on a good 36-inch instrument in a good location. We are not building this observatory for the west coast observatory staffs."

McMath also wrote to the Panel[76] and said he could "see no valid reason for abandoning the 36-inch project." He also said: "Both Dr. Bowen and I appear to be neutral concerning the 16-inch vs. the 12-inch optics. . . . Here Dr. Meinel is free to proceed as he wishes." Whitford[77] was "glad the 36-inch telescope is going ahead" and he favored the 16-inch size for the small telescopes, and in a follow-up letter said he thought there should be two 16-inch telescopes. Bowen wrote to McMath:[78] "I concur completely in your letter of November 13 to the Panel."

McMath's first memorandum to the Panel[79] included his initial thoughts about ownership and management. "It is the writer's opinion that title should remain in the NSF and that the Observatory should be administered by a 'board of directors' chosen from the chairmen of the larger Departments of Astronomy." He thought "the institutions represented by the Board of Directors might well contribute something toward the operation of the telescope." The first Board could come from Harvard, Wisconsin, Princeton, Indiana, and Michigan. Initial terms of one year, two years, etc., up to five years would be drawn by lot. The Board itself would fill the vacancy each year from an agreed-upon list. He suggested: "Ohio State, Minnesota, Columbia, Yale, University of California, Lick, Virginia, and Case, and doubtless many have been overlooked." NSF ownership and institutional financial support were approved by Bowen[80] and Stromgren.[81] Stromgren also felt: "I believe we will have to consider the question, whether or not an astronomer director should be appointed to carry out the policies of the board."

The minutes of the first meeting of the Panel do not reveal why McMath's suggestion was replaced by the legally unacceptable proposal that the observatory should be "operated by a Board appointed by the National Science Foundation on a rotating basis."[82] The questions of ownership and management were not discussed at the second meeting of the Panel, but two months later Seeger came to Ann Arbor to explain the legal facts of life to McMath and Goldberg.[83] It was agreed "that the University of Michigan, in consultation with the NAO–NSF Panel, invite a group of universities to participate in this activity and that each university would nominate its own representative on the proposed board of directors."[84] The target dates were to have a plan ready by June 30, 1956 and to have the new organization take over from the Panel on June 30, 1957.

McMath reminded the Panel in October[85] that the plan agreed to in June "contemplates setting up or nominating a group of Universities which will take over from the Panel at the end of FY57." In early January[86] he suggested that Menzel be chosen as "chairman of a group for the purpose of organizing something similar to AUI or MURA [Midwest Universities Research Association]. Leo Goldberg should be a member of the committee and a direct representative of the Panel." Whitford[87] and Bowen[88] accepted the proposal that Menzel be made Chairman of the organizing group, but Bowen was "not quite so willing to let him remain as chairman of the operating group."

Seeger's ideas about organization were passed on to the Panel by McMath.[89] A suitable group of universities should incorporate as a non-profit organization. The universities would designate the members of the "Trustee Group," and they would appoint the members of the managerial or "Operating Group." The latter group would appoint the Director of the NAO and other personnel with the consent and approval of the "Trustee Group." McMath suggested the following list of universities for discussion at the coming meeting of the Panel: Harvard, Michigan, Wisconsin, Princeton, Indiana, and also Arizona if the site chosen for the observatory were in that state. He was not sure about the possible interest of Chicago and California (Berkeley).

The Panel held its fourth meeting in Pasadena on February 25–26, 1956.[90] The Panel's lengthy discussion of the organization of the NAO was summarized in Hogg's draft minutes. It was omitted from McMath's "official" minutes, which simply said: "The results of this discussion have been summarized and forwarded to the Program Director for Astronomy under date of March 3, 1956."[91] This summary included: (a) Goldberg was unanimously chosen to organize a group of universities into a legal entity which would be competent to receive grants from the National Science Foundation. (b) The question as to whether the group of universities or the NSF should hold title to the observatory was referred to the "Board of Trustees of the Foundation." [Note: McMath consistently failed to use the correct name of the National Science Board.] (c) The "trustee group" should be composed of administrative officers of the several universities. This group would nominate an operating group, to be composed entirely of astronomers. Their appointment would be subject to approval by the NSF. Initial appointments from one to five years would set up the rotation for five year terms. (d) The opening approach should be made to the following universities: Harvard, Princeton, Yale, Michigan, Wisconsin, Indiana, Arizona, Ohio State, Pennsylvania, and Case. Chicago and California (Berkeley) might be interested. The list of those chosen should be subject to NSF approval. (e) The principle of continuous support of the observatory by the NSF is all-important. The National Science Board had accepted the principle of "permanent support" of large-scale facilities at its June 9, 1955 meeting.[92] This was reaffirmed by the Vice Chairman of the NSB at the March 8–9, 1956 meeting of the MPE Divisional Committee.[93]

Goldberg's reaction to this plan was negative. He wrote a strong letter to McMath on March 30 stating his concerns.[94] McMath wrote to the Panel on April 3 and suggested that the organization plan should be reconsidered.[95] He wrote to the Panel again on April 7 to describe conversations he and Goldberg had with University of Michigan officials. "Most of the comments went right along with Dr. Goldberg's letter. Everybody prefers to have the title remain in the National Science Foundation, and everybody thinks that the trustees should be composed of both administrators and scientists."[96] McMath wrote to Hogg[97] on April 28:

> Speaking for the Panel, I would like to withdraw the proposal contained in my letter of March 3 and substitute the following suggestions to the National Science Foundation.
>
> It is the consensus of the Panel that the National Science Foundation retain title to the observatory when and if it is established. The Panel has requested me to enter this as a very strong recommendation to the National Science Foundation.
>
> The Panel recommends organizing a group of universities into a legal entity which will be competent to receive grants from the National Science Foundation. It is now recommended that this group be composed of university administrative officers and scientists, preferably astronomers although not necessarily. It is the Panel's suggestion that one administrative officer and one scientist be appointed by each university which elects to join the trustee group.
>
> The actual operation of the observatory should be in the hands of an "operating group" composed of astronomers. Many members of the Panel feel that this operating group should be chosen from the scientific group on the board of trustees and that the scientists on the trustee board be rotated from time to time in order to be sure that all interested universities and colleges have representation in both the trustee group and the operating group. Details will depend upon the number of universities or colleges joining in the initial non-profit trustee type of corporation envisaged above.

The operating group should be composed of not more than seven astronomers, and five might well be a more practical number. The members of the operating group should definitely be on a rotating basis, and it is suggested that a maximum term of service on this group be five years.

The above operating group would nominate the director of the proposed observatory and such other personnel as might be necessary or for whom the necessary funds can be found. Actual appointment of the director should be by the board of trustees."

McMath ended the letter with a request for a "letter of intent" from the NSF so that a start could be made in approaching universities.[98]

The NSF reply to McMath's request for a "letter of intent" was discussed in a meeting in Waterman's office on June 7, 1956. Those present were: Waterman, C.E. Sunderlin (Deputy Director), W.J. Hoff (General Counsel), Seeger, Hogg, and the author of this book – Hogg's successor as Program Director for Astronomy. Hogg drafted a letter to McMath incorporating the comments of those present. McMath felt he must have the assurance such a letter would give before Goldberg took the first step to form a group of universities, and wrote that this letter did not meet his needs.[99]

Edmondson wrote several drafts, only to have them come to a stop in the office of Hoff, the General Counsel, who was opposed to such a letter for legal reasons. Edmondson expressed his frustration to Klopsteg, Associate Director, during a hallway conversation. A few days later he was asked to submit another draft. This was approved by Hoff with a minor correction and the letter was signed by Waterman and mailed on October 16.[100]

Goldberg started his efforts to organize a group of universities immediately after McMath received the "letter of intent" from Waterman. He asked Professor Laylin K. James, Professor of Law and an expert in corporate law, to draw up a set of articles of association and by-laws which could be used as a basis for discussion with the universities concerned.[101] He proposed to invite a group of universities to meet in Ann Arbor in March to try to reach an agreement on the articles of association and by-laws. NSF would then be told the universities would "form an association in the event that NSF should wish to turn over to them the operation of the Observatory." Goldberg also wanted the Panel's advice concerning the choice of universities that should initially be asked to participate.

The NAO Panel met in Phoenix February 25–26, 1957, with Seeger and Edmondson also in attendance. The criteria proposed for membership were that the number of staff be at least three, plus distinction; size and scope of graduate program; total financial support by university; managerial and operational competence, and geographical distribution. On the basis of these criteria California, Chicago, Harvard, Indiana, Michigan, Ohio State, Princeton, and Wisconsin, were chosen for invitations to Goldberg's proposed April 1 meeting.[102]

A "Meeting of Invited Universities" was held in Ann Arbor on March 29, 1957.[103] Representatives of seven universities attended, Princeton having declined owing to the large amount of staff time required by their heavy involvement in balloon and space astronomy. As a result of the meeting Goldberg sent an "Agreement for Cooperative Research in Astronomy" to the "Representatives of Universities Participating in the Organization of AURA" on April 12.[104]

Goldberg wrote to Waterman on May 3 to report that all seven universities had agreed to sign as soon as they could get formal approval from their Trustees or Regents.[105] In response to this the National Science Board at its May 20 meeting took the following action:[106]

> AUTHORIZED the Director to commence negotiations with the proposed Association of Universities for Research in Astronomy, Inc. looking toward a proposed contract for the construction and operation of the optical astronomical observatory which could be considered by the Board at its September meeting;
>
> Further, the Board SUGGESTED that the Director in these negotiations explore specifically the possibility of "user participation" in the support of the proposed observatory.

Goldberg met with several NSF officials in Waterman's office on June 27 to clarify the steps AURA should take before the September meeting of the National Science Board.[107] This was followed by a meeting of the Organizing Committee for AURA in Ann Arbor on July 1, 1957. Goldberg reported to the Committee that six universities had signed; the Wisconsin Regents had approved and authorized participation subject to the approval of the Wisconsin Attorney General. Goldberg also reported on his June 27 visit to the Foundation, and read the May 20 Resolution passed by the National Science Board. The troublesome question of user participation, raised by NSF, was thoroughly discussed. There followed a full discussion of the proposed by-laws. Goldberg's status was changed to Permanent Chairman of the Organizing Committee for AURA, and it was voted:

> That the Permanent Chairman be authorized to make a Proposal to the NSF and to carry on interim negotiations with the NSF, retain counsel in Phoenix to prepare the necessary documents relative to incorporation and to pay necessary expenses provided that expenditures in excess of $5,000 shall not be made without further authorization from the Organizing Committee.
>
> It was noted at this point that incorporation will be deferred until after NSF's decision on September 6, and that the drafts of the by-laws and proposed contract will be strictly preliminary in character and subject to approval of the corporation after it is formed."[108]

Everything went smoothly at the September 6 meeting of the National Science Board,[109] and the NSF director was authorized to negotiate a contract with AURA along the general lines of the proposal. Goldberg wrote to the Organizing Committee: "This is the green light we have been waiting for, and the way has now been opened for AURA to incorporate."[110] The Phoenix firm of Jennings, Strouss, Salmon and Trask was employed to draw up articles of incorporation,[111] and a meeting to consumate the incorporation of AURA was scheduled for October 28, 1957 in Ann Arbor.[112]

Part II
AURA is created

8

The incorporation of AURA on October 28, 1957.
Actions taken at the second meeting of the AURA
Board of Directors on December 12–13, 1957. The
contract with the National Science Foundation.
Sputnik saves the solar telescope.

Goldberg's reply to Eckhardt's telephone call[1] included: "It occurs to me that the members of the Organizing Committee will undoubtedly wish to have something in writing from the NSF before actually proceeding to incorporate, and I should therefore appreciate a letter from either you or Dr. Waterman which I could present to the Organizing Committee." He also said he was trying to arrange a date for representatives of the Organizing Committee to discuss the proposed contract with NSF staff. Waterman sent the requested letter to Goldberg a week later.[2] The meeting to discuss the contract was scheduled for Wednesday, September 25.

Charles L. Strouss, a senior partner in the Phoenix law firm of Jennings, Strouss, Salmon and Trask, handled the details of incorporating AURA. He wrote to E.A. Cummiskey, the University of Michigan's attorney, to explain the Arizona statutes covering nonprofit corporations.[3] He pointed out that these statutes specify that the business and affairs of a nonprofit corporation are to be managed by a board of "directors" numbering not less than three nor more than twenty-five. He wrote: "I believe that if the 'board of trustees' in the proposed by-laws is changed to 'board of directors' these by-laws will be in conformity with the requirements of our law."

Strouss included with his letter a draft of suggested articles of incorporation for consideration by the Organizing Committee. He pointed out that the incorporators may not be corporations, but must be "persons." Corporations may become members after the incorporation. His advice was: "To accomplish your purpose, not less than three persons should be selected to sign the articles and become the incorporators. As soon as the incorporation is complete the six (sic) institutions may apply for membership. When the institutions have become members, the original directors may resign as directors and as members, and new directors may be elected in accordance with the provisions of the by-laws."

Goldberg suggested[4] that "one member of the Organizing Committee from each of the seven universities should act as an incorporator." He proposed that the meeting to sign the articles be on or about October 10. Menzel and Reynolds, the Harvard members of the Organizing Committee, told Goldberg that seven incorporators were

73

not necessary.[5] Goldberg replied that he had wanted to offer the opportunity to all members of the Organizing Committee, but said "the signing could be performed by Mr. Lee, Mr. Cummiskey and me . . . if the Organizing Committee so desires." Goldberg finally arranged for the articles of incorporation to be signed on Monday, October 14, by Goldberg, Lee, and Edmondson (en route from Indianapolis to Washington).[6] He proposed that the first meeting of the AURA Board be held in Ann Arbor on October 28 at 9:00 a.m.

During dinner on the evening of October 27 the group was told that the articles signed by Goldberg, Lee, and Edmondson had been "lost" and that another copy had been signed by Goldberg, Lee, and Cummiskey.[7] It had been sent air-mail-special delivery on October 25 but had not yet been delivered in Phoenix. The meeting began at 9:00 a.m. on the 28th, but recessed because no formal action could be taken until the articles of incorporation had been filed and recorded in Phoenix. Word was finally received by telephone that the articles had been filed at 9:30 a.m. (11:30 a.m. EST) and recorded at 10:03 a.m. (12:03 p.m. EST).

The three incorporators and representatives of the seven universities convened at 1:00 p.m. During their 15-minute meeting the incorporators voted to admit the seven universities to membership in AURA, Inc., elected the fourteen persons designated by their respective institutions as Directors of AURA, Inc., and then resigned – their function fulfilled.[8] Immediately following this meeting, the first meeting of the Board of Directors of AURA convened and C.D. Shane was elected Temporary Chairman.[9] He stated that Dr. Goldberg, Permanent Chairman of the Organizing Committee of AURA, had appointed as Nominating Committee C.D. Shane (Chairman), W.B. Harrell, and D.H. Menzel. The Board approved the slate of officers, members of the Executive Committee, directors-at-large, and consultants as recommended by the Nominating Committee. McMath took the chair at this time. The terms of Board members were classified later in the meeting as one, two or three years following recommendations of a special subcommittee established by McMath, his first action as President.

The Board next voted to adopt the August 8 draft of the By-laws as written, but with the word "Trustees" changed to "Directors" and the term "Observatory Director" rather than "Director" used to avoid confusion with the Board of Directors. There were also a few technical changes.

Draft No. 1 of the proposed contract[10] between the NSF and AURA was approved with two provisions: (1) That the inclusive dates of the contract be decided by the President after consultation with the NSF. (2) That the word "Trustees" be changed to "Directors" wherever it occurs. It was voted that the President be authorized to sign the contract after these changes were made. A vote of thanks was recorded for the excellent work of the Subcommittee (Goldberg, Lee, Cummiskey, and Peterson) in negotiating the contract with the NSF.

The October 28 draft proposal to the NSF was approved with two minor fiscal changes and a revision to indicate the election of two (not three) directors-at-large. It was voted to initiate steps for the University of Michigan to transfer the unspent NSF funds, responsibility for outstanding commitments, and government property used in the site survey to AURA as soon as possible.

Nearly all of the directors seemed to be in favor of appointing Meinel as Observatory Director, but Kuiper said he had strong reservations about Meinel.[11] He persuaded the Board to delay this action so that members of the Board could evaluate Meinel's work by visiting the Phoenix office and the sites, and by talking with Meinel and the staff. After discussion it was voted: "That the President appoint a group of scientific members of the Board of Directors to visit the Phoenix office, inspect the sites, and make recommendations to the Board at its next meeting concerning the choice of Observatory Director." McMath appointed Shane to organize the group and plan the time of the visit.

One of the most important actions taken at this meeting was the appointment of Reynolds as Chairman of a committee consisting of the administrative members of the Board "to work out a financial plan of operation, including the type of fiscal officer and business manager required, and all other matters relating to the fiscal and financial operation of the Observatory." The awesome array of talent on this committee guaranteed the establishment of a fiscal and business side of the observatory capable of coping with the bureaucracy of government funding.

A Scientific Policy Committee was also established consisting of the nine scientific members of the Board, with the President serving as Chairman. This Committee had been deleted from the draft of the by-laws earlier in the meeting as being too restrictive a requirement.

Responsibility for working with the NSF to prepare a news release about the incorporation of AURA and the signing of the contract was given to Edmondson.

The next meeting of the Board was scheduled for December 12–13, 1957 in Phoenix. It was anticipated that the contract would be ready for signing at that time. The date of the Annual Meeting was scheduled for the first Wednesday in October.

Finally, there were votes of thanks for Dr. Goldberg and the University of Michigan, and for Dr. McMath and the NAO-NSF Advisory Panel.

The meeting adjourned at 4:30 p.m. EST.

The administrative members of the AURA Board began to work on their very important assignment without delay. Reynolds wrote to the other members three days after the Ann Arbor meeting[12] to review the assignments they had agreed to in Ann Arbor. Franklin was to explore banking arrangements with the Valley National Bank, the largest in Arizona. Harrell was to review the organizational set-up of the Yerkes and McDonald Observatories, and J.M. Miller was to do the same for the Lick Observatory and for the Mount Wilson and Palomar Observatories. C.F. Miller was to make recommendations about insurance. Peterson and Lee were to consider the accounting procedure needed for the observatory, the definition of the job of the principal business officer (i.e. business manager or controller), and whether or not a construction manager would be needed during the construction period. Reynolds would get information about the administrative arrangement for the National Radio Astronomy Observatory. He suggested that the group should meet in Tucson the evening before the scheduled December 12–13 meeting of the full Board.

Reynolds also said:

> We all agree on one more point, and that is that each of the observatories
> operated at a distance from our respective Universities under the
> directorship of a member of the Board of AURA should be visited before
> we reach Phoenix. ... Mr. Harrell and Mr. Peterson are willing to visit the

Yerkes Observatory. . . . Mr. Harrell expects to join me in visiting McDonald Observatory. . . . I have asked Jim Miller to invite at least one of the rest of you to join him in a visit to Lick Observatory. . . . The purpose of these visits, of course, is to see how the astronomers, who are our co-directors of AURA, run these stations with so much less in the way of personnel than we all seem to feel will be needed at the new Observatory.

A week later Peterson sent the Committee a suggested organization chart for AURA and the observatory, and a list of five business staff positions and their job descriptions.[13] Reynolds wrote to the Committee:[14] "It is simple and direct, and I think we could all agree with it in principle." He also wrote: "Meanwhile, I had a look at the Harvard College Observatory, which I had never thought of as an outlying station." He included the job titles, job descriptions and an organization chart. He sent the administrative information about NRAO to the Committee after he received it in early December.[15] All of the assigned tasks were performed promptly by the members of the Committee and were reported to Reynolds and the other members.

Franklin was the obvious person to handle the banking arrangements because James E. Patrick, Executive Vice-President of the Valley National Bank was a close personal friend.[16] Franklin first wrote to Patrick the day after the July 1, 1957 organization meeting in Ann Arbor.[17] A graduate of the University of Michigan Law School had been suggested to handle the incorporation of AURA, and Franklin wanted Patrick's opinion of his qualifications. Patrick was lukewarm, and Michigan agreed to delay the selection of a lawyer until Patrick could supply a list of the top lawyers in Phoenix.[18] The firm of Jennings, Strouss, Salmon and Trask was subsequently selected.[19]

Franklin wrote to Patrick again the day after the incorporation of AURA to discuss banking arrangements, including "having an official of the bank act as a treasurer for us."[20] Patrick's reply was positive,[21] causing Franklin to send him a copy of Peterson's organization chart.[22] Reynolds, Harrell, and Lee expressed pleasure with Franklin's progress.[23]

The scientific members of the AURA Board also had their work to do. Shane arranged for a visit to the Phoenix office on November 25–27.[24] In attendance were Shane (chairman), Edmondson, Keenan, and Whitford. Kuiper, who had insisted on this inspection visit, was conspicious by his absence especially since the date had been cleared with him. Bowen (consultant) and Geoffrey Keller, the new NSF Program Director for Astronomy, were also present and participated in all the discussions. Meinel and Baustian attended all but the executive session. C.W. Jones (contractor) and Bruce Rule (consultant) attended during the first two days. The recommendations included: "The committee recommends the appointment of Dr. Aden B. Meinel as Astronomical Director of the National Observatory for a term of years to be determined by the Board of Directors."

November 26 was occupied with a trip to Kitt Peak arranged by the Tucson Chamber of Commerce. A visit to Hualapai Peak had been arranged by the Kingman Chamber of Commerce for November 27 but it had to be cancelled due to the crowded schedule of the committee. The committee therefore recommended that "immediately following the meeting in Phoenix on December 12–13, as many members of the Board of

Directors as possible, and particularly the scientific members, join in a visit to Hualapai Peak."

Transfer to AURA of assets held by the University of Michigan in connection with grant G-2622 and its predecessors was discussed with NSF officials in Washington on December 3, 1957 by AURA Board members Lee and Harrell.[25] The procedure that was adopted would be to have the University of Michigan return to the NSF all physical and cash assets as of December 31, 1957. These assets would then be transferred to AURA by an amendment to the basic contract. A supplement would be added to cover the actual cost of the 36-inch telescope over and above the amount returned by the University of Michigan, as soon as a firm bid could be obtained. J.E. Luton, the NSF Assistant Director for Administration, outlined the understandings that had been reached in a letter to Lee.[26]

Plans for a signing ceremony and news release about the contract were discussed by Edmondson and Lee Anna Embrey, NSF Deputy Public Information Officer.[27] Unfortunately, McMath suffered a serious knee injury on November 9.[28] This ruled out the possibility of having him come to Washington for a signing ceremony and photo opportunity similar to that done when AUI and NSF signed the NRAO contract.[29] Embrey's final version of the press release[30] was scheduled for Noon, MST, on Friday, December 13, the final day of the AURA Board meeting. Edmondson's role was to coordinate the simultaneous release of this story with the Public Information Officers of the AURA member universities, plus Swarthmore and Vanderbilt. There were a few mishaps in the timing in the midwest (Wisconsin and Indiana) due to failure to take into account the one hour difference in time between MST and CST. Frank Carey, of the Associated Press, complained to Embrey[31] and said they felt they were no longer obligated to observe the 12 Noon, MST time. He suggested that she should release United Press (UP) and Inernational News Service (INS) from the time given. It should be noted that Friday, December 13 was also the Ides of December.

The NSF press release said the contract had been signed by NSF Director Waterman and AURA President McMath on December 13. The behind-the-scenes story was quite different. Six unsigned execution copies of the $3,100,000 contract had been sent to McMath in Phoenix by Eckhardt[32] with the explanation: "The Director has been continuously tied up in meetings and has not been available for signing the six copies being sent to you. ... You may be assured that he will sign them if they have been signed in behalf of AURA."

McMath opened the Board meeting on December 12 with a statement that he had received word[33] that Hoff, the NSF General Counsel, had ruled that neither the President or Vice-President of AURA (McMath and Edmondson) should sign the contract to avoid a perceived conflict of interest. McMath had been on the MPE Divisional Committee and Edmondson had been on the NSF payroll.

Discussion led to an action by the Board that "Mr. Harrell was authorized as Contracting Officer of the Corporation for purposes of executing this particular contract on behalf of AURA." The final NSF revision of the contract was reviewed and approved by Counsel Strouss, Harrell, and AURA Secretary Miller.[34] Harrell signed the six copies on December 13 and Miller mailed them to Eckhardt on December 16.[35] The contract was signed by Waterman for the NSF on January 2, 1958 and three of the executed copies were returned to AURA Secretary Miller.[36] AURA was asked to

Second Meeting of the AURA Board of Directors, 13 December 1957, Phoenix, Arizona. Left to right: P.C. Keenan, C.D. Shane, C.K. Seyfert, P. van de Kamp, C.F. Miller, D.H. Menzel, G.L. Lee, Jr., F.K. Edmondson, E. Reynolds, R.R. McMath, A.B. Meinel, M. Moore (Treasurer), J.M. Miller, A.E. Whitford, W.B. Harrell, J.A. Franklin. Absent: G.P. Kuiper, A.W. Peterson. Edmondson photo.

provide NSF with "three certified copies of the AURA Board resolution or minutes, as appropriate, which authorized Mr. Harrell to sign the contract." These were needed to support the NSF copies of the contract.

McMath next reported that Eckhardt had rejected the concept of a separate large solar observatory, but had agreed to plan for a solar telescope at the NAO site. Shane moved and the Board unanimously approved a recommendation: "That solar equipment be located at the site of the National Astronomical Observatory, subject to the suitability of the site."

The final action by the Board in the morning was to appoint Meinel as Observatory Director.[37] The vote was unanimous; Kuiper was the only member of the AURA Board not in attendance.[38] The afternoon of December 12 was devoted to separate meetings of the Administrative and Scientific Subcommittees.

The morning session on December 13 began with the report of the Administrative Subcommittee.[39] It was voted to employ the firm of Jennings, Strouss, Salmon and Trask as the General Counsel of AURA, Inc. The Valley National Bank was chosen as Fiscal Agent and Minton Moore, Executive Assistant to the President of VNB was appointed Treasurer of the Corporation. Approval was voted for the subcommittee's recommendations concerning the staff of the Observatory Director, selection of an audit firm, delegation of signing authority, working capital from NSF, the need for the staff to develop operational business procedures for the observatory, insurance, travel expenses of members of the Board of Directors and Officers, a restudy of the budget, acquisition of land for the site of the observatory, and budget restrictions for the Observatory Director. Edmondson was selected to work with NSF on future negotiations with that agency.[40]

McMath had expressed concern about his physical ability to carry the heavy workload of the President of AURA.[41] The Board approved the Administrative Subcommittee's recommendation: "That the new office of Vice-Chairman of the Executive Committee be created, and that Dr. C.D. Shane be appointed to this office, to assist the President." In addition, the Vice-President and Secretary were "to visit the Phoenix office at monthly or other intervals as experience dictates, to discuss problems with the Observatory Director."

The scientists on the Board would have liked the Administrative Subcommittee to continue in parallel with the Scientific Subcommittee, but the administrators did not agree. After their recommendations had been accepted by the Board, the ad hoc Administrative Subcommittee was discharged.

The morning session concluded with the report of the Scientific Subcommittee. The following recommendations were approved by the Board of Directors. (a) The observatory should be in Arizona. (b) Site testing should be concentrated on Kitt Peak, Hualapai Peak, and Mormon Mountain. Test equipment should be placed on Slate Mountain until Mormon Mountain became accessible. Primary emphasis should be on night seeing and transparency, but there should be occasional checks on solar seeing with the two available 6-inch telescopes. (c) Despite the pressure [from NSF] to make the choice of site by July 1, 1958, the decision should be postponed as long as feasible. (d) A small, temporary optical shop should be established in Phoenix for experimental work. (e) Dr. Nicholas U. Mayall was appointed a consultant. He and Bowen would advise Meinel in connection with telescope design work.

The afternoon session began with a discussion of the transfer of funds from Michigan to AURA. The procedures negotiated by Lee and Harrell[42] were approved, and the President was authorized to sign a contract with C.W. Jones in the event that the Michigan contract could not be assigned to AURA. The present staff of the Phoenix field office would be retained at their initial rate of pay, including medical benefits, etc.

Admission of Yale had been put on the agenda at Edmondson's request.[43] Whitford supported this during the discussion: "Had it been known that Princeton and Cal Tech would not accept invitations, Yale might well have been invited in the beginning." Menzel (Harvard) moved that Yale be invited, and this was approved unanimously.

Counsel Strouss reported that Arizona law requires corporations to file an annual report within 90 days of the close of the fiscal year and pay a $25 filing fee. Approval was given to Secretary Miller's proposed design for the AURA corporate seal, another requirement of Arizona law, and to his proposed AURA letterhead. The meeting then ajourned.

After the contract had been signed and sent to NSF, McMath wrote to Eckhardt[44] asking for $300,000 in operating funds to make it possible for AURA to take over from the University of Michigan. Meinel informed Keller about this by telephone.[45] Of course, NSF could not provide any money until the contract had been signed by Waterman. Eckhardt sent the six signed copies to Luton on December 19,[46] and Luton sent them on to Hoff on Christmas Eve.[47] McMath had phoned Waterman about the status of the contract on December 23,[48] and Luton mentioned this in his memo to Hoff. McMath's phone call probably helped move the contract off Luton's desk.

For the record, McMath summarized his understanding of the details of the phone call in a letter written to Waterman the day after Christmas.[49] Waterman signed the contract on January 2, 1958,[50] and he wrote to McMath about this on January 8.[51] The letter also said "The Department of the Treasury mailed on January 3 a check for $300,000 to Mr. Minton Moore, Treasurer of AURA, in care of the Valley National Bank at Phoenix."

With operating funds in the bank, all that remained to be done was the transfer of assets and responsibilities from the University of Michigan to AURA. On April 3, 1958 Lee provided Luton with an authenticated inventory of items purchased under the grants to the University of Michigan. Unfortunately, the National Science Foundation reverted to its glacial pace mode, and Luton's reply approving the transfer of equipment to AURA was not written until two months later, on June 9, 1958.[52] A sum of $95,000 of unspent funds from the grants remained to be returned to NSF, and this was to be added to the AURA contract.[53] Major changes in the NSF appropriations for FY 1958 and FY 1959 delayed the contract amendment until November 1958.[54]

The first earth orbiting artificial satellite, Sputnik I, was launched by the USSR on October 4, 1957, three weeks and three days before the incorporation of AURA. The public reaction to this event gave the National Science Foundation a wonderful opportunity to go back to the Congress with a request for a supplemental appropriation, actually a restoration of the two $25,000,000 cuts from President Eisenhower's requests for FY 1958 and FY 1959.[55] A supplemental appropriation of $9,750,000 was quickly approved for FY 1958, bringing the total to $49,750,000.[56] The $115,000,000 that had been approved in the FY 1959 appropriations bill was increased to $136,000,000 before passage.[57] Thanks to Sputnik I the National Science Foundation received $30,750,000 in additional funds, an increase of 19.8 percent over the amounts first approved by Congress.[58]

Two and a half weeks after Sputnik I was launched, Goldberg wrote to Waterman:[59] "I have heard rumors that the NSF is considering submitting requests both for a supplemental apropriation for FY 58 and additional funds for FY 59 in response to the Russian challenge." He suggested five astronomical needs that deserved serious consideration, including "the reactivation of the large solar telescope project for the NAO." He also said: "Bob McMath intends to reopen the issues of the solar and X-inch telescopes this week at the MPE Divisional Committee meeting."

The MPE Divisional Committee meeting took place on October 25, only three days before the incorporation of AURA. McMath asked Struve to write a statement concerning the importance of a big solar telescope[60] which he could read at the meeting, and Struve responded by return mail.[61] McMath told Struve he had left the original of his letter with Eckhardt when he wrote to describe what happened at the meeting.[62] He also wrote about the meeting to Bowen:[63]

Dr. Eckhardt, the relatively new assistant-director for the MPE Division, announced repeatedly that the MPE Division would not request funds for facilities involving continuous support by the NSF. He further stated before the full committee that this automatically ruled out the solar observatory.

Now, without any chance to consult with any of my Panel colleagues, I have gone back to Waterman, telling him to take the word "observatory" out of the title and then, that we are requesting a very large solar telescope. This telescope would be put on the same site as the NAO and that, at the most, no more than $10,000 per year would have to be added to the regular operating budget.

This is, of course, a compromise, but not necessarily in a downward direction. I feel sure that American astronomy would be better off with such an instrument than with no big solar telescope at all. On the telephone this morning, Leo concurred.

Even with this change I am not at all hopeful that we can get our big instrument, but at least this change would take the solar project out from under the ban on large-scale facilities with support. I hope very much that you will go along with this. Struve wrote me a special letter in Washington concerning the big telescope and I enclose a thermofax copy.

McMath's letter to Waterman[64] emphasized the semantic problem (solar observatory vs solar telescope) and argued that the proposed plan for a very large solar telescope did not come under Dr. Eckhardt's definition of a new facility. He asked Waterman "to re-establish the original request for, say, 5 million dollars for 'a very large solar telescope' in your FY 59 request." A month later McMath wrote to the Chairman of the MPE Divisional Committee:[65] "Dr. Keller, Program Director for Astronomy, telephoned me on my hunting trip to tell me that the big solar telescope project had been reinstated for FY 59 in the amount of $5.6 million. This was done with the understanding that the big solar telescope would be located on the same site as the NAO." The letter also said he was resigning as a member of the MPE Divisional Committee.

Secretary Miller got approval from the Executive Committee by mail vote[66] for McMath "to prepare and transmit to the National Science Foundation a proposal for the solar observatory [sic], in the amount of $5.6 million, to be located, at least for the time being, adjacent to the stellar observatory." The proposal for $5,000,000 was mailed to Eckhardt on January 27.[67] At the Executive Committee meeting in Pasadena on February 10–11, 1958[68] McMath requested confirmation of the mail vote that had authorized him to submit the solar telescope proposal to NSF. This was approved unanimously.

The increase in the NSF appropriation did not mean that all of the line items in the NSF budget request would be approved, and the funding for the solar telescope was far from being assured. The Senate version,[69] more generous than the House version, included funds for the solar telescope and earmarked $1,890,000 for a road up Kitt Peak to be provided "within the funds available to the Foundation." The bill contained a provision for transferring this money to the Bureau of Public Roads,

Solar telescope. Originally dedicated as the McMath Solar Telescope on November 2, 1962. It was rededicated as the McMath–Pierce Solar Telescope on 7 November 1992. NOAO Photo Archives.

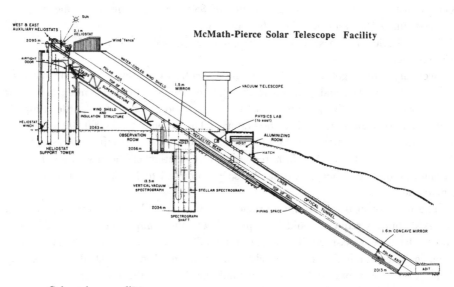

Solar telescope diagram.

Department of Commerce. Waterman reported this unexpected action to the National Science Board.[70] The original intent had been to have the Kitt Peak road included in the Department of Commerce appropriation, but a member of Senator Hayden's staff "let all the Interior, Commerce, etc. bills go through without tacking the road on. Finally there was only the NSF to tack it to, so they have reluctantly accepted it."[71]

The compromise $6 billion budget bill that was passed was vetoed by President Eisenhower "because of an item of $589 million to set up a civil service retirement fund." The revised appropriations bill (PL 85-844) was finally passed and sent to the White House on August 23, 1958.[72]

A telegram from Luton to Ralph Patey,[73] the recently appointed NAO Business Manager,[74] gave the first warning of a problem with the amount of money appropriated for the solar telescope: "Five million recommended by Conference Committee may include funds for both the solar telescope and one million for the road." Senator Hayden and Senator Goldwater believed the Senate-House conferences had "agreed to leave in appropriations bill one million for road and provision for solar telescope which is estimated to cost 5 million."[75] Luton summarized the situation in a memo to Waterman.[76] He said the question had been discussed with Pierce shortly after the appropriation bill was passed, and that he had "indicated without reservation that the telescope could be constructed within the $4,000,000 figure and that Dr. McMath would be quite pleased with that amount." Dr. McMath confirmed this in a later conversation with Frank Sheppard, NSF Controller. Luton wrote: "In view of the conflicting interpretations of Senator Hayden and Mr. Thomas and the relaxed attitude on the part of Dr. McMath I would recommend that we make no further attempt to resolve the issue before Congress returns in January. As you know we are holding $1,000,000 in reserve for the telescope should it prove to be required."

The AURA Executive Committee met in Ann Arbor on September 13, 1958.[77] McMath reported that a revised solar telescope proposal[78] for $4,000,000, the amount appropriated by Congress, had been sent to the National Science Foundation. The Executive Committee voted to approve the action taken by McMath. In the Executive Session McMath recommended that Pierce be appointed Associate Director of the NAO, effective October 1, 1958. Pierce had worked with Meinel, but reported directly to McMath,[79] during the period when the solar site survey was funded by grants separate from the NAO site survey. The Executive Committee approved this appointment.

McMath's failing health and physical inability to cope with the demands of the position of President, led to a major change in the organization. The Executive Committee accepted a recommendation that Shane be appointed President and that McMath be given the new title Chairman of the Board. The necessary changes in the by-laws were to be prepared by counsel and the secretary in time for the annual meeting.

McMath had written to Waterman in July that the funding for the solar telescope should be an amendment to the existing contract, and that renegotiation of the entire contract would not be acceptable to the AURA member universities.[80] Waterman took this up with Luton, who agreed that a contract amendment would be appropriate.[81] Waterman needed the approval of the National Science Board before he could sign the contract amendment, and he brought this to the NSB at the October 13, 1958 meeting.[82] McMath informed the AURA Board the next day that a supplement to the contract of $4,445,000 had been approved.[83] The NSF press release was dated November 9,

1958.[84] The contract amendment was signed for NSF on November 18 by Robert R. Brode, Acting Director, and for AURA on November 22, this time by President McMath.[85]

Kitt Peak had been chosen almost nine months before this, and the design and construction of the large solar telescope now became part of an active construction program that was already well under way.

9 The selection of Kitt Peak on March 1, 1958. Negotiation of a lease with the Papago Tribe following passage of special legislation by the US Congress. The University of Arizona helps with the acquisition of land for the city headquarters.

Whitford's report at the November 1955 meeting of the American Astronomical Society[1] concerning the activities of the NAO Panel described a five-year plan agreed to by the Panel. "In the first two years there would be an exhaustive survey of possible sites in order to find one with the best combination of a good annual yield of clear night hours and a high average of good seeing. As soon as the search had narrowed down to a few promising sites, a 36-inch reflecting telescope would be set up on a temporary mounting in order to provide a more complete test of one site, and at the same time to give an immediate opportunity for photoelectric observing to astronomers who need it. Once the final site was chosen, work on the installation of the first major instrument could begin." In the final three years development of the site, installation of an 80-inch telescope, and creation of an operating organization of universities would be completed. "These universities would be representative of those in the eastern and midwestern sections of the country which would be the most frequent users of the telescope."

The timetable that actually came to pass was not the one proposed by the NAO Panel. The first 60-foot towers were installed in mid-1956, but the first 16-inch telescope was not ready for mountain use until a year later.[2] Whitford's summary of the situation at this time (September 3, 1957)[3] began with the blunt statement "We have adequate astronomical data for something less than half a year on only one site, Kitt Peak." This was just eight weeks before the incorporation of AURA. A description of the five-year plan was in the October 29, 1957 AURA proposal to the NSF,[4] but there was a hint of a speed-up in the fourth paragraph: "On the assumption that a final choice of site can be made by the end of FY58, the first task for FY59 will be the completion of a construction access road." The report of the Scientific Subcommittee at the December 13, 1957 AURA Board meeting[5] made it clear that the speed-up did not originate with AURA: "The Subcommittee recommends that despite the pressure to make the decision as to the choice of site by July 1, 1958, any additional time that can be made available to continue studies should be utilized and the decision postponed as long as feasible."

Site selection was on the Agenda of the AURA Executive Committee when it met in Pasadena on February 10, 1958.[6] Temporary installation of the 36-inch telescope on the most promising site, Kitt Peak, was approved, in agreement with the earlier recommendation by the NAO Panel. The Executive Committee also delegated authority to the Scientific Subcommittee "to act in the matter of site selection." The Subcommittee was scheduled to meet in Pasadena on March 1, and Meinel was instructed to provide in advance of the meeting "a detailed report, including a synopsis to the fullest extent possible of the original observing books." Edmondson was instructed to advise NSF that this action had been taken.

Secretary Miller announced that Yale had accepted the invitation to join AURA, and would be represented on the AURA Board by Professor Rupert Wildt and Charles S. Gage, Treasurer. Wildt would attend the March 1 meeting.

Meinel had written directly to the AURA Board, with a copy to Leo Goldberg, two weeks before this meeting[7] asking the Board to approve the establishment of an ad hoc advisory committee that would advise him. He wrote: "I believe it is necessary to form such a group in order to best accomplish the research objectives of this observatory by basing my recommendations [to the Board] upon the approval of a broad group of active researchers."

The evening before the meeting began he was asked informally to elaborate on what he had written. He said he needed the advisory committee because the scientific members of the AURA Board were not in the main stream of present day astronomy and he couldn't trust their judgement.[8] Needless to say, this generated some heated discussion as soon as it became known to all of the members of the Executive Committee. It also produced a resolution, passed unanimously, in Executive Session, at the end of the first day: "That the Observatory Director shall be responsible to the President of AURA, Inc." The Executive Committee instructed the President to write to the Observatory Director, advising him as to the proper procedures in regard to communications, including visits. Dr. Shane, as Vice-Chairman of the Executive Committee and Secretary Miller were to receive copies of correspondence directed to the President by the Observatory Director, and Dr. Shane would act for the President when so delegated.

Dr. Meinel entered the meeting at 5:30 p.m., and no further record was taken. The meeting adjourned at 6:15 p.m. A short follow-up Executive Session was held beginning at 9:00 p.m. and again no record was taken.

Meinel's 85 page final report[9] on the site survey was discussed by the Scientific Committee[10] in Pasadena on the morning of March 1, 1960. Kuiper was the only member not in attendance.[11] The report included 48 diagrams and five tables. Kitt Peak had the lowest winds among the four sites tested (Hualapai, Summit, Chevalon, and Kitt Peak), and also had lower winds than the Lick Observatory. Microthermal seeing fluctuations were better on Kitt Peak than at Chevalon and Summit; information was not available for Hualapai. Visual records of seeing with the 16-inch telescopes showed little difference between Hualapai and Kitt Peak, but the best on Hualapai never equaled the best on Kitt Peak. Automatic seeing records from the Polaris monitors were available from Kitt Peak on 150 nights. Successful measures were not obtained at the other tower sites "due to a succession of difficulties with the stability of the structure." Photoelectric data about atmospheric transparency had not been reduced in time for this report, but the observers thought the sites were equally good. Meinel gave his assessment of the relative merits of the Kitt Peak and Hualapai sites for 18 items: nine items favored Kitt Peak, four items favored Hualapai, and the two sites were equal for five items. The members of the Scientific Committee then voted unanimously for Kitt Peak, and so did the seven consultants and staff members. "A few qualifications were expressed based on the wish that a longer time could have been available." Kitt Peak was shown to be a good site, but the original goal of the NAO Panel to conduct the most thorough site survey ever done had not been met.

The afternoon session started with a review of the solar telescope project by McMath. This was followed by a discussion of the X-inch telescope. The final session on Sunday morning was devoted to discussion of the plans for the 36-inch and 80-inch telescope mountings. The plans for the 36-inch had been reviewed by a small subcommittee the previous evening and were recommended for approval, which was given. The Committee reviewed the plans for the 80-inch telescope and authorized the engineers to go ahead with the detailed design. The meeting then adjourned for lunch.

Luton had written letters concerning selection of the site to McMath and Strouss[12] on February 4. He told McMath that the National Science Foundation should have "a resolution from the AURA Board of Directors before taking any action with respect to approval or disapproval of a site." The letter to Strouss made a similar statement, and added: "The Foundation would like to work with you very closely with respect to negotiations for the acquisition of the site from the Papago Indian Tribe, either by lease or otherwise. This is particularly true with respect to any legislation which may be required in connection with these negotiations." McMath did not respond to this letter until March 8,[13] after Patey had been in Luton's office on March 3 to report that Kitt Peak had been chosen by the AURA Scientific Committee.

Embrey's draft press release[14] was sent to McMath and Edmondson on March 5. The cover letter to McMath[15] contained a warning: "I believe Ralph Patey has been in touch with you and informed you that the feeling here is that the choice of site must be approved by the National Science Board." She tentatively scheduled the release for March 14, the date of the next NSB meeting.

McMath wrote to Luton again on March 11[16] describing the actions taken at the February 10–11 meeting of the Executive Committee and at the March 1–2 meeting of the Scientific Committee. He said Patey had mailed the minutes of the latter. The letter continued: "I hope very much that this letter, together with Mr. Patey's material, will constitute sufficient material to enable you to go to the NSF Board on the 14th." It was not sufficient, and another letter had to be written.[17] The opening paragraph read: "In accordance with our telephone conversation of this morning, please construe this letter as a formal request from AURA to the National Science Foundation to concur in AURA's choice of Kitt Peak, Arizona as the site for the new National Observatory." McMath sent copies of this letter to Miller, Edmondson, and Shane and wrote:[18] "My own face is somewhat pink, because no one had thought of asking the Foundation for their concurrence in the choice of a site."

Fortunately, McMath's letter reached Washington in time, but just barely, for consideration by the National Science Board at its meeting on Friday, March 14. The NSB unanimously approved the AURA recommendation,[19] and Embrey's news release[20] received nationwide attention.[21]

The Executive Committee met in Phoenix on April 4 and 5.[22] Patey's official minutes of the March 1 meeting of the Scientific Committee were approved, and this constituted AURA's formal acceptance of the recommendation that Kitt Peak be the site. Further discussion obviously was not necessary.

Unfortunately, Charles Strouss died on March 1 in Atlanta, where he had been hospitalized for about a week following a stroke. He had gone to Atlanta to receive an award from the American Bar Association. The legal work for AURA was taken over immediately by Irving A. Jennings, another senior partner in the Phoenix law firm of Jennings, Strouss, Salmon and Trask. He accompanied Meinel and Thompson to discuss the lease of Kitt Peak with the Schuk-Toak District Council on March 5.[23] Jennings presented the proposal signed by Dr. McMath, and the Schuk-Toak District Council passed a resolution recommending that the Tribal Council approve the proposed terms of the lease. The Tribal Council voted approval on March 7.[24] AURA was not represented at this meeting.

The Papago Tribal Council had agreed in early 1956 to allow the testing of Kitt Peak and to negotiate a long-term lease if Kitt Peak proved to be suitable for the location of the observatory.[25] A legal restriction on the leasing of Indian reservation land had been overlooked up to this point,[26] and it turned out that special legislation by the US Congress would be required before a satisfactory lease could be consummated. The restriction was in a bill that had been introduced by Senator Barry Goldwater[27] on January 6, 1955. Its purpose was to liberalize previous restrictions by allowing Indian lands to be leased for a term not to exceed 25 years, and with the possibility of a renewal for one additional term not to exceed 25 years. The bill (S.34) had been signed by President Eisenhower on August 9, 1955.

The death of Counsel Strouss and the move to Tucson made it imperative to hire a legal firm in Tucson. The Executive Committee voted at its April 4, 1958 meeting to terminate the Phoenix law firm as soon as possible, and to employ the Tucson law firm of Boyle, Bilby, Thompson and Shoenhair.[28] This firm was recommended highly by Treasurer Moore, after he had done some investigation. The groundwork for the lease had been initiated by Strouss before his unexpected death,[29] and the first draft was written by Jennings.[30] This work was continued by Theodore K. Shoenhair in Tucson.[31] He played the principal role in drafting the legislation to allow the lease of Kitt Peak for an indefinite period, replacing the legal maximum of 50 years in this one case. He also wrote the final draft of the lease and represented AURA in the negotiations with the Papago Tribe.

Luton informed Patey on June 11, 1958 that the National Science Foundation had received clearance from the Bureau of the Budget for the proposed Kitt Peak legislation,[32] a copy of which was enclosed. Luton said copies had been sent to the offices of Senator Hayden, Senator Goldwater, and Congressman Udall. Patey sent this letter and the enclosure to McMath and the Executive Committee.[33] He wrote: "A verbal explanation of the delay by Luton and Sheppard indicates that this legislation was inadvertently sidetracked in the Department of the Interior. No categorical opposition to the contents of this bill has been indicated to NSF during the travels through the various government agencies."

The bill, H.R. 13499, was introduced by Congressman Stewart Udall in the House of Representatives on July 22, 1958.[34] Senator Barry Goldwater introduced the corresponding bill, S.4167, in the Senate on July 31.[35] Senator Hayden explained to the Senate that the funds for the observatory had been included in the conference report on the independent offices appropriation bill that was adopted on July 30, but these funds could not be spent until S.4167 was passed and signed into law by the President. The bill was "ordered to be engrossed for a third reading, read the third time, and passed." The House passed the bill a few days later, and it was signed (Public Law 85-816, 85th Congress) by President Eisenhower on August 28.[36]

A copy of Shoenhair's proposed lease was sent to Luton on May 16, 1958.[37] NSF reverted to its glacial mode, and gave only tentative approval two months later on July 18.[38] After more changes were proposed by NSF, Patey sent five copies of a revised version to Luton on September 8.[39] Approval of this by NSF came shortly after the September 13, 1958 meeting of the Executive Committee.[40] Mark Manuel, Papago Tribal Chairman, signed six copies of the lease during a meeting of the Tribal Council on October 3, 1958.[41] Those present were Meinel, Pierce, Patey, Shoenhair, Earl Cooper of the Senate Appropriations Committee Staff, 15 members of the Papago Tribal Council, and members of the Indian Agency staff. There were 14 votes to approve the lease and one abstention.

The signed copies of the lease were sent to Washington. Dr. Alan T. Waterman, Director of the National Science Foundation, signed on October 24, and Roger Ernst, Assistant Secretary of the Interior, signed on October 31. Ernst made a technical correction of a word on one line of the lease (East was changed to West) and this was accepted by Mark Manuel and Dr. Waterman.[42] This removed the last obstacle to the start of construction on Kitt Peak.

There was informal contact with the University of Arizona before the start of the site survey.[43] University of Arizona faculty members played a crucial role in persuading the Papago Tribe to allow testing of Kitt Peak as a possible site for the National Astronomical Observatory.[44] President Harvill gave strong support to the effort to get funds for the Kitt Peak road,[45] and he offered temporary office space on the campus during the move from Phoenix to Tucson.[46]

Meinel and President Harvill discussed the procurement of Tucson land adjacent to the university for the city offices more than two months before Kitt Peak was chosen.[47] Secretary Miller reported to the Executive Committee[48] that during his January 9–10 visit to Phoenix Meinel told him the arrangement would be a no-cost lease, provided AURA would construct at the mountain site an observatory facility for the University of Arizona Astronomy Department (i.e. a dome for the Steward Observatory 36-inch telescope). Meinel presented plans for the city office building and described the proposed city land arrangement with the University of Arizona at the Executive Committee meeting on February 10.[49] Meinel, Patey, and Counsel Strouss subsequently met in Phoenix with President Harvill and University Counsel McCormick.[50] The university agreed to submit a draft proposal for consideration by AURA.

Luton wrote to Meinel as soon as he was informed about the Meinel–Harvill discussions:[51] "It will take several weeks for the Foundation to reach a judgement on the proposed exchange of land [sic] with the University of Arizona. There is much more that we will need to know about this transaction before reaching any conclusions. I plan to discuss it with you when I am in Tucson and Phoenix. In the meantime, I believe it would be advisable that AURA hold in abeyance any further steps with

The meeting of the Schuk Toak District Council with AURA representatives on 5 March 1958 at which the District Council approved a resolution authorizing the Kitt Peak Lease to the NSF. Mark Manuel, Chairman of the Papago Tribal Council, and Chester Higman, Administrative Assistant to the chairman, are seated at the table. At extreme left is Harold Thompson, engineer on the site survey, and standing at extreme right is Aden B. Meinel, first director of KPNO. NOAO Photo Archives.

respect to this transaction." He also wrote to McMath:[52] "... This is a somewhat unusual transaction from the standpoint of the government, and I am not at all sure that it can be consummated." Luton thought there was also a policy question: "The Optical Observatory is supposed to be a National Observatory, equally available to scientists throughout the country. In view of this, the question arises as to whether it is wise for a particular university to have facilities in conjunction with those of the National Observatory." Luton had no way of knowing that a year earlier Seeger had suggested to Carpenter that the Steward Observatory telescope be moved to Kitt Peak.[53]

Eckhardt, Luton, Sheppard, and Leigh (Acting Grants Administrator) visited Phoenix, Tucson, and Kitt Peak on March 27 and 28.[54] Shane, Edmondson, Peterson, and Lee represented the AURA Board, and Meinel, Patey, Baustian, and Thompson represented the observatory. The visit included a discussion of the city land and Kitt Peak telescope site question with President Harvill.[55] Meinel summarized the AURA point of view in a letter to Luton.[56] Luton told McMath[57] the visit to Arizona had been fruitful. He also mentioned the policy question about installing the Steward

Observatory telescope on Kitt Peak, and wrote: "I have discussed the matter with Dr. Waterman, Dr. Klopsteg, and Dr. Eckhardt. Our feeling here at the Foundation is that the astronomers should be allowed to decide this question. Dr. Eckhardt has taken responsibility for developing recommendations on this matter."

Shane, Lee, Meinel, and Patey met with Luton, Eckhardt, Sheppard, Leigh, Charles Ruttenberg (Assistant General Counsel), and Keller in Washington on April 18.[58] The main item of business was the financial plan for FY58, 59, and 60. Several other matters were discussed, including the Tucson city land acquisition. "It was agreed by all concerned that the simplest method of acquiring land in Tucson would be to purchase the land from the University of Arizona at a price approximating what it would cost them to erect a telescope on Kitt Peak. . . . If we can purchase the land from the University of Arizona, we would then lease them an area for their telescope on Kitt Peak."

At about the same time, Secretary Miller attended a meeting of the Western Association of College and University Business Officers and reported an interesting conversation:[59] "Regarding the purchase of land from the University of Arizona, I was surprised to find out that Mr. Murphy, the business officer, and Mr. Hull, Vice President of the University of Arizona, know very little of the AURA negotiations. Apparently the President has handled this personally, and has not passed on any information to the other officers. Specifically as to the land purchase, is there any information as to the appraised value of the land we are to buy?"

Luton followed the discussion at the April 18 meeting by sending a letter to Meinel[60] proposing that the University of Arizona acquire the land for the city headquarters and sell it to AURA. This was acceptable to the university.[61] By the time of the September 13, 1958 meeting of the Executive Committee[62] the university had purchased one-half city block across the street from the Steward Observatory (316 × 180 feet and appraised at $86,000), and was willing to buy the remaining one-half block for sale to AURA.

Such a simple and straightforward solution to the city land acquisition problem seemed too good to be true, and it was. Luton became concerned about the cost of improved real estate near the university, and delayed approval of the proposed purchase of the land bought by the university for AURA. Meinel responded to Luton's request for justification of the site with a three-page letter.[63] McMath talked with Luton by telephone on October 24, and followed this with a blunt letter on October 27.[64] Mark H. Klaffter, Real Estate Appraiser, explained the basis of his appraisal to Luton.[65] A month later Secretary Miller telegraphed Luton a strong protest about the continuing delay.[66] Luton's telegram in reply said:[67] "Meeting yesterday with Director and Staff resulted decision that availability other sites must be investigated. Further letter follows." The letter[68] said Meinel's letter failed to make a strong case, and it would not "satisfy a Congressman who raised the question of why the AURA offices could not be located somewhere else in or near Tucson on unimproved property at substantially less cost to the government." The letter also said: "The Director and his staff acted primarily in an advisory capacity and made it clear to me that the final decision is mine as to whether we will purchase the property in question."

Information provided by Meinel and Klaffter[69] ruled out a nearby site that had attracted Luton's attention. He sent a telegram to Meinel[70] authorizing him to initiate negotiations with the University of Arizona with the understanding that the survey conducted by Klaffter shows no suitable unimproved lots available within reasonable proximity to the university. Meinel summarized the background and read this telegram at the December 7, 1958 meeting of the Scientific Committee.[71] He reported that Klaffter had looked at available land within a one, two and five mile radius of the university, and had not found anything that would be suitable. Secretary Miller questioned the wording of the telegram. McMath read a letter he had received from Harvill[72] about the Kitt Peak sublease and the city land purchase. Harvill pointed out that these arrangements had been proposed by Luton in his April 22 letter to Meinel,[73] and that the Board of Regents had approved both proposals for the university.

The annual meeting of the Board of Directors took place the next day.[74] Secretary Miller gave a highly critical review of the "inaction of the National Science Foundation" during the past nine months. He felt Luton's December 4 telegram to Meinel was not sufficient authorization; AURA must have something in writing from the National Science Foundation before negotiations are begun. After more discussion it was agreed that Lee would telephone Luton, and Secretary Miller and Lee left the meeting to do so.

Lee and Miller returned to the meeting and reported that during the telephone conversation Luton agreed that the appraised value of the property plus 5 percent and plus closing costs would not be unreasonable, but he was reluctant to put this in writing without also stating that this price would still be subject to NSF approval. Lee felt it might be to AURA's advantage to allow Luton to clear this proposal through the National Science Foundation first.

A short Executive Session followed for further discussion of the relationship with the University of Arizona. Counsel Shoenhair said the university officials were familiar with and sympathetic to AURA's problems with NSF. Secretary Miller pointed out that title to the property would vest with the National Science Foundation, not with AURA.

Lee left the meeting to take a call from Luton. He reported that Luton told him that a telegram would be sent to him authorizing negotiation on the basis of not to exceed 5 percent over the appraised price plus closing costs. The Board approved a motion that Counsel Shoenhair should consult with President Harvill about a suitable offer to AURA.

Luton's telegram,[75] which was sent to Lee in Ann Arbor on December 8, included a serious error[76] that had to be corrected before negotiations with the University could begin. The telegram read: ". . ., provided direct price not in excess of five percent of appraised price of $86,000." It should have read: "five percent over appraised price." Shoenhair phoned Shane[77] on December 15 to report that the University of Arizona had paid $90,074.81 for the land. He said: "Harvill is very cooperative, but Trustees are most impatient." Shane then called Luton, who agreed to the price and promised to call Shoenhair later in the day. Shane reported this to Shoenhair who was very unhappy and said the deal should be closed immediately because the Trustees are ready to pull out. Shane reminded him that "the U of A would be hurt as badly as we would in that case." Fortunately, there were no more NSF road blocks, and the transaction was concluded.[78] Purchase of the second half of the city block from the University of Arizona for $115,710 was approved by NSF six months later.[79]

The signing of the Kitt Peak lease and the acquisition of the land for the city head-quarters finally cleared the way to go out for bids and to start construction of the National Observatory.

10
The AURA Board of Directors choose a name for the observatory. The start of construction in Tucson and on the mountain. The Fort Huachuca drone aircraft electronics test range threatens Kitt Peak. The March 15, 1960 dedication.

Otto Struve wrote to McMath on June 13, 1955 concerning the proposed organization and name of the NAO.[1] With regard to the name, he wrote: "We could even attach to it the name of the site we shall ultimately agree upon. There is a certain advantage in specifying the location of an observatory in its name." This simple solution, the one finally adopted, was preceded by a series of proposed names, some of them based on bureaucratic considerations.

Goldberg spoke of a "National Observatory" at the Flagstaff Photoelectric Conference.[2] Subsequently the MPE Divisional Committee reacted favorably to Goldberg's discussion of the possibility of the National Science Foundation assisting in the planning of a "National Observatory".[3] The McMath Panel was named "Advisory Panel for a National Astronomical Observatory" when its members were appointed by the Director of the National Science Foundation.[4] This name was used in the FY54 and FY55 NSF Annual Reports.

The terms "Cooperative Observatory" and "Inter-university Observatory" were used in some internal NSF reports and memoranda.[5] Several NSF staff members objected to what they felt was implied in the use of the word "National", and suggested that "American" be used.[6] The July 1955 proposal from the University of Michigan to the NSF used the new word in the title: "Studies Leading to the Establishment of an American Astronomical Observatory".[7] The FY56 and FY57 NSF Annual Reports dropped "National" and changed the name of the Panel to "Advisory Panel for Astronomical Observatory." The initials NAO continued to be used in Panel correspondence, but the minutes of the final meeting of the Panel, written by the NSF Program Director for Astronomy, were headed; Minutes, Fifth Meeting of Advisory Panel for Astronomical Observatory, Feb. 25–26, 1957.[8]

The term "Cooperative Observatory" was once again used in the title of the October 29, 1957 proposal to the National Science Foundation that was approved at the first meeting of the AURA Board of Directors.[9] Hence, it was also used in Waterman's request to the NSB for approval to negotiate a $3.1 million contract with AURA.[10]

The NSF press release about the contract[11] used "Optical Astronomy Observatory" in the heading, and "National Astronomical Observatory" in the text.

Adoption of a name for the observatory was on the agenda of the annual meeting of the AURA Board of Directors in Ann Arbor, Dec. 8, 1958. A month before the meeting Secretary Miller sent the Board a list of five names that had been proposed.[12] McMath added "Kitt Peak National Optical Astronomy Observatory" in a letter to Miller.[13] Following lengthy discussion the AURA Board on December 8, 1958 voted unanimously to adopt "Kitt Peak National Observatory" as the name.[14] McMath forwarded this recommendation to Waterman on December 22.[15] The National Science Board gave its approval at its fifty-eighth meeting on January 23, 1959 and Waterman promptly informed McMath.[16] This was finally reported to the AURA Board during a special meeting in Tucson on March 3, 1959.[17]

Preliminary plans for a city headquarters building had been prepared along with the design of the 36-inch telescope mounting.[18] Meinel presented plans for a 16,000 square foot building on land adjacent to the University of Arizona at the February 10 meeting of the Executive Committee, three weeks before the formal selection of Kitt Peak.[19] The initial estimate of cost was $360,000, but this was trimmed to $225,000, excluding the cost of the land.[20] Budgets sent to NSF on May 13, 1958 included $120,000 for the city facility in FY58 and $296,000 in FY59.[21] The Scientific Committee discussed the plans for the building on July 9.[22] Shane pointed out that a library for 30,000 books could not be built without further appropriations beyond the 1959 funds. Many changes were suggested within the present space limitations, the most important being the plans for photographic dark rooms "because they involve plumbing." The Executive Committee approved the plans for a 15,000 square foot building on September 13.[23] Appointment of E.L. Varney and Associates of Phoenix as architect for the project was approved. Meinel said he would seek architectural services closer to Tucson in the future.

After eight months of frustrating delay, Luton had still not given NSF approval for the purchase of the city land by the time the drawings for the building were ready to go to bid. Meinel reported to the Scientific Committee on December 7 that it had been decided not to go to bid until the land question was resolved.[24] This delayed the opening of the bids until February 12, 1959.[25] The low bid was $365,164 compared to the final budget figure of $270,000, an increase of 37 percent. Meinel's explanation was: "We originally budgeted for 15,000 square feet. By actual count we have 20,600 square feet in the building, an exact increase of 37 percent." Postponement of part of the building until supplemental funds could be obtained seemed to be the best solution. Furthermore, it was known by this time that the solar telescope project would include funds for an addition to the city building.

Construction of the building began in April 1959, shortly after a $327,643 contract was awarded to the low bidder, M.J. Lang Construction.[26] The building was 90 percent finished by Christmas. The move from the temporary Park Avenue headquarters started on February 17 and was completed on March 1, 1960, two weeks before the dedication of the observatory.

The $250,000 grant to the University of Michigan in August 1955 (G-1990) included $49,500 for construction of the 36-inch telescope.[27] It was proposed then "that an order for such a telescope be placed as soon as possible and that it be erected at a temporary site pending the completion of the site survey." The $545,000 grant a year later (G-2622)[28] added $56,300 to the funding for the 36-inch telescope. Planning for this instrument was well along by the date of the incorporation of AURA. During its meeting in Pasadena on February 10, 1958 the Executive Committee chose E.L. Varney Associates, Phoenix, as architects for the building, and approved going to bid for the 36-inch telescope following review and approval of the final specifications for the mounting by the Scientific Committee.[29]

The bids for the mounting and building were opened during the April 4, 1958 meeting of the Executive Committee in Tucson.[30] Boller and Chivens, Pasadena, bid $69,978 for the mounting. Murray J. Shiff Construction Company, Tucson, bid $112,700 for the building. Based on these bids, the total cost was estimated to be $193,028, not including instrumentation or contingencies. This was $26,028 more than the funds available in the budget ($117,000 turn back from the University of Michigan plus $50,000 from NSF). Patey and Meinel were asked to revise the present budgets to obtain a financial plan within $3,100,000, the amount available in NSF Contract C63.[31]

Patey presented revised budgets for 1958, 1959, and 1960 at an evening session. These were discussed from 10:00 p.m. to 11:00 p.m., but action was postponed until the next day. The total amount needed to complete the 36-inch telescope, including instrumentation, was now estimated to be $233,028. The free balance to be returned by the University of Michigan had shrunk to an estimated $90,000. This amount plus the promised $50,000 supplement from NSF was $93,028 short of meeting the need. The revised budget for the period January 1–June 30, 1958 rearranged spending plans to cover the shortage. The Executive Committee unanimously approved $744,929 for this period, and $2,420,985 for fiscal year 1958–59.

The Executive Committee thereupon unanimously approved award of the contract for the 36-inch telescope mounting to Boller & Chivens, and award of the contract for the building, dome and hydraulic platform to Murray J. Shiff Construction Company. Both of these were subject to NSF approval, and the construction contract was also conditioned on the signing of the lease with the Papago Tribe. Purchase of the 80-inch mirror blank for $114,000, FOB Corning was also approved at this meeting.

The contract with Boller & Chivens was approved, and McMath signed it in July.[32] The construction contract was delayed by the slowness of NSF in approving the terms of the Kitt Peak lease.[33] The Shiff Construction Company signed the contract on October 6, 1958 and scheduled work to begin on November 1.[34] The exterior aluminum panels of the building were damaged by violent windstorms on December 24–25, 1959.[35] Except for repairing this damage, the building was ready to receive the telescope when it was delivered on February 26.[36] Installation was completed on March 12.[37] This was just three days before the dedication of the observatory.

Funds for the 80-inch telescope were also provided in the August 1955 and August 1956 grants to the University of Michigan (G-1990 and G-2622).[38] G-1990 provided $11,500 for "80-inch telescope study." The 1956 proposal asked for $229,000 for the 80-inch, including $80,000 for the mirror blank. The money for the mirror blank was not included in the grant because the House had put a restriction on capital funding in the budget bill.[39] As in the case of the 36-inch telescope, planning for this instrument was started before the incorporation of AURA.

There was a brief discussion of the plans for the 80-inch telescope during the Executive Committee meeting in Pasadena on February 10 and 11, 1958.[40] It was pointed out that the new AURA contract with C.W. Jones would add planning the 80-inch telescope mounting to his responsibilities. The Scientific Committee reviewed the plans for the 80-inch mounting on March 2, the day after Kitt Peak was chosen.[41] Purchase of the mirror blank was approved by the Executive Committee on April 5, 1958.[42] The Scientific Committee reviewed changes in the plans for the 80-inch mounting, the design of the mirror, and plans for the observing platform at a meeting in Phoenix on July 9.[43] Final approval of the design specifications was delegated to a small subgroup (possibly Shane, Mayall, Meinel, Bowen and the engineers).

Major progress on the 80-inch telescope was reported in detail to the Scientific Committee on December 7, 1958,[44] and in summary to the AURA Board on December 8.[45] The 80-inch Telescope Committee reported that it had met in Tucson and had approved the design on December 29, 1958.[46] Bid invitations were issued January 15, 1959, and the opening of the bids took place on March 31, 1959.[47] The low bidder for the telescope mounting was Willamette Iron and Steel Company, Portland, Oregon ($365,527 or $363,313, depending on the choice of drives). The Boller and Chivens bid was third ($416,348 or $413,149). The high bid came from Westinghouse Electric Corporation ($661,846 and $655,881). The low bidder for the building, dome, and pier was Murray J. Shiff Construction Company, Tucson ($375,050). The high bid came from Consolidated Western Steel ($467,298). M.M. Sundt Construction, Tucson, was third ($414,145). The FY59 Financial Plan had called for $700,000 for the mounting, and $490,000 for the building, dome, and pier. Even the high bids were lower than these figures.

Secretary Miller telephoned the members of the Executive Committee on April 7, 1959 and received approval for President Shane to execute several construction contracts, including the two for the 80-inch telescope.[48] The pyrex mirror blank, with a diameter of 84-inches, arrived in Tucson on October 7, 1959.[49] The building was about 30 percent complete when the dedication ceremony for the observatory took place on March 15, 1970.[50]

The 36-inch and 80-inch telescopes had been formally recommended as the basic equipment of the observatory at the second meeting of the NAO Panel.[51] The Panel also "believed that a 60-inch solar telescope would be completely justified if the requisite daytime seeing could be found." The Panel voted at its third meeting to include daytime solar seeing in the program of the site survey.[52] The "on again–off again" history of funding for the solar telescope finally came to a satisfactory conclusion when Congress included $4,000,000 for such an instrument in the FY59 NSF appropriation.[53] The National Science Foundation added this to the contract with AURA on November 22, 1958.[54] The appointment of Pierce as Associate Director of the Observatory on September 13, 1958[55] was an administrative change that did not alter his responsibilities in planning the solar telescope.

McMath discussed the solar telescope project with the Scientific Committee on December 7.[56] He proposed that the prestigious firm of Skidmore, Owings and Merrill (SOM) be employed to conduct studies on the solar telescope, dealing mainly with building and structural features, and temperature control. This came to the AURA Board as a recommendation from the Scientific Committee, and the Board approved.[57]

A contract with C.W. Jones in an amount not to exceed $50,000 for engineering studies of the solar telescope heliostat mounting was approved by the Board on March 3, 1959.[58]

The initial proposal by Skidmore, Owings and Merrill was not satisfactory to NSF, but a revised contract for $40,000 was approved by NSF and executed by AURA on April 6, 1959. The contract for $50,000 with C.W. Jones was also executed. These contracts were in force a few days later after being signed by the contractors.[59]

Four Skidmore, Owings and Merrill staff members had previously visited Kitt Peak where they met with Shane, Edmondson, Baustian and Pierce on the morning of March 4 and with McMath in the afternoon in Tucson.[60] These discussions were concerned with evaluating the subsurface conditions of the proposed location of the solar telescope, and with three sets of suggested plans of the building prepared by Skidmore, Owings and Merrill. Review of these plans led to a radical new proposal. The light would still be sent from the heliostat down the south polar axis toward ground level, but instead of being reflected by the 60-inch concave mirror located near ground level, the light would proceed into a 300-foot tunnel underground to the 60-inch concave mirror. It would reflect the light to a focus near ground level. The 80-inch heliostat, located 100 feet above the ground, could be supported by: (1) a full index building; (2) a diagonal tube, plus a vertical tower for support of the north end; or (3) a cantilever as proposed by Skidmore, Owings and Merrill.[61]

The Solar Subcommittee (McMath, Shane and Pierce) met with William Dunlap and Myron Goldsmith of Skidmore, Owings and Merrill in Ann Arbor on June 1, 1959. Baustian and Dr. Orren Mohler, from the McMath–Hulbert Observatory, were also in attendance. Dunlap and Goldsmith presented a complete review of SOM's design and cost studies. Ten possible structures had been investigated and two of them were recommended. The estimated costs were $2,124,000 for the cantilever, and $1,863,100 for the tower and diagonal. The Solar Subcommittee unanimously adopted the tower and diagonal.[62] Following this, Skidmore, Owings and Merrill received a $200,000 contract for the design of the solar telescope building.[63]

Test drilling at the site on Kitt Peak was started in July and completed in August.[64] Three vertical holes were 100 feet deep, and a diagonal hole was drilled along the planned underground portion of the solar telescope. More holes were drilled in September.[65] A total of 730 feet of diamond drilling was completed along the axis of the solar telescope. Seven holes were drilled, five vertical and two inclined, at a cost of $7.53 per foot. Competent rock for foundations was found at depths of 15 to 20 feet. Zones of soft rock several feet thick were found in some of the holes at various depths up to 100 feet. This was not thought to present any serious problems.

A new cost estimate was prepared by Skidmore, Owings and Merrill on December 1.[66] This was discussed with Pierce and Baustian on December 4 in Chicago, and with McMath, Mohler, Pierce and Baustian on December 5 in Pontiac. The estimated total was $2,395,200, compared with the revised budget figure of $1,900,000. $200,000 was saved by eliminating the cover for the heliostat. $50,000 could be saved in the cost of the several hoists. SOM had estimated the cost of cooling to be $501,170. Recent tests of the cooling efficiency of TiO_2 paint suggested that this figure might be too high. It was decided to go to bid with:

(a) Base building including all header, piping, pump, house, etc., but no water cooling surface panels.

(b) Alternate 1 – addition of 50% water cooling panels.

(c) Alternate 2 – addition of 100% water cooling panels.

Bid (a) represented an estimated $200,000 savings. The total possible saving was estimated to be $722,000. April 18 was the planned date to go to bid.

Activities in January 1960[67] included: (1) four new drill holes at the request of SOM to check fault lines; (2) a "wind tunnel test" of the solar telescope building. A 1/16 inch to the foot scale model was placed on top of a car and driven at 70 miles per hour; (3) preparation of bids for surface excavation at the site (4) a revised solar project budget covering the $4,000,000 appropriation was sent to NSF.

A contract for $29,090.00 was awarded to the Tate Mining and Development Company on February 4 for excavation and clearing required by the solar telescope on Kitt Peak.[68] This work included all rough excavation exclusive of the tunnel, at a unit cost of $4.58 per cubic yard. The report on the results of the supplementary core drilling was received from Heinrichs Geoexploration Company on February 11.[69] No serious problems were noted. Tate Mining and Development Company started work on the site on March 14, the day before the Dedication of the Observatory.[70]

The U.S. Army Electronic Proving Ground (AEPG) at Fort Huachuca, Arizona began to develop plans in 1958 for a test range for drone airplanes carrying the latest electronic equipment. The test area was planned to extend 266 miles across southern Arizona from Fort Huachuca to Yuma. Meinel was informed about this by architect Cole, who heard about it during the Fort Huachuca Annual Board Meeting.[71] A telephone conversation with General F.W. Moorman, Commanding Officer at Fort Huachuca, was followed by a letter from Patey to Colonel George F. Moynahan, Director of Combat Developments, AEPG.[72] Patey's letter described the plans for the Kitt Peak National Observatory, and emphasized that electronic interference was a serious concern for the observatory. He suggested that a meeting to discuss the problem would be useful.

Colonel Moynahan's reply[73] said that Fort Huachuca had been chosen as the site for the AEPG "largely because environmental conditions here are ideal for testing radiating electromagnetic systems." A map of Arizona showing the planned installations was enclosed. He proposed that a meeting should be held at an early date.

Meinel and Patey met with Colonel Moynahan and members of his staff at Fort Huachuca on January 7, 1959.[74] They learned that the track for the drone airplanes would be in close proximity to the north side of Kitt Peak. Kitt Peak itself would be used as a location for a microwave relay and a tracking radar. "Colonel Moynahan repeatedly impressed upon us that the comparatively small amount of money invested in our project does not balance against the money and national defense factors with which he is concerned."

Meinel reported to Shane[75] that he had talked to Chester Higman, Business Manager for the Papago Tribe, on the afternoon of January 7. Higman said the Tribe had no knowledge of the AEPG plans. He believed that no one, not even the military, could enter and use parts of the reservation without the consent of the Tribe, and he planned to explore the situation. He felt the Tribe would honor the terms of the Kitt Peak lease limiting the use of the mountain to scientific work related to astronomy.

The January 9 news story released by AEPG said: "Since much of the land between Fort Huachuca and Yuma is owned by either the federal or state governments, only a few acres of private land will have to be acquired."[76] The AEPG officials who were quoted in the press release did not know that Indian Reservation land is not government land!

Meinel wrote to Shane a week later:[77] "The Bureau of Indian Affairs confirms our previous information that the AEPG cannot enter and use reservation land without consent of the Tribal Council." Meinel also reported that a representative of the AEPG had approached the Papago Tribal Council on January 12 "to request approval for access to the sites on the reservation, one of which was Kitt Peak. He was advised to submit the request at the next meeting of the Tribal Council for action, but was informed of the expected refusal regarding Kitt Peak."

The sneaky attempt to get access to Kitt Peak only five days after the meeting at Fort Huachuca caused Meinel to tell Colonel Moynahan to send all future communications to Dr. C.D. Shane, the President of AURA.[78]

Following the meeting at Fort Huachuca, Meinel described the situation to Keller at NSF and Dr. James B. Edson, Assistant Secretary for Research and Development, Department of the Army.[79] In addition to being a "close personal friend" of Meinel,[80] Edson was the brother-in-law of the discoverer of Pluto, Clyde Tombaugh. He had also been on the staff of the Lowell Observatory. Meinel told McMath on the telephone that he thought he had "headed off the Army problem through his friend Dr. Edson."[81] He told the Board of Directors on March 3: "The proposed Army encroachment on the Kitt Peak site was prevented with the help of the Indians."[82]

Colonel Moynahan paid no attention to the opposition that was developing and asked for 10 additional copies of the Kitt Peak site topographic map. Meinel refused to send them, and told Colonel Moynahan:[83] "We have been informed by the National Science Foundation that the proposed usage of Kitt Peak by the AEPG will be opposed at high level." Twelve weeks later the AEPG had yielded to higher authority and had developed a new plan.

General Moorman described the new plan in a full dress presentation to Shane, Meinel and Patey at Fort Huachuca on April 14 following lunch.[84] A chart was displayed showing the major change that had been made in the drone track. It would now pass south of Baboquivari, about 20 miles south of Kitt Peak. This eliminated the danger to the observatory and was a highly satisfactory conclusion to a difficult problem.

Following the presentation General Moorman accepted an invitation to visit Kitt Peak in May. He and three members of his staff, Colonel Moynahan, Lieutenant Donald K. Adams and Dr. John T. Myers, arrived on Kitt Peak by helicopter on May 7.[85] They landed on the clearing at the solar site, and were given a tour of the mountain before driving to Tucson to visit the city offices. It was a friendly conclusion to the encroachment issue.

Shane telephoned Waterman on April 15 about the meeting with General Moorman at Fort Huachuca, and he followed this with a letter and a copy of the map showing the new drone range.[86] Keller told the Executive Committee on June 2 that Waterman intended to make further inquiries about the Army's interest in the general area.[87] Waterman finally wrote to the Secretary of the Army, Wilbur H. Brucker, on June 22, 1959.[88] The letter included: "The purpose of this letter is to tell you that these difficulties which we foresaw now appear to have been solved.". . . I have been advised by Drs. Shane and Meinel that this revised proposal is satisfactory to them with regard to the protection it would afford the Kitt Peak National Observatory from serious radio disturbances." He added: "We very much appreciate the cooperation of the Commanding General of Ft. Huachuca in this instance and will be grateful if this situation is kept in mind in any future planning of Army activities in the neighborhood of Kitt Peak."

A ground-breaking ceremony for the 36-inch telescope building had been discussed in the October 6, 1958 weekly staff meeting, and the decision was made not to have one at this time.[89] The idea was revived, apparently on Patey's initiative, when word was received that Robert B. Brode, Klopsteg's successor as NSF Associate Director, and Geoffrey Keller, Program Director for Astronomy, planned to visit Tucson on November 4–5, 1958. Patey made an off-hand remark during a telephone conversation with Embrey[90] that "they might set off a stick of dynamite or something" to mark the ceremonial beginning of the project. She thought this was an excellent idea because it could get some national publicity for the observatory and NSF. A few days later Patey saw Luton in Washington[91] and "discussed some of the problems that have arisen in connection with this ceremony." The guest list was turning into a major problem, and Luton felt that Meinel should be told that NSF would not be offended "if AURA decides that the simple arrangements initially planned by AURA are getting out of hand and should be called off." Keller used these very words in a letter to Meinel.[92] Embrey's reply to Luton[93] was defensive, and blamed Patey for going too far without AURA approval of his proposal. The minutes of the October 20, 1958 weekly staff meeting read:[94] "Because of the lack of time for adequate planning this proposal was abandoned in conference on October 25th with Edmondson, Shane, and Miller in consultation." Nevertheless, Patey mailed a copy of the proposed guest list (39 names) to Edmondson the next day.[95]

Patey again raised the question of a ground-breaking ceremony, this time in connection with the March 2, 1959 meeting of the NSF Advisory Panel for Astronomy in Tucson and the March 3 Special Meeting of the AURA Board. Edmondson wrote to Meinel:[96] "I told Ralph Patey on the phone that I do not look upon the coming Tucson Board Meeting as being anything as formal as a ground-breaking ceremony. We are meeting at that time primarily because the NSF Astronomy Panel is also meeting in Tucson." He said he thought AURA should try to persuade NSF to schedule a meeting of the National Science Board in Tucson, and use that as the occasion for a dedication of the 36-inch.

Ground-breaking ceremonies were discussed by the AURA Board during the March 3, 1959 meeting.[97] Edmondson suggested to the Board that AURA and the University of Arizona should invite the National Science Board to meet in Tucson in January 1960, and that a ceremony could be planned for that time. The Board agreed, and appointed a Special Committee on Ground-Breaking Ceremonies: Edmondson (Chairman), Carpenter, and Patey.

Edmondson wrote an exploratory letter to Waterman[98] about the possible interest of the National Science Board in holding a meeting in Tucson in connection with a ceremony at the Kitt Peak site. March 1960 was suggested as a good date, owing to the possibility of bad weather on the mountain in December or January. If the Board agreed, a joint invitation would be extended by AURA and the University of Arizona. Waterman's reply[99] said that "probably the December or January meeting would be more convenient from the point of view of the Board than the March meeting." He said he would present this suggestion to the Board at the May 24–25 meeting.

Carpenter informed Harvill about the appointment of the Special Committee. Harvill told Edmondson that he and McMath had some conversations prior to the AURA Board meeting about a joint invitation to the NSB, and he reiterated his desire to cooperate with AURA.[100]

Edmondson consulted with Shane and the staff in Tucson before sending a reply to Waterman:[101] "We are all agreed that we would like to have the National Science Board come in December or January, if those dates would be more convenient than the March meeting." McMath objected:[102] "A meeting in December or January would be very awkward from my standpoint, as one month is the end of the year month and the other the beginning of the new year–both exceedingly busy ones for me." Edmondson replied[103] that he had pushed for March in his first letter to Waterman, and had accepted December or January on the basis of Waterman's response.

McMath got his way on the date of the meeting by going straight to Waterman.[104] The NSB accepted McMath's personal invitation to visit the Kitt Peak National Observatory in connection with their March meeting rather than the previously approved January meeting. The scheduled NSB meeting dates were March 14 and 15, Monday and Tuesday.

McMath's letter to Waterman had bypassed both the Special Committee and the University of Arizona. This led Patey to raise the question of who would be the official host.[105] McMath undertook immediate damage control by writing to Harvill:[106] "I enclose a copy of a very recent letter from Dr. Waterman for your information. As I remember our last conversation concerning this, it was agreed that your University and AURA would act as co-hosts to the National Science Board." Harvill thereupon wrote to Waterman:[107] "I am very pleased to renew the invitation to hold meetings at the University of Arizona." Harvill acknowledged McMath's letter:[108] "The proposal meets with our complete approval. We shall take great pleasure in acting as co-hosts and arranging for meetings of the National Science Board." After receiving Harvill's letter, McMath wrote again to Waterman:[109] "I note that in my previous correspondence with you concerning the March meeting of your Board in Tucson, I totally neglected the fact that the University of Arizona would be co-host." . . . "I am afraid I just took it for granted, in writing to you, that everyone would understand that the University of Arizona would be co-host." A copy was sent to Harvill.

A letter from Patey[110] gave the Chairman of the Special Committee, Edmondson, his first word about the changed dates for the NSB meeting. Eventually an apology was received from McMath:[111] "I have been very remiss and I apologize. My correspondence with Alan Waterman was spontaneous, and, in a way, personal. I just plain forgot that you were Chairman of the 'Host Committee'." The annual meeting of the AURA Board of Directors was tentatively re-scheduled for March 11 and 12.

A brochure about the observatory and the roles of AURA and the National Science

Foundation was well under way by this time.[112] Edmondson was the coordinator with draft copy being supplied by Lee Anna Embrey for NSF and by Meinel, Pierce and Patey for the observatory. Edmondson sent copies of all the material he had received to Embrey, Patey and Shane on November 7, 1959 and said he hoped the text could be put in final form when he planned to be in Tucson December 7–11.[113] Unfortunately, at this time Embrey was relieved of her public-relations responsibilities with respect to astronomy projects.[114] The NSF Public Information Office (PIO) was now fully staffed, and no longer needed the assistance she had been giving on an interim basis. Her successor in the PIO was Roland D. Paine, Jr.[115]

The University of California Press estimated two months prior to the March meeting as the time needed to print the brochure.[116] Dr. and Mrs. Shane accomplished a miracle in having the brochure ready for distribution at the Dedication. The AURA Board passed a resolution thanking Mrs. Shane for the excellent layout and her "splendid work on the brochure." Secretary Miller sent her a copy of the resolution and added his personal thanks.[117]

Formal planning for the meeting and dedication began when Edmondson asked Harvill to designate someone from the university to work with the AURA Special Committee.[118] Harvill responded by designating Dr. David L. Patrick, Vice-President of the University of Arizona, to chair a small committee on arrangements for the meeting.[119] The university made hotel arrangements for the NSB and NSF staff, and also provided a large meeting room on the campus for the National Science Board.

Edmondson, Carpenter and Patey met in Tucson on December 8–11 to discuss plans for the meetings of the NSF and AURA Boards, and for the dedication ceremony. The report of the Committee[120] included a list of guests to be invited in addition to the NSF and AURA Boards and staff. The report also proposed a dedication program for Tuesday, 15 March, 1960. Shane would preside, and the speakers would be Meinel, McMath, Waterman, and Dr. Detlev Bronk (Chairman of the NSB). Dr. W.W. Morgan was the unanimous choice of the Committee to give the dedication address. Edmondson's cover letter to McMath[121] asked for suggestions about additions to the guest list. He also reported that he and Mrs. Shane had made substantial progress on material for the brochure in Tucson on December 7. He met with President Harvill and Vice-President Patrick to report on his December 2nd discussion with Vernice Anderson, Executive Secretary of the NSB, about the logistics of the NSB meeting.[122]

McMath had been in bed with a sinus infection, and his reaction to the report of the Committee was transmitted in a letter written by Orren Mohler.[123] McMath felt that Waterman should give the dedication address and that "Waterman could feel injured if an invitation to deliver this important address goes to a nearly complete outsider." McMath's second choice was Struve. Edmondson sent thermofax copies of this letter to Carpenter, Patey, Meinel, and Secretary Miller.

Edmondson's reply to McMath[124] explained why Morgan had been chosen for the dedication address. The Committee felt the dedication address should be given by an astronomer on an astronomical subject. Morgan was chosen for the following reasons: "a) He is an outstanding research astronomer working on the frontiers of modern astronomy. b) He is an enthusiastic speaker who would be capable of giving an address that would be interesting to all of the members of the NSF Board and the AURA Board. c) Dr. Morgan will be the first Consulting Astronomer on the staff of the Kitt

Peak National Observatory and this seemed to make him an especially happy choice."

Edmondson told McMath that discussions regarding protocol, seniority, and the dedication address had been held with Clyde Hall, Roland Paine, Lee Anna Embrey and Vernice Anderson at NSF on December 2. They told him that if NSF were to be responsible for the dedication address it should be given by Dr. Bronk, because the Chairman of the NSB is senior to the Director of the NSF. There was a very favorable reaction to having an astronomer give the address. The program proposed by the Committee had been approved in detail by Dr. Waterman, according to Vernice Anderson in response to an inquiry from Edmondson that followed receipt of Mohler's letter.

McMath mailed two letters to Edmondson on December 22,[125] before he had received Edmondson's letter explaining why Morgan had been chosen. "Frank–Disregard!" was written in longhand at the bottom of the first letter. The PS at the end of the second letter said: "This supersedes all other and previous letters." This letter listed eight suggestions about the program made by Leo Goldberg, and the final one was: "8. Dr. Morgan will then make the dedication address." McMath wrote to Morgan on January 14, 1960 formally inviting him to "be the Dedication speaker."[126] Morgan sent his acceptance on January 20.[127]

The guest list initially suggested by the Special Committee had 35 names.[128] Among those added were the new Commanding General of Fort Huachuca, Dr. Wernher von Braun, and Lewis W. Douglas (a personal friend of Waterman). Keller and Gerard Mulders (Associate Program Director for Astronomy) proposed Admiral Rawson Bennett (Chief of Naval Research, ONR) and Lt. General B.A. Schriever (Commander, Air Research and Development command).[129] Waterman felt that if these two persons were invited, five other high ranking military people would have to be invited for protocol reasons. They probably would not come, but would send low ranking representatives. The two names were not added to the list.[130]

The final list of guests that was sent to Waterman for approval contained 48 names, not including members of the National Science Board and the AURA Board.[131] In addition, Waterman would extend special VIP invitations to Senator Carl Hayden, Senator Barry Goldwater, Representative Stewart Udall, Representative John J. Rhodes, Representative Albert Thomas, and Dr. T. Keith Glennan (NASA Administrator).

Waterman quickly gave his approval to the proposed list of guests.[132] He also wrote: "The proposed program is satisfactory with Dr. Bronk and me. I hope Dr. Morgan will be able to give the dedication address." This approval cleared the way to finish planning the detailed arrangements.

On March 6, 1960, nine days before the dedication of the observatory, Mrs. Joseph H. Perue, Jr., the wife of a visiting scientist, found a cache of miniature Papago pottery in a small crevice near the summit of Kitt Peak.[133] She took these to the mountain headquarters where they were seen by Meinel. He immediately told her to put them back where she had found them. On March 9 he called Dr. Emil Haury, Director of the Arizona State Museum and Professor of Anthropology at the University of Arizona, and told him what had happened.[134] Haury accompanied Meinel to Kitt Peak on March 10 to examine the material and to decide on a course of action. Enos Francisco, Chairman of the Papago Tribal Council, was told about the cache on March 12. Permission was requested and obtained to remove the specimens to the Arizona State Museum at

The first five National Science Foundation Program Directors for Astronomy. From right to left in order of service: Peter van de Kamp, Helen S. Hogg, Frank K. Edmondson, Geoffrey Keller, Gerard F.W. Mulders. Taken at the dedication of KPNO on 15 March 1960. NSF Photo Archives.

the university for cleaning and restoration, and to hold them until the museum on Kitt Peak was completed. A small party from the Arizona State Museum came to Kitt Peak on March 13, 1960, two days before the dedication, to recover the entire collection of whole and fragmentary vessels. This was a potentially explosive situation, and the dedication could have been disrupted by Papago protests. Meinel and Haury acted in a very careful and diplomatic way to defuse the situation before it got out of hand. The pottery is now in a very attractive display at the Kitt Peak Visitors Center.

The schedule of meetings and the dedication filled four days.[135] The AURA meetings began on Saturday, March 12, with a meeting of the Scientific Committee in the new city headquarters building. The Executive Committee met on Sunday, March 13, also in the headquarters building. Monday was devoted to the meetings of the two Boards. The AURA Board met in the headquarters building, and the National Science Board met in the Arizona State Museum on the campus. The University of Arizona was host at a reception and dinner at the Pioneer Hotel, which also provided rooms for the NSB during the meeting period. The AURA group stayed at the new Del Webb HiWay House.

The National Science Foundation group paid an inspection visit to the city headquarters on Tuesday before proceeding to Kitt Peak. The dedication ceremony began at 11:00 a.m. and lasted about 45 minutes. The Vice-Chairman of the NSB, Paul Gross, substituted for the Chairman, Detlev Bronk, who was prevented from attending at the last moment. The other speakers were: McMath, Waterman, Meinel, and Morgan. The new Chairman of the Papago Tribal Council, Enos Francisco, was invited to speak at the luncheon.[136] Tours of the mountain were conducted between 2 and 4 p.m. The final scheduled event was a cocktail hour at the HiWay House at 6:30 p.m., sponsored by the Valley National Bank.

An excellent article about the dedication was published in *Sky and Telescope*.[137] The author was Charles A. Federer, Jr., at that time Editor of the magazine. He had received a special invitation to attend because he was the senior person in his field.

The meetings of the Scientific Committee, the Executive Committee and the AURA Board of Directors will be reviewed in Chapter 12. It should be mentioned here that Meinel's resignation as Observatory Director was accepted by the AURA Board on March 14.[138] He was appointed Associate Director in charge of the Stellar Division and the proposed Space Division of the Observatory.

The dedication was a great success by all accounts, and Shane sent a note to all employees expressing AURA's appreciation for the part they had played and a job well done.[139]

Part III
New directions for AURA

11 The Kitt Peak space program

The launch of Sputnik I on October 4, 1957 had the beneficial effect of restoring funding for the Kitt Peak solar telescope, as described in Chapter 8. The creation of the National Aeronautics and Space Administration in 1958[1] provided AURA with a new opportunity. A telegram from Lloyd V. Berkner, chairman of a National Academy of Sciences group on space satellite problems, was read by Shane at the July 9, 1958 meeting of the Scientific Committee.[2] He wrote to McMath:[3] "I introduced this subject to get the Scientific Committee thinking about it since I feel it is something we cannot ignore, though there seems no way at present for us to contribute to it."

Six months later Meinel told Shane there was a way that AURA and KPNO could contribute.[4] He wrote: "We are rapidly nearing the completion of creative work on the stellar instruments and my attention has been directed to pending projects[5] in the placement of optical telescopes in space vehicles. This field looks very attractive, inasmuch as my interests lie in the planning and design of new observing techniques." He said he would like to "submit a proposal for the development of a large aperture pointable satellite telescope," and he hoped that AURA would permit him to direct a major portion of his efforts toward such a program. It might be necessary to add an astronomer to the staff to be in charge of the stellar division in order to provide him with time for the new activity in addition to his duties as Observatory Director.

Shane wrote about this to McMath:[6] "After phoning you yesterday I talked to Code about the Space Vehicle and Princeton matter. . . . We have to get into this business or be left behind, and if we can get Princeton to come in and have N.S.F. support we should accomplish great things. But <u>this must not be at the expense of the projects now under way</u>. It may however replace the X-inch [telescope]." He also said: "The only persons who know about the Space Vehicle and Princeton Business are McMath, Code, Shane, Meinel, Patey, and the latter only because he occupies the same office with Meinel and necessarily overheard telephone conversations. After January 30 we can increase slightly the number who know about it, i.e. if all goes well."

Vice-President Edmondson was soon brought into the loop and joined Shane and Meinel in a visit to the National Science foundation on January 29 and Princeton on

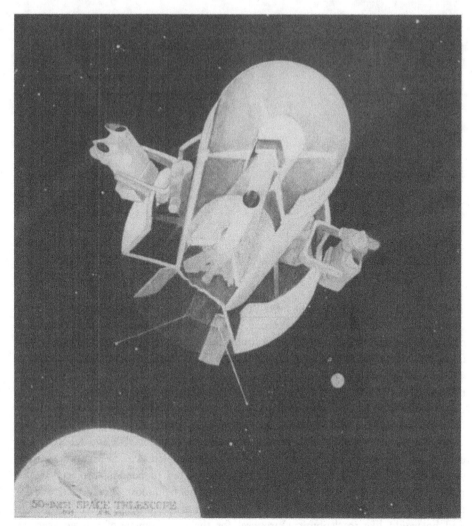

50-inch (1.25 meter) space telescope, proposed by Aden B. Meinel in 1959. NOAO Photo Archives.

January 30.[7] All did go well on both of these visits. NSF was "very much interested in the plans for development of space astronomy at KPNO."[8] The focus of the discussion at Princeton was the possibility that Princeton would join AURA if Meinel's proposal were accepted by the AURA Board.

Princeton had declined the invitation to be a founding member of AURA[9] because Lyman Spitzer and Martin Schwarzschild had heavy ongoing commitments. Spitzer was half-time with the Princeton Plasma Physics Laboratory (controlled nuclear fusion) in addition to his duties as Chairman of the Astronomy Department, and Schwarzschild was fully occupied with Project Stratoscope (a balloon-borne astronomical telescope). "Martin and I discussed it at some length and decided that while we approved of AURA and would like to support it, we didn't feel that our potential contribution for either of us as a board member was that central to the organization, to override the very strong personal arguments against our permitting that much time. So that is essentially why we didn't join."[10]

Spitzer's activities also included "plans for ultraviolet space astronomy, which led a few years later to the Orbiting Astronomical Observatory (OAO) program."[11] He had been interested in space research since 1946.[12] During the January 30 meeting Shane proposed "a program of minimum participation" that Spitzer felt he could live with. He would be expected to attend the annual meeting of the Board of Directors,[13] and he would be Chairman of a sub-committee of the Scientific Committee to deal with the problems of observations from space vehicles. The committee would include Code, Meinel and Shane. The Princeton administrative member of the Board would serve on the Executive Committee.

This paved the way for discussion of the proposed space program by the Scientific Committee in Tucson on March 1, 1959.[14] Kuiper and Menzel were absent. Spitzer, Keller, Robertson and Luton were in attendance. Robertson said the National Science Foundation looked favorably on the program after some preliminary review. Spitzer discussed the Princeton projects, neither of which would become a part of the AURA program. Keller suggested that AURA might consider undertaking a feasibility study of the space program. It was agreed that further discussion would be deferred until the next day. The Committee adjourned at 11:45 p.m.

The Committee reconvened on Kitt Peak at 11:00 a.m. on March 2, with Menzel also in attendance. The Committee reviewed the previous evening's discussion and then unanimously approved a recommendation to be presented to the Board. "That AURA undertake studies in the nature of preliminary conceptual design and experiments, leading toward the development of telescopes for making astronomical observations from outside the earth's atmosphere; the studies are to be supported by funds especially provided for this purpose; the studies shall not reduce in any degree the emphasis on existing programs of the Observatory; the facilities for space observations that may result from the studies shall be available to astronomers on a similar basis to the Observatory's surface instruments."

Code presented the recommendation from the Scientific Committee at the Board meeting the next day.[15] Meinel and Edmondson left the meeting at the beginning of the discussion. When they returned Edmondson said they had gone to the airport to meet with Robertson, who was on his way back to Washington, regarding the information on the Kitt Peak space program to be released to the press. He read the proposed press release to the Board, and it was approved. The *Tucson Daily Citizen* carried the story on page one the next day with a banner headline: "Kitt Peak Astronomers Planning Space Scope."[16]

During the absence of Meinel and Edmondson the Board approved the recommendation of the Scientific Committee regarding the space program. Authorization was granted to the AURA staff to make an application to NSF for a grant in the amount of $160,000 to fund the feasibility study on the space program. The Board also approved the appointment of Mr. Russell A. Nidey as Associate Engineer for the space program, subject to the availability of funds.

In Executive Session the Board unanimously agreed to invite Princeton University to become a member of AURA. Secretary Miller was requested to write to the President of Princeton in this regard; the letter was mailed on March 19.[17] The response from

President Goheen, including a check for $2,500, was dated April 8.[18] Miller forwarded it to McMath who acknowledged Goheen's letter.[19] He wrote: "I am very pleased that you have designated Dr. Lyman Spitzer, Jr. for a three-year term and Mr. Raymond J. Woodrow for a two-year term. Both of these gentlemen were very helpful as guests at the last meeting of the Board of Directors held in Tucson, Arizona, on March 3, 1959."

The proposal for $160,000 was sent to NSF in April.[20] Waterman signed the grant letter on June 9 (Research Grant NSF-G9145).[21] Meinel had already reported the approval at the June 2, 1959 meeting of the Executive Committee.[22] Meinel also reported that AURA's interest in the space telescope had been presented to several Government agencies including the Army, Advanced Research Projects Administration (ARPA), Army Ballistic Missile Agency (ABMA), and the Air Force. He also reported that there was "some degree of reluctance on the part of the National Aeronautics and Space Administration to agree to the AURA program." There had been a conference to discuss the AURA program at NASA on May 26 attended by Dr. Nancy Roman, Head, Observational Astronomy Program (NASA), Dr. Gerhardt Schilling, Chief, Astronomy and Astrophysics Programs (NASA), and Dr. G. Keller, Program Director for Astronomy (NSF). The signed memorandum summarizing the understandings that were agreed to was read to the Executive Committee by Keller. The Executive Committee voted to accept this statement, and thanked Keller for his efforts.

In 1982 Roman recalled some of the reasons for NASA's reluctance:[23] "Aden and I think Russ too, surprisingly, were just completely naive as to what was involved in trying to build a 50-inch telescope. They had no background of engineering in the field, except for Nidey's very restricted guidance experience. . . . All the problems that are involved in designing space instrumentation were completely foreign to them and, as far as I could tell by talking to them, they not only had no conception of what these problems were, they thought I was being very foolish in feeling that the problems existed." Spitzer agrees[24] that he "and probably many others underestimated the size and amount of the engineering activity that would be needed."

Meinel found some positive feedback from the Fort Huachuca problem.[25] He wrote to Shane[26] that "Dr. Edson will set up a meeting with ARPA and ABMA groups for Friday, April 24, at the Pentagon." He said that Keller was concerned about the politics of the "civilian" and "military" space programs, and that NSF would prefer to work with NASA "if they get realistic about sharing the field."

Meinel wrote a report of the April 24 meeting for the AURA Board[27] and sent a copy with a letter to Shane.[28] The letter ended: "To sum up the negotiations, I feel that our objectives were attained and that both AURA and NSF are achieving a position where the ebb and flow of the missile struggle will balance."

Meinel sent Shane two letters[29] dated May 28, 1959. The second letter described meetings at NASA and with Air Force representatives at the Pentagon on May 22. The first letter told about a May 18 visit in Tucson from Mr. C.D. Swanson and Dr. Rudolph Festa of ABMA. "The chief topics of discussion were technical and political aspects, should our telescope be designed for the specific capabilities of the Saturn rocket. The Saturn booster is a cluster rocket, basically consisting of seven Jupiter rockets, and develops a thrust of 1,500,000 lbs. It is expected to fly in 1960 and be operational in 1963. It is an ARPA project being developed by ABMA. . . . At the present time ABMA missiles, when developed, are deployed by the Air Force." He told Shane that "ABMA would like to invite us to come to Huntsville to meet with the staff and see their installation."

The Huntsville (Redstone Arsenal) visit was scheduled for Wednesday, 12 August 1959.[30] Edmondson represented AURA, Meinel and Nidey represented KPNO, and Mulders represented NSF. Edson represented the Pentagon (US Army). The name of Dr. Wernher von Braun headed the list of eleven names on the distribution list. The meeting began at 11:00 a.m. with short presentations by Meinel on payload characteristics, H.H. Koelle on Saturn performance, and O.H. Lang on ABMA funding. The afternoon included a conversation with von Braun in his office, and watching a static test firing of a Jupiter rocket from the safety of the blockhouse. On exhibit was a 3 foot tall model of the Saturn rocket with the AURA telescope in the nose cone. An artist in the Graphic Engineering and Model Studies Department had prepared a picture of the "Large Orbital Telescope by AURA" with a portion of a space station in the foreground and the Earth in the background. 8 × 10 inch copies were given to all participants.[31]

The Satellite Telescope Subcommittee presented a report at the October 9, 1959 meeting of the Scientific Committee.[32] Meinel said the satellite telescope would have a 24-hour orbit with five tons payload, and the vehicle would be a Saturn rocket.

There were also some negative developments in 1959. The National Science Foundation was not happy with some of Meinel's business practices. A letter from Luton[33] told Meinel to shape up or expect trouble with GAO (General Accounting Office) or NSF auditors. "The contract currently requires that no expenditures, purchase, or commitment is to be made for an amount in excess of $100 unless it is reduced to writing. It is my understanding that you have already expended several hundred dollars for engineering services without a written document to cover the transaction and also without having obtained prior approval from the Foundation as specified in your contract. Steps should be taken immediately to bring this transaction in line with the provisions of the contract." The letter included new authorization to arrange for certain engineering services without prior written approval from NSF "provided the amount involved does not exceed $5,000 for any one contractor in a given fiscal year." Luton warned Meinel: "The authority is not to be used to cover a transaction in excess of $5,000 by issuing several orders each of which is within the $5,000 limitation."

Shortly before the March 1960 dedication of the observatory, it was discovered that Meinel had overspent his FY 1960 budget by almost $100,000.[34] Following the discussion by the officers McMath asked for Meinel's resignation as Observatory Director, and this was presented to the Executive Committee in Executive Session.[35]

> The Executive Committee unanimously agreed to present the following recommendation to the Board of Directors in Executive Session: (a) That the resignation of Dr. A.B. Meinel as Director of the Kitt Peak National Observatory be accepted; (b) That C.D. Shane perform the functions of the Director of the Kitt Peak National Observatory pending appointment of a new Director; (c) That under the acting Director there be four major divisions: Stellar, Solar, Space, and Administration; (d) That Dr. A.B. Meinel be appointed Associate Director in charge of the Stellar and Space

Divisions; (e) That A.K. Pierce be continued as Associate Director in charge of the Solar Division; (f) That a committee consisting of the following astronomers be appointed to seek a candidate for the position of Observatory Director: Drs. Menzel (Chairman), Edmondson, Code, Morgan and Herbig; (g) That a committee composed of the following administrative officers be appointed to seek a candidate for the position of Associate Director – Administration, in consultation with the President: Messrs. Harrell (Chairman), Lee and Peterson; and that the President be authorized to make the appointment, upon approval by the National Science Foundation as to salary.

The Board of Directors approved these recommendations in Executive Session after lengthy discussion of each item.[36] The Board agreed that the effective date of Meinel's resignation would be March 31, 1960[37] and his appointment as Associate Director in charge of the Stellar and Space Divisions would start on April 1, 1960.

The reorganization of the observatory and the appointment of Dr. Nicholas U. Mayall as Observatory Director, effective October 1, 1960, will be described in the next chapter.

Luton wrote to Mayall on October 14, 1960 asking for clarification of several items in the FY 1961 financial plan, including the "plan of operation for the development of the space telescope."[38] Mayall's reply[39] emphasized that the present studies were not aimed at the *immediate* construction of a 50-inch orbiting telescope. The Space Division had nine full time employees and eight unfilled positions. The staff would grow to 32 by the end of FY 1965. Launch would occur after FY 1966, and the total cost up to that time was estimated to be $29,620,000.

The Space Subcommittee met in Berkeley on November 10, 1960.[40] Spitzer, Shane, Code, and Meinel were all present. Edmondson, Mayall, Ziemer (NASA), and Nidey were in attendance. The chairman named Nidey as Secretary. Technical reports were given by Meinel and Nidey, and a resolution was passed:

> Resolved: That it would be highly desirable for KPNO to collaborate closely with the OAO program; this collaboration might include the development of an instrumental prototype which could be flown in an OAO; and (that) other types of collaboration should also be explored.

Shane included this resolution in a letter to Keller.[41] He began with a summary of the history of the space program from the AURA perspective. He then wrote: "The close cooperation with NASA initially envisaged has not come about for two reasons. First, there seems to have been some hostility to our project on the part of certain NASA personnel because we were not under their control and insisted on following our own ideas, particularly with reference to the 24-hour orbit. Second, the lack of any official connection with the NASA Orbiting Astronomical Observatory project was not conducive to the free exchange of information, though no deliberate obstacles were placed in the way." Next he pointed out that one of the most serious problems confronting the KPNO project is the $30 million cost of the orbiting telescope, not including the cost of launch. This is beyond the foreseeable financial resources of NSF, and an additional source must be found. He wrote: "The obvious source is NASA." Roman recalled:[42] "It was quite clear they were going to come to NASA for the money to do the project."

Shane read a statement from Meinel offering his resignation as Associate Director of the Stellar and Space Divisions at a joint meeting of the Executive Committee and the Scientific Committee in Executive Session on March 11, 1961.[43] Meinel had become involved in a number of activities related to the orbital telescope and had been away from the observatory a considerable amount of time during the past year. Mayall said: "As a result, the work of the Stellar and Space Divisions has suffered. His staff has been uncertain as to what was expected, and there has been a lack of communication between Dr. Meinel and his staff which has been especially aggravating in budget matters." He felt that Meinel should be relieved of his administrative responsibilities "for the good of the Observatory." It was unanimously voted to present the following recommendation to the Board of Directors in Executive Session: "That the statement of Dr. Meinel in which he requests relief from his administrative responsibilities in the Kitt Peak National Observatory be accepted, and that Dr. Meinel be continued in the rank of Astronomer with no reduction in salary." Mayall said that Nidey should be designated Acting Associate Director of the Space Division until a permanent head could be found, and this was approved.

There was an extended discussion of the future of the space program and its relation to KPNO in the open session that followed. Code, a leader in the OAO program, felt that the space program should be a large one, and that it should be under the Kitt Peak National Observatory. A small program would not be a very imaginative approach. McMath called the two philosophies "go for broke" and "play it cozy." Code felt that AURA should "go for broke." Harrell's view was that the space program should be a separate organization, not an integral part of the Kitt Peak National Observatory, and we should not run the risk of wrecking the land-based astronomy program. Appointment of a working committee to study the AURA space program was approved. The Chairman and President jointly appointed: Spitzer (chairman), Mayall, Woodrow, Harrell, McMath and Shane *ex officio*.

The Board of Directors met two days later.[44] The report of the working committee was presented in Executive Session, as follows: "That a new research organization, separate from Kitt Peak National Observatory and with its own scientific and administrative staff and its own facilities, be established at the earliest practicable date. It was felt that the location of the space center should be decided at a later time. It is also recommended that the Committee be authorized to recommend a director for the new organization." This recommendation was unanimously approved by the Board.

The Board also approved the recommendation presented by Shane "that Dr. A.B. Meinel be relieved of his responsibilities as Associate Director of the Space and Stellar Divisions and that he continue to hold the rank of Astronomer at no reduction below his present salary." Mayall later received a note from Meinel's doctor saying that Meinel had been under his medical care from March 10 to April 3, and that Meinel was in a very nervous, upset condition and could not receive visitors. The doctor said he should be allowed to return to work when he felt able.[45] Meinel resigned from KPNO effective at the end of August to accept a faculty appointment at the University of Arizona.[46]

Shane telephoned Waterman to inform him about the change in the space program that the AURA Board had approved.[47] McMath also telephoned.[48] He told Waterman that he had no knowledge of these plans prior to the presentation of the resolution by Shane. He felt it had not been considered as carefully as it should have been by the members of the Board, and he was opposed to it. McMath asked Waterman to invite Miller, KPNO Associate Director for Administration, to attend the March 26 meeting at NSF so that his views would be represented. Miller's notes are the only written record of the meeting (References 49 and 50).

AURA representatives (Shane, Mayall, Miller) met with NSF staff members (Keller, Luton, Mulders, Robertson, Ruttenberg, Scherer) in Washington on March 26, 1961 to discuss and defend the AURA Board action on the space program.[49] Robertson pointed out that NSF had explained to Congress that KPNO is a national observatory including a space facility. He said Congress might not be willing to fund it separately. He said he would personally prefer reorganization under KPNO and not to set it up as a new institute. Luton wanted to know the basic reason for AURA's action. The lengthy response by Shane did not answer the question. Other questions included money, personality problems, telescope operation after launch, and Meinel's time away from the observatory during the past year. Miller said OAO consulting with NASA had taken him away 25 percent of his time.

Waterman joined this group at 11:00 a.m.[50] He opened the meeting by reviewing the development of the space telescope program. He said: "Having developed this plan with the Bureau of the Budget and Congress, he would like to carry it out with a minimum of questions on their part for, if given opportunity, Congress has a great ability in questioning." The discussion that followed was diffuse and sometimes at cross purposes. Waterman said the AURA action had happened too fast. He would prefer thinking it over. Shane said he had expressed the AURA position, and if NSF doesn't approve we must reconsider it in AURA. The meeting ended with Waterman again suggesting reconsideration of the situation, and AURA to review it with the NSF staff.

McMath was anxious to know what had happened at the meeting. He wrote to Mayall[51] and mentioned that Miller had phoned him and said "things had gone 'my' way." Miller sent him the transcripts of his notes after they were typed, and McMath's reply[52] said he would take Miller's advice to "lay low" for the time being. Mayall wrote to McMath:[53] "What it all boiled down to was a polite, but firm, no from NSF for a separate set-up for space astronomy. As a corollary, we were given a very clear mandate to find, as soon as possible, a suitable Associate Director for our Space Division."

The AURA Board of Directors held a special meeting at the National Science Foundation headquarters in Washington on June 14, 1961. The first action taken at the meeting was to rescind the March 13, 1961 action to have a separate organization for AURA's space astronomy activity.[54] The full resolution was in five parts. Parts 2 to 4 insisted that the original action was correct, but that "circumstances relating to the National Science Foundation make such an organization impractical at the present time." Part 5 stated that "under either form of organization, the headquarters of the space telescope activities should be in Tucson in order to maintain a close relationship with the ground-based astronomy." Robertson said this resolution took care of the concern NSF had about how Congress might react to a separate space organization under AURA. The vote of the Board to approve was unanimous.

The next item was the status of the search for a new Associate Director–Space Division. Menzel reported that the Space Subcommittee had met in Philadelphia in late April, and had agreed on a list of seven candidates, headed by Dr. James A. Van Allen of the University of Iowa. Mayall offered the position to Van Allen orally during a visit to Iowa City on June 12, 1961, and confirmed it with a letter.[55] Van Allen took the offer seriously, but finally decided to stay at the University of Iowa.[56]

Herbert Friedman and Richard Tousey of NRL were the next names on the list of seven. Mayall consulted the members of the Space Subcommittee and several other Board members before setting up an approach to both Friedman and Tousey.[57] Friedman would be Associate Director–Space Division and Tousey would be Astronomer–Space Division. Waterman had approved the salaries that would be offered. Mayall and Menzel met with Friedman on July 21, and Mayall sent the formal offer in a letter a week later.[58] He told Friedman that, with his permission, an offer would be made to Tousey if he accepted. Mayall told McMath that he was not sure we would be successful.[59] "His NRL division operates on a budget of 5 million, and he seemed concerned about NASA not wanting more competition from another organization."

Shane, Menzel and Mayall were to meet with Friedman at NRL on the morning of September 25. The four of them were to meet with NSF staff members (Scherer, Robertson, Keller, and possibly Waterman) in the afternoon.[60] Mayall told Friedman:[61] "We think it highly desirable that all of these persons jointly hear your views about how the AURA space astronomy program should be prosecuted." He said the meeting with Waterman might have to be on Tuesday, and he hoped Friedman could be present.

The discussion led to an astonishing proposal.[62] AURA would take over Friedman's present group intact, and let them continue with their present line of research in their present location for the time being. Support from the Navy for Friedman's rocket program would continue but the funds would be channeled through NSF, which would add funds for the space telescope project. A new laboratory for Friedman's group would be constructed in the Washington area as soon as possible. The big problem would be to sell this to the Navy.

Mayall presented this proposal at a special meeting of the Board of Directors in Tucson on October 27, 1961, one day before the fourth anniversary of the incorporation of AURA.[63] He said that "neither Friedman nor his staff was anxious to move to Tucson, as they preferred to live and work in the East." He also reported a discussion with Dr. F.J. Weyl, Chief Scientist of the Office of Naval Research, and said that Weyl "was quite against the proposal." Several members of the Board (Slettebak, Carson, Code, and Edmondson) expressed opposition to the proposal. Code felt "that if AURA was no more ambitious than to take over another organization, it should not be in the space program at all." Shane said: that "the opposition of the members of the Board had not been anticipated, but that it should be expressed now and not delayed, lest much embarrassment be caused." Edmondson moved and Code seconded a motion: "It is the sense of this meeting that the Associate Director–Space Division and the space organization should be located in Tucson." The vote was eleven "Aye", four "No", and the motion carried. The Board also voted that a new ad hoc committee be appointed to continue the search.

Mayall communicated the action of the Board to Friedman,[64] and said: "I strongly hope you will reconsider your wish, expressed orally to me and to some others in the AURA organization, to remain in the east to work on this program."

Shane appointed the new ad hoc committee:[65] Menzel (chairman), Carson, Edmondson, Goldberg, Leighton, Mayall, Shane, and Spitzer, and a meeting was held in Washington on November 15th. The morning was spent touring Friedman's laboratory and shops at NRL, and the afternoon was devoted to discussion at the National Science Foundation. Spitzer's summary flatly stated: "There is a conspicuous lack of enthusiasm for the administrative arrangements proposed by those favoring the 'Friedman package.' "

Following the meeting, Edmondson sent a three-page memorandum to those who were in attendance, with copies to the AURA Board.[66] The memorandum began with a one page historical summary, pointing out that the present ad hoc committee is the one "the Board of Directors voted for in June and didn't get." The next two pages described the meeting, and said that after 5:00 p.m., when Keller had to leave, the committee began to talk about other candidates. "In other words, we finally got down to doing the job the Board had instructed us to do." The memo ended: " The function of this Committee is very clear. It is to search in good faith for an Associate Director of the Space Division of KPNO who will live in Tucson and head up an organization located in Tucson. It is not the function of this Committee to try to subvert the policy of the Board which has been clearly and overwhelmingly affirmed on two different occasions."

McMath was in the hospital while all of this was taking place, and Mohler was taking care of his communications on AURA business.[67] Mohler read Edmondson's memorandum to McMath and reported:[68] "As nearly as I can remember, he said, 'That's a damn good letter. I certainly hope that Frank continues to tell the boys what's what.' " Mohler said that McMath was scheduled to leave the hospital on November 26, but that his health continued to be precarious. He died at the age of 70 on Tuesday, January 2, 1962.[69]

During lunch on November 15 Friedman had reaffirmed his position that he would like to have AURA take over his entire organization and set them up in business in the east, preferably in Washington. He was not asked to give a yes or no answer to Dr. Mayall's offer if the Board held to its policy about headquarters in Tucson. Finally, on November 30, he wrote to Mayall:[70] "First, I want to state that I do not wish to move to Tucson to undertake the development of the Space Science Division there. If that requirement remains firm, I must decline the position of Associate Director of KPNO." Two and a half more pages followed this statement in which he re-stated all of the old arguments in favor of a Washington location for the AURA–KPNO space program. The last paragraph said: "I believe that an AURA Space Science Division, developed on the present NRL activity, can be established within a framework of joint Navy–NSF sponsorship. I would feel highly privileged to direct such a program." The beginning of the letter seemed to say he was declining Mayall's offer, but the end said he hoped AURA would reconsider and hire him on his terms.

Mayall sent copies of Friedman's letter to the members of the AURA Board. Mayall's letter to the AURA Board says:[71] "This decision by Dr. Friedman means that the ad hoc Committee appointed to consider the AURA Space Astronomy Program now has a clear mandate to proceed with the search for someone else." He expressed hope that some arrangement could be worked out for close cooperation of AURA and KPNO with the Friedman group.

Nancy G. Roman, Head, Observational Astronomy Programs, NASA, phoned Edmondson on November 20, 1961, following informal conversations with Hiltner and Wildt about the AURA–KPNO space program.[72] She felt the AURA Scientific Committee should meet with NASA officials and tour the Goddard Space Flight Center (GSFC), and Edmondson said he would ask Menzel to get in touch with her about this. Menzel called her and then wrote to Code[73] about the conversation. He said she had two points to make: "One was to express the interests that NASA had in an early appointment by AURA of an Associate Director for Space. . . . The second proposition that she raised, however, dealt with the whole NASA space program. She felt that the Scientific Committee should visit the Goddard Space Flight Center, probably sometime in January." Menzel proposed a two-day meeting in Washington, starting with a visit to Goddard on the morning of Friday, January 5 or 12. This would be followed in the afternoon by a special meeting of the AURA Board with representatives from ONR, NSF, and NASA in attendance. He wrote: "This frank meeting and exchange would enable us to explore and evaluate the long-range support of the AURA space program by NASA and NSF, and ONR also, if Friedman should be brought into the picture." The meeting would continue on Saturday morning for further discussion and possibly some decisions by the Board.

The final plan that evolved was a combined meeting of the Scientific Committee and the ad hoc committee, but it did not include a meeting of the full AURA Board.[74] The two committees met on Sunday, January 14, 1962.[75] The visit to the Goddard Space Flight Center by the Scientific Committee took place on Monday, and the Tuesday morning meeting was held at NASA headquarters. The Executive Committee met in the afternoon also at NASA headquarters. A schedule conflict prevented attendance at this meeting by NSF officials, and a meeting with them was arranged for January 26.

The Sunday meeting began at 10:00 a.m. in the conference room of the Hotel Lafayette in Washington and adjourned at 6:20 p.m. Menzel opened the meeting with a reference to Edmondson's November 22 communication,[76] which mentioned the fact that the AURA Board had voted twice that the headquarters of the AURA space effort should be in Tucson. He then reported a telephone conversation with Keller, and said Keller had "stressed the advantage of the space operation being separated from the KPNO. . . . The basic reason was that this would keep the budget separate and the projects less in competition for the same funds. . . . Waterman had not been consulted but will probably be realistic and go along with the recommendation."

Menzel next directed discussion to "the type of space program desired in or under AURA." Edmondson, Code, Spitzer, Goldberg, Mayall, Wildt, and Carpenter all had something to say, and Menzel invited Friedman to speak. Menzel summarized the consensus of the group: "(1) That there should be a space effort within AURA; (2) That the program should start from a relatively small beginning and include intermediate objectives, then the large orbiting telescope and perhaps eventually a telescope on the moon; (3) That AURA should proceed as rapidly as possible to the appointment of an Associate Director for the Space Division; (4) That an interim person in charge of the space program should be designated; (5) That this objective should be presented to the Executive Committee and to the Board of Directors for approval." He believed this action would persuade NSF to release the $946,000 now being held for the space program.

After a long lunch break (12:40–4:05 p.m.) the meeting reconvened to discuss the candidates for Associate Director – Space Division. Menzel presented a list of 25 names that had been suggested, and each person on the list was discussed in detail (except for the four who were participants in the meeting). A series of votes reduced the list to six names; the list did not include the four who were participating in the voting.

There was a short discussion of candidates for Associate Director – Stellar Division before the meeting adjourned at 6:20 p.m.

The Scientific Committee visited the Goddard Space Flight Center on Monday, January 15, 1962.[77] They left Washington at 8:30 a.m. accompanied by Dr. Roman and Dr. Jocelyn Gill from NASA Headquarters, and arrived at the GSFC at 9:20 a.m. Dr. Meredith, Head of the GSFC Space Sciences Division, briefed the group on the organization and functions of the GSFC in relation to NASA Headquarters. He also described the functions of the Marshall Space Flight Center at Huntsville, Alabama, and the Jet Propulsion Laboratory at the California Institute of Technology in Pasadena, California. The GSFC budget for FY 1962 was $160,000,000, and would be more than $200,000,000 in FY 1963. He said that NASA Headquarters determined which experiments would go on which satellites. He also mentioned the use of sounding rockets to test the performance of equipment before experiments were placed in satellites. This was followed by a tour of the Space Control Center, the Test and Evaluation Building, and the Space Sciences Building. Next there was a briefing on some current NASA programs. Three spacecraft called Orbiting Astronomical Observatory (OAO) were to be launched with instruments provided by Code (Wisconsin), Whipple (Smithsonian) and Spitzer (Princeton). The Orbiting Solar Observatory (OSO) with an instrument provided by Goldberg (Harvard) was also described.

After lunch Roman described the NASA astronomy program. She said that the mission of NASA, as she saw it, was "to provide the scientific community with the means of obtaining scientific data from space." She concluded with a discussion of how a scientist could get an experiment in a rocket or satellite. The group returned to Washington at 4:30 p.m.

The Scientific Committee held an evening meeting at the Hotel Lafayette starting at 9:50 p.m. Several items not related to the space program were discussed first. Hiltner said he would like to discuss the space program further, based on what we had seen and heard at the Goddard Space Flight Center. He asked: "Was AURA trying to replace NASA in providing facilities for astronomers that already existed?" Shane said that NSF was anxious to get into the space field, and Menzel said that AURA was "missing the boat" by not taking over the Friedman operation. Menzel then read his notes of his recent telephone conversation with Keller.[78] Keller favored having a separate organization for the space program, and said "the endorsement of Dr. Waterman was expected." Mayall and Shane joined Menzel in singing the praises of Friedman and urging that AURA reconsider its position. Shane was very sure that "the National Science Foundation was disappointed in the earlier decision of the Board of Directors that the space program should be located in Tucson." He felt that "if this view persisted, AURA's stock in the eyes of the National Science Foundation would be lowered." Code said he thought the actions of the AURA Board had been proper.

Edmondson suggested waiting until the January 26 meeting to learn directly from Waterman what his position was.

The joint meeting of the two committees reconvened at 9:20 a.m. on Tuesday, January 16, 1962 in Room 415 at NASA Headquarters.[79] Dr. John F. Clark, head of the Geophysics and Astronomy Branch, and Dr. Nancy G. Roman, Astronomy and Solar Physics (NASA Headquarters) were present at this meeting. Shane discussed the history of the AURA space effort, including the offer to Friedman. He said there was now "a chance that the Board of Directors might be convinced to leave the space effort in the East, and the National Science Foundation was apparently ready to agree to a space program separate from the Kitt Peak National Observatory and located in the East." Clark reviewed these Minutes with Edmondson nineteen years later.[80] His comments about Friedman include: "I was rather surprised that AURA viewed the possibility as real that Herb Friedman could be enticed away from NRL. . . . I was personally very much of the feeling that the best interests of space astronomy would be served by leaving Friedman's group where the astronomers in the Navy had grown up. . . . It's very hard to see that taking over that group and transplanting it to another location could do other than reduce the total effort in the field of space astronomy." His comments about the AURA space telescope include: "I see that I expressed some concern about going immediately to the large space telescope; that was an understatement. This was at a time that I was wondering if the Orbiting Astronomical Observatory itself might not be too large a step. Indeed, as matters turned out, the first OAO was the only spacecraft in which the Goddard Space Flight Center attempted to take too large a step and fell flat on its face." In response to questions about continuing annual support from NASA for the AURA space program, NASA Administrator Webb's step funding system was briefly described. A different plan for AURA would require a policy decision at the highest level. "I saw no reason to believe that the national astronomy program would have benefitted by such a transfer of responsibility and I tried to make that clear to the group without giving offense." Homer Newell transferred from NRL to NASA in the same month that NASA opened for business, October 1958, and was NASA's Associate Administrator from 1967 until he retired in 1973. He recalled:[81] "We found it quite distressing when NSF chose to fund rocket astronomy at NRL and Kitt Peak, and to fund the AURA satellite telescope; but for the good of the field of astronomy we chose not to make a fuss about it."

The joint meeting adjourned for lunch at 12:25 p.m., and the Scientific Committee reconvened at 1:45 p.m. Code said he felt the morning meeting had been profitable. Shane repeated his belief that NSF had changed its mind about a separate space organization, and Edmondson insisted that the real attitude of the National Science Foundation was not known as yet. Menzel felt the group deferred too much to NSF, and that AURA should decide what it wanted to do in this area. Additional comments were made by Spitzer, Shane and Dean Whaley. A resolution presented by Shane was modified by omitting a reference to Friedman, and then was passed: "That the officers of the Corporation and Dr. Menzel be authorized to negotiate with the National Science Foundation and other Government agencies with the object of AURA's undertaking a major space effort at a location most suitable for the program, and that any proposed arrangement be reported to the Board of Directors for approval."

Shane suggested that nothing further be done about the Associate Director – Space

Division until after the January 26 meeting with NSF representatives. Menzel said that the number one candidate on the final list had been approached by another organization. It was agreed that Mayall should make an informal approach to him before January 26. The meeting adjourned at 3:10 p.m.

The Executive Committee convened in the same room at NASA Headquarters at 3:15 p.m.[82] The first action was to appoint Lee as a member of the Executive Committee for this meeting to fill the vacancy caused by McMath's death. After two non-space items the Executive Committee discussed the space program, and the resolution proposed by the Scientific Committee was approved. Next, Menzel one more time reported his telephone conversation with Keller. Shane said that the National Science Foundation was willing now to approve a space organization separate from Kitt Peak National Observatory, although Dr. Waterman's approval had not yet been obtained. Edmondson said he hoped the NSF representatives would make statements as positive on January 26 as the NASA representatives had made, and he felt conclusions about the attitude of NSF had been drawn from interpretation and not from Keller's own words. Woodrow inquired about NASA's attitude, and Shane said it was stand-offish. Miller said the funds for the AURA program were equivalent to the $12,000,000 NASA had available for all university support. Menzel downplayed the importance and authority of Clark and Roman.

The Executive Committee endorsed Mayall's request to NSF for the release of the withheld $946,000 of space telescope funds. Also approved was a resolution authorizing Mayall to negotiate with the candidates on the list for Associate Director – Space Division and to present his recommendation to the AURA Board for approval. Several other matters were discussed before adjournment at 5:20 p.m.

Nidey phoned Francis Johnson on Sunday, January 21,[83] and learned that Johnson had offers from NCAR, Aerospace Corporation, NASA (under Newell) and the Graduate Research Center in Dallas. Mayall called Johnson on Monday, January 22.[84] He declined, saying that his interests were more in planetary atmospheres than in space astronomy.

The January 26 meeting with NSF was attended by nine AURA representatives (Shane, Menzel, Hiltner, Miller, Pierce, Harrell, Woodrow, Mayall, and Whaley) and eight NSF representatives (Waterman, Robertson, Keller, Mulders, Scherer, Hoff, Schoen, and Sheppard).[85] Menzel reviewed the history of the KPNO space program, and then called on Shane for comments. Shane expressed the opinion that Friedman was the only solution to the AURA space program. The questions and comments from Waterman, Robertson and others showed little enthusiasm for a space program outside KPNO, and no enthusiasm for AURA to take over the Friedman operation. Hiltner wrote:[86]

> In my evaluation of this meeting, there is no question whatever what NSF wishes in the AURA space program. – Astronomy done under the KPNO administration. It was obvious at all times that Waterman and his staff did not care to move Friedman from his present location. Friedman receives much support from the Navy, and NSF, and NASA will (and does now) help when required. AURA should have a separate program under KPNO. KPNO must appoint someone with astronomical research ideas and proceed

with their astronomy. It is interesting that Whaley supported Shane and Menzel in their attempt to sell Friedman to NSF.

An immediate consequence of the AURA–NSF meeting was the decision by the Nominating Committee to nominate Edmondson as President of AURA to succeed Shane, replacing the planned nomination of Menzel.[87] The ad hoc committee finally did what it had been instructed to do, and nominated a candidate for Associate Director – Space Division.[88] Edmondson was elected President for a three-year term and Joseph W. Chamberlain was appointed Associate Director – Space Division at the Annual Meeting of the AURA Board in Tucson on March 12, 1962.

The beginning of the space program under Chamberlain's direction and the March 1962 meetings of the Scientific Committee and the AURA Board of Directors will be reviewed in Chapter 12. The transformation of the space program from planning for a large orbiting telescope to an active planetary research program using rockets, the change of name from Space Division to Planetary Sciences Division in 1969, and the termination of the rocket program in 1973 will be discussed in Chapter 20.

12

The observatory is reorganized on March 14, 1960, and a new Observatory Director is appointed. Planning begins for a 150-inch telescope. Reorganization of AURA is discussed in 1961–62, and new officers are elected in 1962. The November 2, 1962 dedication of the McMath Solar Telescope.

The need to strengthen the business and administrative side of the observatory had been under discussion for several months prior to the March 1960 AURA Board meeting and the dedication of the observatory. In February 1959 Luton had criticized Meinel for spending several hundred dollars for engineering services without a written document to cover the transaction, in violation of the terms of the AURA contract.[1] Seven months later Shane learned that an unexpected statement from architects Blanton and Cole for museum design work they had done on their own without a written order had been sent "at this time" only because Meinel had suggested it.[2] The Audit Committee of the AURA Board reported to the Executive Committee on March 13 that "budget controls were quite inadequate, and that nothing had been done about overdrafts."[3]

Lee, the Chairman of the Audit Committee brought his assistant, Harold E. Bell,[4] to Tucson a few days before the Executive Committee meeting to straighten out the KPNO accounts in different categories, and to recommend the fund transfers needed to place adequate funds in each category to cover the overdrafts. After three days of hard work Bell found overdrafts in several accounts totalling $87,068. This was the situation that confronted the officers when they arrived in Tucson.[5] Funds to cover the overdrafts could be transferred from the General Reserve Contingency Fund on approval by NSF. Shane, Lee and Miller met with Luton the evening before the Executive Committee meeting, and they reported that Luton seemed willing to give his approval. The Executive Committee voted to authorize President Shane to request NSF to approve the transfer.

Shane presented two plans for reorganizing the observatory, and two more plans were proposed during the lengthy discussion by the Executive Committee. The plan that was finally agreed to was Shane's Plan II, and seven recommendations to implement this were presented to the AURA Board of Directors.[6] After full discussion the Board voted approval separately for each item:

> a) That the resignation of Dr. A.B. Meinel as Director of the Kitt Peak National Observatory be accepted, effective March 31, 1960.

b) That the President, Dr. C.D. Shane, perform the functions of the Acting Director from April 1, 1960, pending appointment of a new Director.

c) That under the Director, there be four major divisions: Stellar, Solar, Space, and Administration.

d) That Dr. A.B. Meinel be appointed Associate Director in charge of the Stellar and Space Divisions.

e) That Dr. A.K. Pierce be continued as Associate Director in charge of the Solar Division.

f) That a committee consisting of the following astronomers be appointed to seek a candidate for the position of Observatory Director: Menzel (chairman), Edmondson, Code, Morgan and Herbig. [It was understood that McMath and Shane would be ex officio members.]

g) That a committee composed of the following administrative members be appointed to seek a candidate for the position of Associate Director – Administration, in consultation with the President: Harrell (chairman), Lee and Peterson; and that the President be authorized to make the appointment, upon approval of the National Science Foundation to the extent required by the contract.

McMath wrote to Mrs. Walter S. Adams, a close friend of the McMaths and widow of the second Director of the Mount Wilson Observatory:[7] "I ran into a terrible mess in Tucson and worked very much harder than I wanted to – or felt like, for that matter. We reorganized, and Dr. Meinel is no longer Director, having been demoted to Associate Director. Dr. Shane, at writing, has the title of Acting Director." He wrote to another friend,[8] and added: "I believe [Shane's] wife told him that he should have his head examined."

The committee composed of administrative members of the Board promptly proposed AURA Board member J.M. Miller to be the Associate Director – Administration. As AURA Secretary he had, perforce, been performing some of the duties now assigned to the new position owing to the weakness of the business and administrative side of the Observatory. He accepted, and Shane wrote to Waterman on April 5 requesting approval of the salary.[9] He was expected to begin his new duties in Tucson about the third week in May.

Menzel sent the Search Committee a list of 16 potential candidates for the position of KPNO Director on March 28, 1960.[10] He asked the Committee to list these names in rank order, and also to propose additional candidates. Edmondson ranked four and crossed out seven, leaving five unranked.[11] Shane ranked five, with Willie Fowler first and Nicholas Mayall second.[12] All the votes were in by April 6,[13] and Menzel reported that Mayall was the marginal front runner with two first place votes (McMath and Menzel), one second place vote (Shane), and one third place vote (Edmondson). Mayall, a member of the National Academy of Sciences, was a highly respected research astronomer who had been appointed a consultant to AURA in 1958 because of his experience in planning the Lick Observatory 120-inch telescope. He had no experience in administration, but Shane, McMath, and Menzel felt strongly that Jim Miller's appointment as Associate Director for Administration would take care of this problem.[14]

On April 15, 1960 the Search Committee named Nicholas U. Mayall of the Lick

Observatory as their unanimous choice to be the new Observatory Director.[15] Mayall wrote to McMath[16] that he felt "like a person struck by lightning." He said: "At present I am still undecided, but more than tempted, and possibly favorably inclined." He decided to accept the position in early May,[17] but there were problems with his University of California pension that would have to be worked out. The AURA Board approved the appointment of Mayall, effective October 1, 1960, at a Special Meeting in Ann Arbor on June 6, 1960.[18] Shane had received verbal approval from Waterman before the meeting; he wrote to Waterman on August 1 to request formal approval of Mayall's appointment and the proposed salary.[19]

The Board agreed that Miller should continue as Secretary of the Board for the duration of his present term and that he should be eligible for reappointment, and that Miss Elliott should continue as Assistant Secretary. James H. Corley, Vice-President – Governmental Relations and Projects, was designated by University of California President Clark Kerr to complete the unexpired term of Miller on the AURA Board.

The Board also voted to transfer to the Associate Director – Administration the responsibilities previously assigned to the Business Manager. Miller replaced Meinel in the financial resolution regarding check signing authority, and he was given the responsibility, previously assigned to the President, to act in behalf of AURA in the disposal of surplus equipment.

A factual, albeit sanitized, account of the reorganization of KPNO is in the 1960 annual report to the American Astronomical Society.[20]

The reorganization was a traumatic experience for members of the KPNO scientific staff[21] who were bewildered by the apparent suddenness of the change; Mayall's arrival on October 1st provided a stabilizing influence. Mayall began to participate in AURA–KPNO activities immediately by attending a meeting of the Executive Committee in Tucson on June 30,[22] and a full-day meeting about the Chile Observatory project on August 10.[23] The first AURA meeting after he took office as Director was a meeting of the Executive Committee on November 9, 1960 in the library on Kitt Peak.[24]

Mayall's first AURA Board meeting was the annual meeting on March 13, 1961. At this meeting Meinel was relieved of his duties as Associate Director of the Space and Stellar Divisions,[25] and he left KPNO five months later to take a faculty position at the University of Arizona.[26] The divisive discussion of the future of the AURA space program that began at this meeting has already been described in Chapter 11.

The joint meeting of the Scientific Committee and the Space Subcommittee[27] prior to the Board meeting was not limited to the space program. The discussion included proposals to initiate a program to design a 150-inch telescope, and a program to study the physics of seeing. The Chile Observatory project, which had become AURA's responsibility in 1960, was also discussed; the history of this project will be reviewed in the next two chapters.

The AURA Board instructed the Chairman and President to appoint a 150-inch Telescope Review Committee. The Board also voted that a request for funds for the physics of seeing project should be sent to NSF. A motion by Menzel to abolish the Scientific Committee triggered a lengthy discussion, and the motion was defeated. The Board approved an amendment to the by-laws, proposed by Shane, that removed the restriction that the Vice-President had to be a member of the Board of Directors.

Edmondson's term as Vice-President ended at this meeting, and Mayall was elected Vice-President for a three-year term.

Before Shane's departure, McMath and Shane had a face-to-face disagreement the morning after the Board meeting about the proposal to separate the Space Division from KPNO.[28] McMath and Miller had a similar disagreement with Mayall during a meeting at noon. That evening Miller also expressed his views to those at a party at his home, including NSF representative Mulders. The next day McMath phoned Luton, and after expressing his views he handed the phone to Miller and ordered him to give his views to Luton. Five days later, McMath called Waterman to express his views, and to ask that Miller be invited to the forthcoming AURA–NSF meeting about the space division as McMath's representative. Copies of Miller's minutes of this meeting were sent to Shane, McMath, and Mayall on April 15, 1961.[29] Shane ordered Miller not to distribute them to the AURA Board, because Shane would inform the Board.

Shane's first reaction to this opposition was to undermine Miller's relationship with Mayall,[30] and to cut off relations with McMath. In a letter written to Harrell on September 5,[31] McMath said he had not heard from Shane since May 4. The President had been in frequent communication with the Chairman prior to this.

Shane's second reaction was to propose an administrative reorganization of AURA, which he presented in a short Executive Session at the beginning of the June 14, 1961 Special Meeting of the AURA Board.[32] The Board voted: "That Harrell (chairman), Edmondson, Menzel, Peterson, Reynolds, and McMath and Shane *ex officio*, be appointed members of a committee to recommend an organizational plan appropriate to AURA's present and foreseeable responsibilities, after thorough study of recent problems that have arisen." The Reorganization Committee was directed to report to the Board at a special meeting in the Fall of 1961.

Harrell, Reynolds and Peterson met with McMath on August 1 in his Detroit office.[33] The Committee met in Tucson on September 8 and 9, and conducted private interviews with Shane, Mayall, Moore, Miller, and seven other KPNO staff members.[34] The Committee concluded "that the difficulties, though serious, were not insurmountable and that a full reorganization is not necessary at this time."[35]

A preliminary report of the Reorganization Committee was presented at the special meeting of the AURA Board on October 27, 1961 by Peterson, owing to Harrell's sudden illness.[36] He summarized the work of the Committee, and stated some tentative conclusions: "That McMath be appointed Chairman Emeritus; that another Vice President of AURA be appointed; that four scheduled meetings of the Executive Committee be arranged each year; that a clear definition of the duties of the Observatory Director and Associate Director – Administration be set forth; and that no change in the organization of AURA and no important changes in the By-laws, except as necessitated by the above, be made." Menzel added that the new Vice-President would be chosen from the administrative members of the Board. Following lively discussion, with several motions made and withdrawn, Peterson said that under the circumstances there would be no formal recommendation from the committee at this time.

Shane had included five revisions of the by-laws in the agenda. Ignoring the report of the Reorganization Committee that had just been given, he called for discussion to get a sense of the meeting before having Counsel draw up proper phraseology for presentation to the Board for adoption at the annual meeting in March 1962. There was no discussion after Peterson moved that Shane's proposed changes in the by-laws be referred to the Committee on Reorganization for consideration and a report in time for action at the annual meeting.

Peterson, Menzel, and Edmondson[37] had several conversations with Shane, Mayall and Miller, and one with Pierce during the two days of the meeting. They wrote summaries describing the conclusions they had reached, and there was no disagreement.[38] They made it clear that Shane's success in undermining the working relationship between Mayall and Miller had done serious damage to the administration of the observatory. Menzel wrote that Shane "had driven a wedge between Nick and Jim to the point where conditions are impossible for the latter." He also said that we learned from Pierce that it was widely believed that Shane was the *de facto* Director of KPNO.[39]

Pierce drove the three members of the Committee to and from Kitt Peak on October 26, and this provided an opportunity for the Committee to hear his views in private. He said that Mayall was leaning too much on Shane and not enough on Miller. Mayall had been trying to do all of the minor administrative chores that were properly part of Miller's job, and Shane had been doing those parts of Miller's job that Mayall was not doing. Menzel had a long session with Miller on the mountain, and the things Miller told him were consistent with what Pierce had said. The Committee met with Mayall that afternoon and Menzel spoke for the Committee. Mayall's statements in reply to Menzel's comments were not consistent with the information the Committee had gotten from Pierce and Miller. Menzel wrote: "The voice was Mayall's but the words very clearly were Shane's."

To try to understand and resolve the inconsistencies, the Committee scheduled a breakfast meeting with Mayall and Miller in Peterson's room on October 27. Menzel made an opening statement, and then asked Mayall and Miller to talk to each other about the issues each had raised. It soon became clear that each one did not understand what the other was saying. Members of the Committee had to explain to each one what the other had said. At the end Mayall and Miller seemed to be understanding each other much better, and the Committee felt that major progress had been made.

The Committee had a short meeting with Shane before lunch, and Peterson had a 30-minute private discussion with Shane following the Board meeting. After this the Committee met with Shane, Mayall and Miller. Peterson wrote: "Menzel reviewed briefly the discussions that had been held with each of the three separately and suggested that since everyone independently appears to be in agreement on the basic division of duties and responsibilities the Committee believes that the three principals concerned are now ready to resume the harmonious and effective working arrangements which existed prior to March 1961. Shane, Mayall and Miller agreed." If something seems too good to be true it is usually is, and this was no exception.

Menzel's summary, written five days after the meeting, reported that Shane had called him on October 31 and said that Mayall was extremely unhappy. Menzel checked with Mayall and found that this was not the case. He was happy about the decision concerning Miller, but unhappy about the Board's decision about the space program. Shane made dire predictions about Miller's already moving in to "take over" AURA, and told Menzel that Mayall needed protection against Miller. Mayall asked Menzel: "What do I do, when Shane tells me to keep Jim out of things and the Board wants Jim to act?" Menzel told him his responsibility was clearly to AURA, which had never taken away Miller's responsibilities as Associate Director–Administration. The cover letter with Peterson's summary said: "I am very much disturbed by Menzel's report that apparently Dr. Shane has not accepted the decisions of the Committee, which I thought had been clearly explained to him following the Board meeting."

Reynolds read the three summaries and replied:[40] "I am personally sorry that I was unable to attend the October 26–27 meetings in Tucson, but perhaps it was just as well. I am not as patient as the rest of you and might not have been able to restrain my sense of outrage at the neglect and distortion of our conclusions at the end of our Tucson visit in September." He agreed with Edmondson's conclusion that "Shane is responsible for 95% of our <u>present</u> troubles." He said he had talked to Whitford during the October 26–27 AUI meeting at Green Bank, and that Whitford had a pretty good understanding of the situation the Committee had found in September. He told Whitford that "our Committee accepted with some relief the fact that Shane's term as President expires next March and that we had made it clear to Shane that there would be no attempt to extend his term in that office." He told Whitford the AURA Board would welcome his nomination by the University of California to succeed Shane on the Board. Finally, he thought our Committee should communicate this view, if we agreed the Board would support it, to the nominating authority in the University of California.

Menzel wrote to Harrell again following a stopover in Tucson on November 10–11:[41] "I am glad to report that the situation, though not perfect, seems to be well in hand as far as Jim and Nick are concerned. The antagonism of Donald Shane toward Jim is something that Jim is willing to put up with, until March at any rate. He accepts it with good grace, if not with good humor." With respect to Shane's charge that Miller was "taking over," Menzel wrote: "From my open and frank discussions with Nick, I get no feeling at all on Nick's part about Jim's "taking over." Nick tells me that Shane has continually warned him that he (Nick) should not let Jim take over. And Nick expressed some mystification as to the reason for Shane's philosophy." Menzel also wrote: "Nick is turning more to me for advice. Calls me every few days. Although I hope this transfer is a salutary one, I am trying to use it as a means of getting Nick to stand on his own feet and make his own decisions." Harrell was still in the hospital at this time, December 1, and would be there for another week or ten days.[42]

An important change had taken place at the National Science Foundation two and a half weeks after the Special Meeting of the AURA Board, as reported by Keller:[43] "On November 13, 1961 Dr. Randal M. Robertson, formerly Assistant Director for Mathematical, Physical and Engineering Sciences here at the Foundation was appointed to the position of Associate Director (Research). I have succeeded him in the position of Assistant Director for MPE.

The new Program Director for Astronomy is Dr. Gerard F.W. Mulders. We will be pleased to hear of possible candidates for the position of Associate Program Director for Astronomy which is now vacant."

The 1961 Annual Meeting of the AURA Board turned out to have been McMath's last AURA meeting. Mayall's frequent long newsy letters were almost his only source of information about the observatory, starting with a letter about the March 27, 1961 meeting at the National Science Foundation.[44] In a separate letter written the same day Mayall invited McMath to serve on the 150-inch telescope Committee,[45] and McMath accepted[46] after several other letters had been exchanged. He told Mayall his health problem was a depressed EKG (electrocardiogram) curve caused by a lack of potassium, and he needed a lot of rest. His health continued to deteriorate, and in August he told Mayall how serious his problems were.[47] He was in the hospital a month later,[48] and died at the age of 70 on January 2, 1962.[49]

The Nominating Committee had made a tentative decision to nominate Menzel to succeed Shane as President.[50] The Shane–Menzel maneuvers about the Space Division, reported in Chapter 11, had raised some doubts about how Menzel might perform as President. For this reason, Edmondson wrote to Harrell:[51]

> Menzel certainly deserves a lot of credit for the part he has played in helping to ease the Tucson problems. It is clear that Nick will "take it" from Menzel with better grace than from anyone else. It also became clear to me during the October meeting that Menzel is his own man, and not Shane's. This has a bearing on the recommendation of the Nomination Committee for the next President of AURA. I felt it was my duty to inform Carson and Hiltner about this, and I did so. I told them that I had no fear that Menzel would be a stooge for Shane, and that his personal relationship with Nick would be a real advantage, in working out the problems that are still with us. I am telling you this because of the nature of the correspondence I showed you on the plane last September.

Menzel's nomination seemed to be secure until the January 26, 1962 meeting at the National Science Foundation. Waterman's strong restatement of the NSF position about separating AURA's space program from KPNO demolished the credibility of Shane and Menzel.[52] It also caused the two other members of the Nominating Committee (Hiltner and Carson) to withdraw their support from Menzel and to nominate Edmondson to be the next President of AURA, as reported at the end of the previous chapter.[53]

The Reorganization Committee met in Chicago on February 15, 1962 to prepare a final report.[54] Edmondson wrote to Harrell after the meeting[55] to express his concerns about some of the actions that were taken by the Committee: "I believe that Nick, like Shane, does not understand the difference between 'disloyalty' and 'disagreement'. . . . I do not feel that any statement, however frank, made to any member of our Committee . . . should be considered disloyal." The final paragraph included: "If I really do become President of AURA, I shall endeavor with all my might to try to bring about effective working relations among the members of our existing team in Tucson."

Paul Scherer, who had been elevated to the position of NSF Associate Director – Administration after Luton's appointment had been terminated, was openly hostile to Miller. Edmondson phoned him on February 19[56] to verify a comment he had made about Miller that Keller had passed on to Spitzer, and Spitzer had passed on to Edmondson. After this was discussed, Edmondson briefed him on the status of the Committee's work. An addendum to Edmondson's notes says: "Sometime during the conversation Scherer mentioned the disagreement between Nick and Jim, Keith, and Nidey about the space budget. This took place in Washington on January 26. I asked Scherer if he was trying to say they were doing their staff work at NSF and not at home. He agreed this might be one way to look at it."

The annual meeting of the AURA Board of Directors on March 12, 1962[57] began with an informal meeting at 9:00 a.m. to give Whaley an opportunity to read a statement. It proved to be a sermon about AURA's recent problems, and his prescription for solving them. This was followed by a brief discussion of the future of the Reorganization Committee and the problems of the space program. During the discussion Whaley used the Chile project as an example of a future problem. He predicted that "complex problems related to the national interest regarding Chile and the whole Latin American area would arise, and if these problems were approached with procedures used in dealing with the Kitt Peak National Observatory or space program problems, an awful mess would result." The informal meeting adjourned at 10:15 a.m.

The Board meeting began at 10:20 a.m. in Executive Session. Miller reported that Harrell, Code, Gage, Spitzer and Lee had been redesignated by their universities as directors for three-year terms, and that Orren C. Mohler would replace McMath and Albert E. Whitford would replace Shane. Albert P. Linnell of Amherst College was elected a director-at-large for a three-year term. Edwin F. Carpenter, University of Arizona, was elected a scientific annual consultant; his three-year term as a director-at-large ended at this meeting.

Menzel had reported to the Reorganization Committee that Mayall was not comfortable about being Vice-President of AURA because he felt there was a conflict of interest in his dual role of Observatory Director and Vice-President.[58] Mayall submitted his resignation to Shane,[59] and Shane presented it to the Board. It was accepted, and this paved the way to nominate Harrell as Vice-President, the beginning of the custom of having an administrative member of the Board hold this office.

The report of the Reorganization Committee was presented by Harrell. The Committee had proposed a new position, Deputy Observatory Director, but the Board was not convinced that this position was needed, and voted to delete it from the report. The position of Deputy Director was also in the proposed by-laws, and this presented a legal problem that required advice from Counsel Shoenhair, who should have been in attendance. He was playing golf, and his displeasure at having his game interrupted was obvious when he came to the meeting in the afternoon. This delayed final consideration of the report of the Reorganization Committee until the Executive Session reconvened at 5:45 p.m.

The report of the Nominating Committee was presented by Hiltner, owing to Menzel's absence caused by a serious medical problem. The nominations, as given in the call to the meeting, were President Dr. F.K. Edmondson; Vice-President, Mr. W.B. Harrell; Chairman, Scientific Committee, Dr. R. Wildt; Treasurer, Mr. Minton Moore; Assistant Secretary, Miss Julia Elliott; Executive Committee, Dr. O.C. Mohler and Dr. A.E. Whitford. The nomination for Secretary had been left open for discussion at the request of the Reorganization Committee, although a majority of the Nominating Committee supported Miller for this position. Harrell said this request was part of the total effort to clarify the actions of the Board, and that Miller was his personal choice. The Board unanimously elected the list presented by the Nominating Committee, and also unanimously elected Miller as Secretary of AURA. Edmondson took the Chair at this point.

Hiltner read a statement of appreciation to Shane, who responded that he would always treasure the experiences he had on the AURA Board and he thanked the directors for this expression of appreciation.

Members of the 1963 Nominating Committee and several consultants were elected before the Executive Session adjourned for lunch at 12:45 p.m.

The Regular Session convened after lunch. Reports were presented by the Audit, Patent, and Scientific Committees. The Scientific Committee had met on March 9 and 10,[60] and their report included several items requiring Board action. The most important was the appointment of Dr. Joseph W. Chamberlain as Associate Director – Space Division.

Mayall had offered the position to Chamberlain in a telephone call on January 26, following the meeting at the National Science Foundation, and had sent a confirming letter on January 30.[61] Chamberlain visited Tucson on February 14 and wrote a conditional letter of acceptance on March 8.[62] He made it clear that he was interested in research as a primary goal and that the planning and development of hardware and instrumentation would be secondary. The Scientific Committee reported to the Board that it endorsed this statement, and Keller told the Board that NSF concurred. Chamberlain entered the meeting and answered questions by Code, Shane, Hiltner and Pierce. With regard to the large orbiting telescope, he said he hoped the staff would not feel bound to the 50-inch orbiting telescope program as a goal. Keller said the orbiting telescope could be postponed for as long a period as seemed wise and necessary. Chamberlain left the meeting and the AURA Board unanimously voted to offer the position of Associate Director–Space Division to him.

A memorial resolution for McMath had been prepared by Edmondson, Lee and Pierce, and this was read by Edmondson. The Board then stood in a moment of silence. The resolution said: "It will be proposed by the AURA Board to name this telescope the ROBERT R. McMATH SOLAR TELESCOPE at the time of its dedication." Subsequent to the meeting Waterman gave NSF approval, and copies of the resolution were sent to Mrs. McMath and daughter Madeline Sloan.

Shane reported on the Chile project, Baustian reported on construction projects, Mayall gave the Observatory Director's Report, Moore gave the Treasurer's Report, and the fiscal year 1964 Budgets for Kitt Peak and Chile were discussed.

Edmondson had reorganized the Scientific Committee into Subcommittees for Stellar, Solar, and Space. He asked the Board to approve appointment of a committee for the Chile project with members: Hiltner, Rutllant, Smith, Clemence, Whaley, Shane (Consultant), Harrell, and Edmondson (chairman). The Board voted approval.

The Regular Session of the Board meeting adjourned at 5:35 p.m., and the Board reconvened in Executive Session at 5:45 p.m.

The Board in Executive Session unanimously approved the report of the Reorganization Committee as amended (Attachment II to the minutes). The Board also unanimously approved the revisions to the by-laws as amended by omitting the proposed position of Deputy Director. Counsel Shoenhair had advised that if the by-laws proposed in the call to the meeting were amended it would be necessary to obtain a waiver of notice and vote of approval from three-fourths of the members of the Board, including those who were not present at the meeting. The Board instructed Miller to do the required paperwork.

The Board also appointed Gage to be a member of the Reorganization Committee replacing Reynolds, and decided that the President and Vice-President should also be members.

Hiltner pointed out that the President and Vice-President had been elected in the morning session under the old by-laws that provided for three-year terms. The Board's intention to conduct the election under the new by-laws that provided for annual election had been frustrated by the absence of Counsel Shoenhair from the morning session. The Board agreed that the President and Vice-President had been elected for three-year terms. The Executive Session adjourned at 6:35 p.m.

A meeting was held at 9:00 p.m. to discuss the space program and the possibility of supplementation for FY 1964 with Keller and Chamberlain. Chamberlain also requested and received approval for the appointment of four astronomers for three-year terms. The meeting adjourned at 11:00 p.m.

McMath had proposed at the 1961 Annual Meeting that AURA start planning for a 150-inch telescope.[63] The Chairman and President appointed a 150-inch Telescope Committee consisting of Mayall (chairman), Shane, Bowen, Code, Hiltner, Mohler, Rule, and Stromgren. The Committee met in Tucson on September 23, 1961.[64] Following discussion of programs, it was agreed that a general purpose telescope was needed, and a discussion of instrumentation was followed by a discussion of the optical parameters. The afternoon session was devoted to seeing and site selection, mechanical design of the telescope, automation, mirror material, dome and building, cost estimate, time schedule for design and construction, and needed staff organization and support facilities. The meeting adjourned at 5:30 p.m.

A month later Mayall reported the conclusions that were reached at this meeting to the Executive Committee.[65] He said the $200,000 authorized by the Board for preliminary design and location studies was being requested in the FY 1963 budget. Shane felt that outside discussion at this time should be limited to the statement that the possibility of constructing a 150-inch was being studied.

Mayall gave a detailed report to the Scientific Committee on January 15, 1962 in Washington.[66] Artists renderings of the 150-inch telescope building and mounting had been prepared by Skidmore, Owings and Merrill. Following Bowen's advice, the telescope was to be 150 feet above the ground in order to eliminate the poor seeing at ground level, and this added $1,000,000 to the cost. The total would be $10 million with a fused quartz primary mirror, or $9 million with a pyrex mirror. The Scientific Committee voted to recommend that funds for the mirror blank and the design study be included in the FY 1964 budget. Edmondson suggested that two 150-inch telescopes be recommended, one for KPNO and one for Chile. The committee voted unanimously that the importance of having a 150-inch telescope in the southern hemisphere should be recognized, and that the Executive Committee should transmit this recommendation to the National Science Foundation. These recommendations by the Scientific Committee were approved by the Executive Committee the next day.[67]

The artists renderings of the 150-inch telescope and dome had been left in Washington after the January meeting of the Executive Committee for display at the National Science Foundation. They were returned to Tucson for display at the March 12, 1962 AURA Board meeting, and the plans that had been developed by the Committee were summarized in Mayall's report to the Board.[68] The appointment of David Crawford as Project Manager for the 150-inch telescope, and other events leading up to the signing of contracts for mirror blanks, mountings and domes for two 150-inch telescopes will be described in Chapter 15.

The dome for the 80-inch telescope was 30% complete at the time of the dedication of the observatory on March 15, 1960, and it was accepted from the contractor in June 1961.[69] The mounting was delivered to Kitt Peak in October, and assembly was completed on December 21, 1961.[70] The pyrex mirror blank, with a diameter of 84-inches, had arrived in Tucson on October 7, 1959,[71] but the optical shop was not ready to start working on it until February 20, 1961.[72] Steps were taken to insure the blank when it arrived, but the initial proposal from the insurer contained some words that Counsel rejected as inappropriate. It was not necessary to say that "flavor changes" and "moth damage" would not be covered.[73]

The figuring of the mirror in the optical shop was finished in May 1962,[74] and it was taken to Kitt Peak in August for testing in the telescope.[75] The mounting was plagued with a series of problems. It was out of commission for six months, starting in October when metal pickup damaged the right ascension gear.[76] Problems with the back support system for the 84-inch mirror caused further delay.[77] A problem with the oil-film pads for the polar axis bearing rendered the mounting inoperative for nearly a month.[78] Finally, with all the problems apparently taken care of, an announcement was published in several journals stating that the telescope would be available for visiting astronomers on September 15, 1964.[79] The first regularly scheduled visiting astronomer to use the 84-inch telescope was Dr. G. van Biesbroeck on his 84th birthday.[80] Four visitors, including van Biesbroeck, used the telescope for 13 nights in October.[81]

The official size of the telescope had been "80-inch" because that is the size that had been in the NSF budget request, and that Congress had approved as a line item in the appropriation bill. The mirror blank had been ordered with a diameter of 84-inches as insurance that it would have 80-inches of clear aperture even if the outer two inches could not be satisfactorily figured. By the time of the June 12, 1962 meeting of the Executive Committee the optical work had progressed to the point where it appeared that the full 84-inch aperture would be satisfactory. Mayall called this to the attention of the Executive Committee and said "it would probably be safe to refer to the telescope now as an 84-inch telescope."[82] This met with general approval from those present including the NSF representative, Mulders, and the sign outside the dome was promptly changed.

Site preparation for the solar telescope started on March 14, 1960, the day before the dedication of the observatory.[83] Final drawings were received from Skidmore, Owings and Merrill on May 2, followed on May 4 by the specifications.[84] Eight qualified bidders were invited to submit proposals, and six responded.[85] Western-Knapp Engineering Company and Robert E. McKee, Inc. were the low bidders, and were also considered to be the best qualified. Detailed bids from these two were opened on June 21, and Western-Knapp was the low bidder by five percent. This was $240,000 over budget, so the $395,000 heliostat cover was eliminated.

Bids on the heliostat, No. 2 and No. 3 mirror mountings were received on June 28 from seven companies. The low bidder was Westinghouse Electric Company of Sunnyvale, California, the high bidder earlier for the 80-inch telescope mounting. The good news this time was that this bid was $193,701 under budget.

The Executive Committee on June 30 authorized the President to execute a contract for $1,956,000 with Western-Knapp and a contract for $215,268 with Westinghouse.[86]

The Western-Knapp project engineer, S.R. Hurdle, was on site in early July.[87] His professional heavy construction experience included Hoover Dam, Parker Dam, Shasta Dam and several other big Western-Knapp dam contracts. Western-Knapp assigned the Kitt Peak job to him because of the underground nature of the solar telescope.[88] About the time the construction of the solar telescope was nearly finished the position of Kitt Peak Mountain Superintendent became vacant, and Hurdle was hired to fill it. He was on the KPNO payroll when the solar telescope was dedicated.

The underground part of the solar telescope presented some difficult problems, but Western-Knapp and Hurdle were more than equal to the task. During October a mining crew made substantial progress on the horizontal tunnel and the incline, and the foundation and first 15-foot section of the cylindrical heliostat tower were poured.[89] The tower was scheduled to be completed by March 10, 1961, and the 300-foot inclined tunnel by May 1, 1961.[90] The final 12-foot section of the tower was all that remained to be done in February,[91] and it was finished on schedule. The tunnel was behind schedule, but it was finished by the end of May. During June 250 tons of structural steel (75% of the total) was put in place.[92] The crown block (8500 pounds) was installed at the top of the heliostat tower on October 23 in a "flag raising ceremony."[93] The finished building was accepted on March 1, 1962, with a few minor details yet to be completed.[94]

The first meeting of the Executive Committee under the new officers was held in Tucson on June 12, 1962.[95] The question of a dedication ceremony for the McMath Solar Telescope was introduced by Harrell, who had been told by Pierce that a date in early November would be feasible. Friday, November 2, 1962 was chosen after consultation with NSF Director Waterman by telephone. A major public address by Waterman in Tucson in the evening would be the closing event of the dedication, following the daytime ceremony on Kitt Peak. The Scientific Committee also requested the President to invite the National Science Board to meet in Tucson, either at the time of the solar telescope dedication or in April, 1963. Pierce reported at the next meeting of the Executive Committee[96] that 226 invitations to the dedication had been sent out, and 40 acceptances had already been received.

The dedication ceremony on Kitt Peak[97] was attended by about 150 people, including representatives of AURA and NSF, and invited guests. The weather was pleasant and the sun's image was on display in the observing room in the late morning. Lunch was served in the horizontal spectrograph room using a long table, and folding screens were used to separate the dining area from the observing room. KPNO Director Mayall was master of ceremonies for the dedication ceremony following the luncheon. The speakers were: Dr. R.M. Petrie, Dominion Astrophysical Observatory; Dr. Helen Dodson Prince, McMath–Hulbert Observatory; Mr. Enos Francisco, Chairman of the Papago Tribal Council, who spoke in Papago and English; Mr. Morris Udall, member of Congress; Dr. A. Keith Pierce, and Dr. Alan T. Waterman. Edmondson read the Resolution of the AURA Board, and a letter from President Kennedy was read by Waterman following the unveiling of the bronze plaque by Mrs. McMath and Madeline McMath.

Waterman's evening lecture was given in the Liberal Arts Auditorium on the University of Arizona campus. Edmondson spoke a few words of welcome and then introduced President Harvill who introduced Waterman. His topic was: "The Changing World of Science."

The dedication received excellent press coverage, national as well as local, including a column in the science section of the November 9, 1962 issue of TIME. This was gratifying because the dedication took place only two weeks after the start of the Cuba missile crisis.

The Scientific Committee met the day after the dedication, November 3, 1962.[98] Possibly the most important item on the long agenda was the vote to approve offering the position of associate Director – Stellar Division to Dr. Arthur A. Hoag. Candidates for Associate Director – Chile were also discussed.

Exactly three weeks to the day after the Dedication of the McMath Solar Telescope, Cerro Tololo was chosen by the Chile site survey team as the location for the southern hemisphere counterpart of Kitt Peak. The origin of the "Chile project" in 1958 and its transfer from the University of Chicago to AURA in 1960 will be described in Chapter 13. Chapter 14 will continue through the dedication of the Cerro Tololo Inter-American Observatory on November 6–7, 1967. The funding of 150-inch telescopes for Kitt Peak and Cerro Tololo will be described in Chapter 15.

13 The Chile project. Professor Rutllant's June 1958 visit to the Yerkes Observatory. Initial support by the US Air Force. Transfer of responsibility for the project from the University of Chicago to AURA, with continuing funding by the National Science Foundation. Relations with the University of Chile.

The first step toward the establishment of the Cerro Tololo Inter-American Observatory was taken during the May–June 1958 visit to the United States by Professor Federico Rutllant, Director of the Observatorio Nacional of the Universidad de Chile.[1] He attended the annual meeting of the Astronomical Society of the Pacific in Los Angeles, the meeting of the American Astronomical Society in Madison, Wisconsin, and the sessions of a Consulting Committee called together by the Organization of American States in Washington to provide information about the development of science in American countries. He was seeking help to build up astronomy in Chile, and visits to major US observatories were an important part of his trip. He was received politely at the west coast observatories, but did not arouse any interest in what he was seeking until he met with Gerard P. Kuiper, Director of the Yerkes and McDonald Observatories, and W.A. Hiltner at the Yerkes Observatory.

Kuiper wrote to Shane, Chairman of the AURA Scientific Committee and Vice-Chairman of the AURA Executive Committee, on July 7, 1958 to say that he would not be able to attend the July 9 meeting of the Scientific Committee.[2] He made comments about five items on the agenda, and suggested an additional item for discussion. He told Shane that Dr. Rutllant had met with the Yerkes staff and had presented his views on a proposed collaboration between AURA and the University of Chile for the building and operation of a moderate sized observatory. Rutllant told Kuiper that Keller had told him that NSF support for such an observatory would depend on the interest of AURA. Kuiper's letter concluded: "Dr. Rutllant would, I believe, appreciate to know whether AURA is willing to: (1) send him an expression of interest in any such plan; (2) would consider an exploratory trip to Santiago, Chile, by a small delegation from AURA, possibly next winter (their good season is October–April)."

Twenty years later Shane recalled his reaction to Kuiper's letter:[3] "The suggestion was not brought before the Committee probably because AURA was just newly established and should not at that time consider extending its field of operation." Eight months after Shane decided not to respond to Kuiper's letter, a brief discussion of a southern hemisphere telescope was reported in the minutes of the AURA Executive Committee:[4]

> 7. Dr. Shane stated that he had received a letter dated February 6 from Dr. Irwin of Indiana University, containing a plea that AURA attempt to secure funds and authorization for an 80-inch southern hemisphere telescope. In his reply to Dr. Irwin on February 11, Dr. Shane had stated that all astronomers agreed that a southern hemisphere telescope would be desirable, but questioned whether this would be the time to ask for it, since he understood Congress had excluded fiscal year 1959 funds for the southern hemisphere astrograph from the budget. Dr. Robertson stated that this was correct, but that it was hoped that money could be obtained in 1960. Dr. Hiltner stated that Yerkes Observatory expected to receive funds for a southern hemisphere telescope from the Government, and in response to an inquiry from Dr. Keenan, stated that there was no reason to believe that visitors would be excluded from using this instrument. Dr. Shane stated that he would inform Dr. Irwin of this.

A few days before Rutllant's visit to the Yerkes Observatory, Kuiper had submitted Hiltner's proposal for a research project entitled "Galactic Field Studies" to the Air Force Cambridge Research Center, Geophysics Research Directorate.[5] Hiltner was named as the project director, and W.W. Morgan, H.A. Abt, and R.P. Kraft were listed as participants in different aspects of the project. The Air Force contract that funded this research would be used later as a convenient mechanism for funding the southern hemisphere telescope that Hiltner mentioned during the AURA Executive Committee meeting.

Twenty-one years later Rutllant's visit was recalled by Hiltner:[6] "The three of us, that is Kuiper and Rutllant and myself, were sitting in Kuiper's office. And Rutllant at that time, at least it was my impression, was very discouraged about getting any cooperation from American institutions to build an observatory in Chile. And I can so well recall in our discussion my remarking to Kuiper, 'Well, how about Chicago trying it?' " Kuiper wrote to Vice-President R.W. Harrison:[7] "I considered that with our plans for McDonald we would have our hands full and should support such a venture through AURA; I promised Dr. Rutllant to present a plan to AURA. I did so, but apparently no action has been taken (I was unable to attend Directors' meetings recently).[8] Meanwhile our department got more and more excited about the opportunities a southern station would offer. . . . I therefore inquired with the Air Force whether they might be interested in the proposal to establish and support such a station." Kuiper's letter to the Air Force GRD[9] was answered by E.L. Eaton, Director (Act'g) GRD:[10] "We are really no less interested in the proposal to support astronomical observations in South America. In January 1958 we proposed to Hq ARDC the establishment of an Air Force Observatory in the Southern Hemisphere, primarily for planetary research. Your interest in and corroboration of the scientific value of such a facility are encouraging to us." Kuiper immediately wrote to Rutllant:[11] "I have now found an agency which appears to be willing to support such a station. We are thinking of starting with a 40-inch reflector with the hope that later on larger equipment might be added." Rutllant discussed Kuiper's letter with the Rector of the University of

Chile, who accepted in principle the project of cooperation with the University of Chicago.[12]

The thirty-year contract between the University of Texas and the University of Chicago for operation of the McDonald Observatory by the University of Chicago was due to expire in 1962, and in late 1958 Kuiper was already negotiating an arrangement that he thought would safeguard Chicago's future participation.[13] Joint operation of the McDonald Observatory would start September 1, 1960, following the creation of a Joint Astronomy Department in the fall of 1958; formal operation of the department did not begin until by-laws were adopted on June 28, 1960.[14] These plans were approved by the University of Texas Board of Regents at their December 12–13, 1958 meeting, and the Chile observatory now became a Chicago–Texas–Chile project.[15]

For administrative convenience the Air Force GRD decided to fund the Chile telescope through an amendment to Hiltner's Galactic Studies project because this would be quicker than writing a new contract.[16] The proposed contract amendment was submitted on January 8, 1959.[17] Kuiper initially proposed construction of a 40-inch reflecting telescope, and got preliminary estimates from a Dutch engineer, B.G. Hooghoudt.[18] This was increased to 60-inches after Hooghout estimated a cost of $80,000 for a 40-inch telescope and $160,000 for a 60-inch, and because the Air Force had anticipated spending $120,000.[19] Kuiper sent Miczaika a detailed explanation of the reasons for increasing the aperture to 60-inches,[20] and a month later Miczaika told Kuiper this change would require a formal amendment to the contract.[21] More than five months passed before Kuiper responded by including this in a request for extending the Galactic Studies contract beyond April 30, 1960.[22] The request worked its way through the Air Force bureaucracy, and the 60-inch telescope received formal approval on December 3, 1959.[23]

Kuiper, Miczaika and Slavin (Eaton's successor) visited Chile March 10–19, 1959.[24] They had three meetings with the Rector of the University of Chile and several with Rutllant during which an agreement between the Universities of Chicago, Texas, and Chile was drafted. Three prospective mountain sites north of Santiago were selected during several flights in the US Embassy DC-3, made available by the Air Attache. They also did some ground-level reconnaissance by automobile and on horseback.[25]

The news stories that appeared following this trip[26] did not mention the USAF Cambridge Research Center (GRD) as the source of funds for the telescope, and Miczaika complained about the New York Times story to Kuiper:[27] "It would be appreciated greatly if in the future steps will be taken to insure reference to the role of the Air Force Cambridge Research Center in an appropriate form in this important project."

Testing of the three mountain sites began in April and May 1959 when Dr. Jurgen Stock spent three weeks in Chile working with astronomers from the University of Chile. Stock had spent two days with Kuiper at the McDonald Observatory in late November 1958 to discuss his possible interest in employment as Resident Astronomer in charge of development and maintenance of the scientific equipment.[28] Kuiper proposed Stock's appointment at the December 12, 1958 faculty meeting, and it received a favorable response.[29] Stock accepted the offer,[30] and Kuiper's reply confirmed that the appointment was effective July 1, 1959.[31] Kuiper had discussed the Chile telescope project with Stock during their meeting at McDonald, and he persuaded Stock to go to Chile to start the site survey three months before his formal appointment began.[32]

Stock wrote four longhand letters to Kuiper during his May 1959 sojourn in Chile.[33]

Stock and Rutllant visited Kuiper in Williams Bay on May 24,[34] following Rutllant's attendance at the Second Astrometric Conference held at the Cincinnati Observatory on May 17–21, 1959.[35] Kuiper's letter to Miczaika said: "I feel very pleased about the competence with which Dr. Stock has carried out his assignment in Chile." Kuiper took Rutllant to Chicago on May 25 to discuss new business arrangements for the site survey with Irene Fagerstrom, Assistant Business Manager of the University of Chicago.

Stock took up his position as Resident Astronomer at the McDonald Observatory on July 1, 1959 as scheduled, but a second trip to Chile was already being planned.[36] He was in Chile August 24 – September 18, and decided that the highest site, Cerro Colorado, 11,000 feet, should be abandoned because of its inaccessibility during the winter. Two new sites were added to make a total of four under investigation, all of them between 7,000 and 8,000 feet.[37]

Two recommendations of the OAS Committee meeting that Rutllant had attended in 1958 attracted the attention of the National Science Foundation.[38] Seeger called Merle Tuve, Carnegie Institution of Washington, who was a member of the OAS Committee, "to ascertain what steps, if any, the NSF should take to assist in implementing the recommendations." They agreed that a meeting of interested astronomers would be useful, and such a meeting was held on October 19, 1959.[39] Keller sent invitations to 12 people.[40] The meeting actually was a consequence of Kuiper's discussion of the progress of the Chicago–Texas–Chile operations with Shane (AURA) and Keller (NSF) in Toronto on September 1, 1959 at the meeting of the American Astronomical Society. Shane remembered:[41] "It is my recollection that he felt the proposed telescope was underfunded and needed broader support. Further interest in South American Astronomy on the part of Clemence, Brouwer and Merle Tuve resulted in calling a conference in Washington for October 19." The discussion covered astrometry, radio astronomy, the Chicago–Texas–Chile project, and Air Force plans to put a large Baker–Nunn Super-Schmidt satellite tracking telescope, with the mounting modified for dual use as an equatorial or alt-azimuth, at the University of Chile Observatory. A resolution was passed "that a Subcommittee of the U.S. National Committee of the IAU be formed to study Inter-American Astronomy problems."

A follow-up meeting to discuss the 60-inch Chile telescope was held in Chicago on November 9, 1959.[42] The Air Force Cambridge Research Center was represented by Miczaika and Gerlach, who presided. The University of Chicago was represented by Vice-President Harrell, Dean Zachariasen, Vice-President Johnson, and Drs. Kuiper, Morgan, and Stock. AURA was represented by Shane, and NSF was represented by Keller and Mulders. Gerlach began by stating that "the main reason for the meeting was to review the background and the financial status of the project before Dr. Miczaika leaves the Air Force." Following the discussion, Dean Zachariasen proposed to set up a panel advisory to the University of Chicago on the Chile project, composed of representatives from the Air Force, NSF, AURA, and the staff of Yerkes and McDonald. Keller suggested that Kuiper should convene a meeting of the Advisory Panel as soon as the members had been appointed, to look into the problem of how much additional financing would be needed for the site survey. He said NSF could supplement the Air Force funds in future fiscal years.

A few days after this meeting, the staff of the Yerkes Observatory voted that Kuiper's term as Chairman of the Astronomy Department should not be renewed when it expired on August 31, 1960.[43] He was summarily removed from all of his administrative posts by the University of Chicago on January 4, 1960,[44] and by the end of January he had accepted a professorship at the University of Arizona, effective July 1, 1960.[45] The University of Chicago appointed W.W. Morgan[46] as Kuiper's successor as Chairman of the Astronomy Department and Director of the Yerkes and McDonald Observatories.

The first meeting of the Policy Advisory Board – Chilean Observatory Project (PAB-COP) took place in Chicago on January 5, 1960.[47] Vice-President Warren Johnson presided. Kuiper's changed status was not explicitly mentioned in the report, but its impact can be felt. It was decided that the next meeting should be held after the March 12 meeting of the AURA Scientific Committee;[48] March 30 was the date that was finally chosen.[49] Stock was not present at this meeting, but his next trip to Chile had already been scheduled by Kuiper to be for five months, starting on February 8, 1960.[50]

The AURA Scientific Committee made a careful study of the plans for the proposed Chilean telescope, and approved three recommendations:[51] (1) the fork type mounting should be abandoned in favor of one that would allow ample room for a Cassegrain focus and a 4-mirror Coude focus; (2) the primary focal ratio of 5 should be changed to one as short as 3; (3) the type of mounting planned for the KPNO 36-inch telescope was recommended. Shane sent these recommendations to Vice-President Johnson on March 21.[52]

The March 30, 1960 meeting of the PAB-COP[53] was attended by Vice-President Johnson, Vice-President Harrell, Dean Zachariasen, Dr. Kuiper, and Dr. Morgan, for the University of Chicago; Dean Whaley, for the University of Texas; Dr. Rutllant, for the University of Chile; Dr. Shane, for AURA; Dr. Mulders, for NSF; and Dr. Dieter, for the Air Force Cambridge Research center. Johnson presided.

Whaley could not be present until noon, and Kuiper arrived late, so it fell to Rutllant to begin the report on the site survey; Kuiper finished the report after he arrived. Next, Shane read the Report of the AURA Scientific Committee, and the Policy Advisory Board approved adoption of these recommendations. Morgan moved and the Board approved the appointment of Meinel and Baustian as Consultants. The discussion that followed was mostly about costs, funding, and logistics.

Vice-President Johnson opened the discussion of the organization that would now be needed to carry out the project by noting the debt owed to Kuiper, Rutllant, and Miczaika for initiating the project. Kuiper's departure from Chicago presented a difficult problem, and Johnson proposed that AURA be asked to take over the project. This suggestion was received favorably by Dieter, Mulders, Shane, Kuiper, Rutllant, Zachariasen, Morgan, and Whaley. However, Whaley pointed out that there was a problem because "the University of Chicago is a member of AURA, while Texas is not", and Zachariasen said that a joint-membership with Chicago would be a possible solution. Vice-President Johnson stated that "if this transfer of responsibility occurs the Policy Advisory Board should be dissolved. In addition, it would be necessary to abrogate the tri-University agreement. The arrangements would then be made through AURA and the University of Chile." He proposed "that an attempt be made to effect the transfer by September 1, the date of Dr. Kuiper's departure from Chicago."

The following motion was made by Mulders, seconded by Whaley, and approved unanimously: "that the committee recommend that the University of Chicago, the University of Texas, and the University of Chile enter negotiations with AURA to arrive at an arrangement under which AURA shall proceed with the construction and operation of the joint-Chilean Observatory."

Kuiper wrote a long letter to Stock following the meeting.[54] He said he had told Rutllant that transfer to AURA would be the best solution, and that Rutllant had proposed this to Vice-President Johnson the day before the meeting. Johnson welcomed this, but was worried about two possible difficulties: (a) the position of the Air Force; (b) the position of Dean Whaley, representing the University of Texas. Identifying the 60-inch telescope as an Air Force project satisfied Dieter. Whaley accepted the logic of Shane's argument that the Chicago–Texas Joint Department of Astronomy would justify a joint Chicago–Texas membership in AURA. Kuiper told Stock he felt "the meeting ran surprisingly well." He also described the reorganization of KPNO that had taken place on March 14 (see Chapter 12). Stock thanked Kuiper for his letter[55] and said: "I am happy that the Chilean Project is in good hands in spite of all the changes, although I do not quite see where I fit into the picture." Kuiper forwarded a copy to Shane,[56] and Shane replied by telephone that he would like to have Stock continue to serve the project.[57] Kuiper then advised Stock "to look in that direction rather than a solution at McDonald or Austin." Stock's reply said he would be interested.[58]

The question of AURA taking on the responsibility for the Chile telescope was brought to the AURA Board of Directors in Executive Session on June 6, 1960.[59] Morgan summarized the history of the development of the Chile program (with a few understandable mistakes since he had not been involved in the project). He reported that the PAB-COP had passed a motion asking AURA to take over the planning, construction and operation of the Chile telescope. Shane stated that AURA would soon receive a formal request from the University of Chicago, and he read a draft of the proposed letter.

Shane described the Chicago–Texas–Chile agreement, and said it must be taken into account if AURA is to take over the project. He mentioned Dean Whaley's concern about the status of Texas, and asked for an expression of opinion from Board members about a joint Texas–Chicago membership. Edmondson read a draft resolution that he had written as a basis for discussion.

The Board unanimously adopted two resolutions: "(1) It is the sense of this meeting that AURA undertake the Chile project with due consideration to the political and economic factors; and further that the Executive Committee be empowered to act. (2) It is the consensus of the Board of Directors that, by reason of the existence of a joint department of astronomy of the University of Chicago and the University of Texas, the membership of the University of Chicago is hereby changed to be a joint membership of the University of Chicago and the University of Texas. Such joint membership is a direct consequence of the formation and continuing existence of the joint department. This action does not affect the status of the University of Chicago as an organizing member of AURA. Chicago–Texas shall be represented on the AURA Board of Directors by one scientist from the joint department and by one administrator from either Chicago or Texas."

The formal request from the University of Chicago was dated June 17, 1960, but an item about transfer of Air Force funds from the University of Chicago to AURA had to be deleted at the request of the Air Force.[60] Corrected copies were sent to Shane prior to the June 30 meeting of the AURA Executive Committee in Tucson. Johnson's cover letter[61] pointed out that the main reason the transfer to AURA would be difficult was because "the present contract between the University and the Air Force actually lists Mr. Hiltner as the investigator and is in support of two programs, one for Mr. Hiltner's work proper and the other for the Chilean project." Johnson also said that he had been told that Dr. Gordon Wares would be taking Mrs. Dieter's place as the Air Force representative.

Discussion of the Chile program at the June 30, 1960 meeting of the Executive Committee[62] was far from perfunctory because several members were hearing the details for the first time. Two actions were taken: "(1) The Secretary was instructed by the Executive Committee to obtain approval of the Board of Directors on the joint Chicago–Texas membership by mail vote. (2) the Executive Committee voted to accept the University of Chicago proposal, subject to approval by mail vote of the Board of Directors of the joint Chicago–Texas membership, and that the officers be authorized to negotiate with the Air Force, the National Science Foundation, the University of Chile and the University of Chicago, looking toward the establishment of the Chile Observatory."

The next major event was a conference on the Chile Observatory in Tucson on August 10, 1960.[63] The 19 people in attendance represented AURA (7), University of Chicago (4) University of Texas (1), University of Chile (1), Air Force GRD (1), NSF (1), Yale University (1), Columbia University (1), Lick Observatory (1), and the US Naval Observatory (1). Shane presided.

The conference began with Kuiper's detailed and authoritative review of the history of the Chile Observatory project. Shane discussed the AURA action on the request from the University of Chicago for AURA to take over the project, and Johnson discussed the project from the perspective of the University of Chicago administration. Rutllant said that Chile was pleased to have American astronomers interested in its plans for development of astronomy in the southern hemisphere.

Shane called on Stock to report on the site survey, and mentioned his enjoyment of the letters and reports that Stock had written.[64] Stock recalled his previous experience during the 1957 European Southern Observatory site survey expedition to South Africa. The site survey in Chile had given him an opportunity to use the experience he had gained in South Africa. He listed eight sites, four in the Santiago area and four to the north. The sites near Santiago were visited in April and May 1959 because of funding limitations and because they were near existing roads and were accessible. His August and September 1959 visit was coordinated with the Yale–Columbia group who were seeking a site for the Southern Hemisphere Astrograph.[65] The third visit, from February to August 1960, concentrated interest in the four northern sites because conditions had been found to improve to the north. He described the interferometer-type instrument used for testing seeing, consisting of two lenses about 60 to 65 inches apart which produce two images moving independently in the same eyepiece. Two of these instruments were now in use. He displayed maps, and showed slides and movies. He said Tololo appeared to be the most promising site at present. Morado is larger, but is downwind from Tololo.

Schilt discussed the Yale–Columbia survey for a site for the Southern Hemisphere Astrograph. Four Danjon cameras had been used in Argentina, with encouraging results. Two were still operating 130 kilometers north of San Juan in the same latitude as Vicuna; and the other two were being offered to Stock for use in Chile.

Stock finished his discussion of the site testing after lunch. He felt the program should continue for at least one more year, with the possibility of looking further north. He discussed the four northern sites in detail, and repeated that Tololo and Morado were his first choices.

The plan for the observatory was presented by Shane in the categories: (a) Instruments, (b) Buildings, (c) Facilities, (d) Staff, and (e) Supporting Staff. The group expressed general agreement with the plan.

Shane stated that the purpose of the observatory would be "to provide visiting astronomers with observing facilities; to train graduate students (Latin American, with emphasis on Chilean students); and to permit research by the resident staff." The group agreed with these purposes.

The organization of the observatory was described in the categories of (a) Name, (b) Authority, (c) Use of KPNO shops, and (d) Exchange with KPNO astronomers. Several names were suggested, and the group accepted without much enthusiasm Shane's second suggestion: The Inter-American Observatory of Chile – A Division of the Kitt Peak National Observatory. It was agreed that the astronomer in charge in Chile would report to the Director of KPNO.

Other topics on the agenda were: relations with the University of Chile, Southern Hemisphere Astrograph, visit by Mayall and Shane to Chile in 1960, financial problems, program of construction, and Russian application to Chile.[66]

Following this meeting the initial negotiations with the Air Force for funding the 60-inch telescope were difficult and progress was very slow, and there were also problems with NSF funding. A preliminary proposal for $150,000 to complete the site survey was sent to NSF on October 17, 1960.[67] This included $107,400 for partial development of a road to site; this could not be paid from the available NSF research funds,[68] and some quick changes had to be made in the budget. A revised proposal for $81,000 was sent to NSF on November 30.[69] The National Science Foundation was being very cautious about its relations with the Appropriations Committee, and was asking the Bureau of the Budget to include only $50,000 for the Chile project in the budget for FY 1962.[70]

The joint Chicago–Texas membership was approved by the sign-out vote of the Board of Directors, and this was reported to the Executive Committee on November 9, 1960.[71] Dean Whaley was appointed a consultant by the Executive Committee, and Hiltner and Harrell would continue on the Board as the representatives of Chicago–Texas.[72] Miller said that NSF will ask for $800,000 for the Chile project in FY 1963, if the $50,000 request for FY 1962 is successful. He will meet with Wares and others on November 21 at the Air Force Cambridge Research Center to discuss the proposed FY 1961 contract with AURA.

Shane and Mayall visited Chile December 1–17, 1960.[73] They flew from Santiago to La Serena on December 3, and were driven to Vicuna by Stock. A pilot of the La Serena flying club flew them around the tops of Tololo, Morado, and Blanco on December 4. The next day Shane and Mayall rode horses, Stock and four others rode mules (mulas in Chile), and there were two pack animals. They reached Los Placeres about 5 p.m., and were greeted by Don Rogelio Ramos and his family. Los Placeres is located at a spring near the base of Tololo, and has been occupied by the Ramos family for two hundred years. The group, accompanied by Don Rogelio, rode up Tololo on December 6, and spent the night in a stone hut near the summit. On December 7 they rode to the top of Morado, and then back to Los Placeres for an overnight stay. After a restful day and a dinner party hosted by Stock in Vicuna on December 9, they rode to the top of Blanco on December 10 and spent the night there in a mud hut. They returned to Vicuna on December 11, and to La Serena and Santiago on December 12.

The morning of December 13 was spent at the old National Observatory discussing administrative and financial problems with Rutllant and Stock, and the operating arrangement when AURA takes over with Rutllant. In the late afternoon Rutllant took them to see the new National Observatory which was under construction on Cerro Calan, a low hill east of the city.

Arrangements had been made through the US Embassy for a flight to the Copiapo area in a US Air Force plane on December 14. However, Mayall had picked up a cold and stayed at the hotel. Stock and Shane were accompanied by Carlos Torres who had been working on the site survey with Stock. They spotted two suitable looking mountains, and Carlos Torres was planning to visit the Copiapo area starting December 17 to get more information.

Shane and Mayall spent the morning of December 15 at the Observatory and looked at the instruments that were to be moved to Cerro Calan. In the afternoon Shane, Stock, and Carlos Torres rode with Hugo Moreno, who was also participating in the site survey, to Alto del Toro near the ski resort of Farellones. This site was no longer under serious consideration.

Before lunch on December 16 Rutllant, Shane, and Mayall called on the Rector of the University of Chile, Juan Gomez Millas, and on Professor Carlos Ruiz Bourgeois, head of the legal department of the university. They also called on Miss Malena Saavedra, Counsel for the US Embassy. After lunch they prepared a statement for the press, and then called on Dr. Heilmaier of the Catholic University. He took Shane and Mayall to Cerro San Cristobal to see the 36-inch reflector that the Lick Observatory had sent to Chile in 1903 to measure radial velocities in the southern hemisphere.[74] It was given to the Catholic University when that program was completed. A banquet in the evening at the Union Club was hosted for a party of eleven by Carlos Mori Ganna, Dean of the School of Physical and Mathematical Sciences. With Rutllant's help, Shane and Mayall did some shopping on the morning of December 17 before taking the flight to Miami.

When the AURA Board of Directors met in Tucson on March 13, 1961 the solution that was proposed to solve the KPNO space program problem (see Chapter 11) overshadowed everything else. Discussion of the Chile project was limited to the bare essentials, and the December 1960 trip to Chile was summarized in eleven lines in Mayall's Report to the AURA Board.[75]

Miller made a trip to Chile a month after the Board meeting to set up administrative procedures for the site survey, now being financed by NSF, and to inspect site survey operations in the field.[76] Accompanied by his wife Jane, Miller travelled directly to Copiapo, arriving on May 1. They had decided to visit Cerro Checo de Plata before coming to Santiago. This revised itinerary was in a letter to Rutllant mailed from Lima on April 26. It arrived in Santiago the next day, but was not delivered until six days later. Stock was in Santiago, and when Miller phoned from Copiapo on May 3, he was told that a trip to Checo could not be made on such short notice. Hence, the Millers came to Santiago on May 4.[77] On May 5 Miller visited the Observatory in the afternoon and had a long discussion with Rutllant. Other business was conducted on the morning of May 5 and on May 6 and 7. A trip to Alto del Toro was made on the afternoon of the 6th.

The Millers flew to La Serena with Stock on May 8, and spent the night in Vicuna. A five hour horseback ride the next day took them to Los Placeres, and they rode to Morado on the 10th. Jim was not feeling well, but Jane rode from Morado to Tololo with Stock on the 11th. She decided it would be more comfortable to walk to Los Placeres from Tololo, and this took about an hour and a half. Jim had ridden his horse from Morado, and was feeling much better. They returned to Vicuna on the 12th, and to La Serena on the 13th to take a flight to Copiapo. May 14 was devoted to an all day trip to Cerro Checo de Plata, with an hour at the summit. They returned to Santiago on May 15. May 16 and 17 were spent on business at the Banco de Chile, and in discussions with officials from the US Embassy and the University of Chile concerning how legislation could be initiated that would enable equipment for the observatory to be brought in free of import duties.[78] The Millers visited Cerro Calan on the morning of 17th, and left on an afternoon flight to Buenos Aires.

Miller reported on his trip to Chile at a meeting of the AURA Executive Committee on June 14, 1961.[79] He described his discussion with US Embassy and University of Chile officials regarding customs legislation for the Observatory. He also presented a draft of the proposed AURA agreement with the University of Chile that would replace the Chicago–Texas–University of Chile agreement. This was discussed in detail, and the Executive Committee voted unanimously authorizing the President "to execute a cooperative agreement between AURA, Inc. and the University of Chile for the establishment of the Southern Hemisphere Astronomical Observatory substantially as set forth in Attachment II, subject to approval by the University of Chile and the National Science Foundation."

He also reported that the House had passed the National Science Foundation budget at $250,000,000 with no objection to the southern hemisphere observatory item. The first installment of Air Force funding for the telescope had finally come through with a grant of $100,000 from the Air Force Office of Scientific Research.[80] This was for the design of the mounting, purchase of the 60-inch mirror blank, and general planning. By the next meeting of the Executive Committee on October 26, 1961, Congress had approved the $50,000 line item for the southern hemisphere observatory[81] and the future of the project was assured. A financial plan for the $50,000 was unanimously approved.

In December 1961, Shane paid a second visit to Chile to study the results of the site survey and to visit the sites that had been added since his 1960 trip.[82] He arrived in Santiago on November 29, and met the new Ambassador, Charles Cole, former President of Amherst College, the next day. He and Rutllant discussed the proposed customs legislation with Ruiz Bourgeois on December 1, and they compared the Spanish text with the English translation. The three of them called on the Rector of the

University of Chile, Juan Gomez Millas, who had written to the Minister of Education in October urging passage of the law. Stock and Shane drove to Vicuna on December 2, arriving about 9:30 p.m. They started for Tololo at 9:30 a.m. on the 4th in the Chevrolet pickup truck and changed to horses at the corral at El Zapallo. They reached Los Placeres after an hour and forty-five minute ride. They continued on to Tololo after lunch and arrived at the summit about 5:15 p.m. That night Shane thought the seeing was good, but Stock said it was somewhat below average. They returned to Vicuna on the 5th, with a stop for lunch at Los Placeres.

It was determined that Tololo offered the best possibility in the Vicuna area, so Shane had directed Stock to terminate operations on Blanco and Morado in order to concentrate more on the Copiapo area. Stock and Shane drove from Vicuna to Copiapo in the Chevrolet on December 6, arriving at 8:30 p.m. Paul Kuiper[83] had driven the jeep to Copiapo, and he gave his room to Shane because the hotel had failed to hold three rooms; he stayed at a less desirable hotel. On December 7 Stock, Shane and Kuiper were joined by an employee, Fernando Richards, for the ride to La Peineta, elevation a little over 10,000 feet. They reached the top about 12:30 p.m. and stayed until 2:15. This mountain is a few miles to the southeast of Checo, and seems to have better conditions. Stock, Shane and Richards spent the night of December 8 on Checo, and were back in Copiapo at 9:30 a.m. on the 9th. They called on the Intendente of the Province before lunch, and Shane took the afternoon plane to Santiago. He was invited to have dinner with the Moreno family on Sunday. There was a fire on the slope of Cerro Calan, and Hugo Moreno was at the Observatory until late afternoon. Shane visited the US Embassy on Monday morning, December 11, and spent the afternoon at Cerro Calan looking at the cloudiness records based on the observations made by Carlos Torres. He went to the embassy again on Tuesday morning, December 12, and left for Buenos Aires at 1:30 p.m.

Shane felt that "La Peineta was a site worth studying."[84] In this case there was a choice between using "mulas" or the construction of a road. Stock investigated the feasibility of a road on the Monday after Shane left, December 11. Upon his return to Copiapo, he went to see Luis Gonzales, an engineer of the Empresa Nacional de Mineria, a government agency, who was in charge of their road construction equipment. He gave Stock an acceptable estimate of the cost, and started to work on the road on December 15. A crew of ten men with a bulldozer and a tractor did all the work, and the road was open for traffic on January 24. The cost was $1,792.50.

Stock left for Venezuela on February 19, 1962 and expected to return to Chile by the end of March.

14 Completion of the site survey, and the selection of Cerro Tololo on November 23, 1962. The start of construction in La Serena and on the mountain, following purchase of the land. The November 6–7, 1967 dedication of the Cerro Tololo Inter-American Observatory.

In the absence of Stock, who was in Venezuela, a preliminary report on the status of the site survey was presented by Rutllant at the meeting of the AURA Scientific Committee on March 10, 1962.[1] Miller pointed out that the budget for the southern hemisphere telescope had been based on the anticipation that the site would be chosen by October 1962, and Shane thought this schedule could be met. A 24-inch telescope had been included in the fiscal year 1963 budget as a back-up to the 60-inch Air Force telescope, but several members of the Committee thought this was too small. A motion by Shane, seconded by Hiltner, to increase the size of this telescope to 36 inches was unanimously approved.

The new AURA President, Edmondson, was in the chair when Shane reviewed the history and present status of the Chile site survey at the meeting of the AURA Board on March 12, 1962.[2] Shane described his recent trip to Chile[3] and said he was greatly impressed with the progress that had been made since his 1960 trip. He said Stock was continuing observations on Tololo, intensifying investigation of Checo, and starting observations on La Peineta. He told the Board that a site decision could be made in October, and Miller said that funds were available to continue until March 1963. He also reported on the progress of the proposed Chilean legislation which would allow observatory equipment to be imported into Chile duty-free. This was very important because the duty on industrial machinery was 100 percent. Finally, Rutllant summarized Stock's seeing data for the Board.

Miller presented the Chile project budget for fiscal year 1964. He pointed out that there would be a city headquarters similar to the Kitt Peak arrangement, and mentioned the substitution of a 36-inch telescope for the previously requested 24-inch telescope. In response to a question, Keller said the National Science Foundation felt it would be best "to keep the Southern Hemisphere Observatory [contract] separate from the Kitt Peak National Observatory [contract] in case Congress should turn against expenditures outside the United States." The Board unanimously approved the $1,404,000 budget as presented.

Six days after the Board meeting Shane, Mayall, Stock, Edmondson, and five other astronomers were in Caracas, Venezuela for a week.[4] They had been invited by Dr. Jose Abdala, Director of the Observatorio Cagigal, to advise him concerning the best location and use of $4 million worth of modern telescopes. These had been ordered by the previous director shortly before his death, and belonged to the Government of Venezuela. Edmondson had two or three extended conversations with Shane about the Chile project, and these convinced him that he should make a trip to Chile before the June meeting of the Executive Committee. The only possibility for this in his very full schedule was the third week of April, from Friday, April 13 through Saturday, April 21. At the start of the visit he met with the Rector of the University of Chile, Dr. Juan Gomez Millas, the Dean of Science, Dr. Carlos Mori, and other University officials. He also met with the US Ambassador, Charles W. Cole, who had been President of Amherst College, Miss Malena Saavedra, the legal advisor of the US Embassy, and Madison Monroe Adams III, the Science Advisor. Rutllant had not yet returned from the United States.

During the week Edmondson and Stock travelled all the way from Santiago to Copiapo and back in the AURA jeep. Arrival in Copiapo was at 3:00 a.m., about 500 miles and 19 hours after leaving Santiago. The next night was spent on La Peineta, where the temperature dropped to 40° F. The sky was clear and the images in the telescope were very sharp and very steady. The final night in Copiapo had an unexpected bonus, a fine concert by the Orquesta Filarmonica de La Serena and the Coro Atacama, de Copiapo. It was interesting to hear the Halleluiah Chorus sung in Spanish. The concert was sponsored by the Copiapo Club de Leones and the Municipalidad de Copiapo.

The return to Santiago was made without an overnight stop, but it included a side trip from La Serena to Vicuna. Departure from Vicuna was at 9:00 p.m., and arrival at the Hotel Carrera was at 5:00 a.m. on Saturday, April 21. Rutllant called at about 9:00 a.m. to arrange a date for lunch.

After lunch Rutllant drove Edmondson around the city for some sight seeing. They also discussed the proposed visit of AURA Board members to Chile. Rutllant, Hugo and Adelina Moreno, and Carlos and Regina Torres were at the airport to see him off on the flight to Miami. Father Theodore Hesburgh, President of the University of Notre Dame, was on the evening northbound plane. He was at that time a member of the National Science Board and Chairman of the Board Subcommittee on International Science, so Edmondson took advantage of the opportunity to brief him about the status of the Chile project and the expectation that the site would be chosen before the end of the year.

Edmondson's report, which was presented at the June 12, 1962 meeting of the Executive Committee,[5] said that during the week in Chile:

> I became increasingly convinced that it would be necessary for other members of the AURA Board to visit Chile. Too much is at stake in this project, both in money and in human effort, for us to make the site selection and the evaluation of other aspects of the project on the basis of recommendations by only one or two individuals. Dr. Stock and I prepared a tentative one-week schedule before dinner in Vicuna, and I showed this to Dr. Rutllant on Saturday afternoon. He gave approval to the suggested calendar, and professed considerable enthusiasm for a visit by some of the

> AURA directors to Chile. I feel that the optimum would be for the entire
> Board to go but this would certainly lead to a charge of "junketing".
> Perhaps we should have the courage to face this charge. If the whole Board
> does not go, then we have the question of how to choose those who do go,
> and I think this question is something that must be discussed by the
> Executive Committee today.

He asked the Executive Committee for suggestions about who should participate in a visit to Chile in October or November, and presented a tentative schedule for October 20–27 or November 16–24. It was agreed to make November the target date for the trip.

Edmondson and Mayall represented AURA and KPNO at the NASA Space Science Summer Study at the University of Iowa, starting June 24. Mayall stayed to the end on July 7, but Edmondson left on July 3 because he was leaving for France on July 7.

Miller had replaced Shane's name with Whitford's on the list of AURA Board members after the March meeting. This inadvertently led to Shane no longer receiving the Stock Reports, and he and Mayall complained about this to Edmondson.[6] Before leaving for France, Edmondson wrote to Shane that he had instructed Miller to restore his name to the mailing list.[7] Edmondson also sent a memorandum about the distribution list for the Stock reports to Harrell (AURA Vice-President), Mayall, and Miller.[8] After criticizing both Miller and Mayall for errors of judgement in the way they had handled the Shane problem, the memorandum said: "This episode does help to point out the fact that communications between Nick and Jim need to be further improved. . . . Both Nick and Jim have called me more than once since March 12 to ask questions that they should have been asking each other." Harrell responded:[9] "Agree with your position. Applaud the suggestions contained in your memorandum to Messrs. Harrell, Mayall and Miller dated July 5, 1962." Mayall and Miller each thought the criticism of the other was justified, but each also felt that he had been criticized unfairly. Edmondson's attempt to be evenhanded finally bore fruit in September, with help from Menzel. There will be more about this in connection with the September meeting of the Executive Committee.

Edmondson also instructed Miller to go to Chile as soon as possible[10] to find a Chilean legal firm to represent AURA, check up on the status of the customs legislation, and do whatever else needed to be done before the November AURA site selection visit. Miller wrote to Stock and Rutllant that he would be arriving in Santiago on July 18.[11]

At the time of the June Executive Committee meeting Edmondson, Harrell, Clemence, Mulders and Whaley agreed that it would be advisable for AURA representatives to meet with State Department officials to discuss the Chile project. Harrell urged Edmondson to arrange for such a meeting after he returned from France.[12] Mulders wrote that Waterman and Scherer wanted this meeting to be an internal AURA–NSF matter, and they did not favor inviting State Department or National Academy of Sciences representatives.[13] Edmondson replied[14] that he could understand the position taken by Waterman and Scherer on internal AURA–NSF discussions; however, there were some matters where AURA wished to solicit assistance from the State Department and the NAS. He suggested that the internal AURA–NSF meeting should start at 9:30 a.m. on Monday, August 13, and that State Department and NAS representatives be invited to join the meeting at 11:00 a.m. This format was accepted by NSF.

The first part of the meeting in the NSF Boardroom was attended by seven AURA representatives, four from KPNO, six from NSF, and one from the National Academy of Sciences.[15] Stock reported on the site survey and the coming visit to Chile by representatives of the European Southern Observatory (ESO) to make independent observations at several sites, and Mulders described the Yale–Columbia survey in Argentina. Miller summarized the results of his July trip to Chile[16] in the following categories: Legislation, Audit, Counsel, Architect, Construction, Land Acquisition, Water Development, and Historical Record.

There was a brief recess before continuing the discussions at 11:10 a.m. with State Department representatives, Dr. John Rouleau (Science Advisor's Office) and Mr. Ralph Richardson (Chilean Desk), and Mr. Frank Skelding of the NAS. Aaron Rosenthal, NSF Controller, entered the meeting at this time. NSF Director Waterman joined the group to describe the expected congressional action on the FY 1963 budget. There was a short recess before Waterman left for another meeting.

Edmondson asked Harrell to take the chair at this time. In response to questions from Whaley, Richardson said that the State Department was "trying to establish itself in science in Chile and that the AURA–NSF project is most satisfactory." He said that from the State Department point of view AURA's relations in Chile were "basically good." Miller described AURA's present legal backing from the University of Chile and the US Embassy in response to a question from Richardson, and Richardson described the newly established Regional Science Office in Rio de Janeiro. This office included two NSF representatives, Dr. Harlow Mills (from the University of Illinois) and Max Heilman. This part of the meeting ended after Clemence and Stock described the Russian plans for astrometric observations in Chile. The afternoon AURA–NSF meeting was concerned with AURA–NSF–NASA relationships.[17]

The discussion of the Chile project was resumed at 9:25 p.m. in the West Room, Lafayette Hotel. Mulders represented NSF, and Woodrow was the only member of the morning AURA–KPNO group not in attendance. Acceleration of the 60-inch and 36-inch telescope was approved, based on availability of funds. It was voted to recommend to the Executive Committee that the Santiago legal firm of Puga, Puga, Pascal and Pacheco be employed as AURA counsel, based on Miller's discussion and recommendation. Edmondson read the names of the group who had been invited to make the site selection trip to Chile in November. The meeting adjourned at 11:25 p.m.

The Chile project was the major item on the agenda of the September 22, 1962 meeting of the Executive Committee.[18] The Committee on Organization had recommended that "the responsibility for administering the Chile Program be assigned to the Director of KPNO." Mayall said he planned to operate the Chile observatory as another division within KPNO with an associate director in charge. There should be a staff astronomer and a bi-lingual administrator in residence in Chile. This was unanimously approved by the Executive Committee; Mulders and Scherer entered the meeting at this time, and expressed their agreement.

The Executive Committee unanimously approved employment of the Santiago legal firm that had been recommended by Miller and endorsed by the Chile Committee. Sr. Enrique Puga would have direct responsibility for AURA affairs. His father, the founder of the firm, had been Ambassador to the United States.

Other actions and reports about the Chile project included: establishment of bank accounts and delegation of authority, status of Air Force and NSF financing, status of customs legislation (it had not yet been passed), architectural services, the Russian astrometric expedition to Chile, the November site selection trip, establishment of radio communication between Tucson and Chile, and the name of the observatory. Mayall reported the good news that NSF had appropriated $3.75 million for KPNO and $1.0 million for Chile in FY 1963.

This was Menzel's first meeting since his recovery from the effects of the blood clot that prevented him from attending the annual meeting of the Board in March. He queried Edmondson about his motivation for writing the July 5 memorandum to Harrell, Mayall, and Miller, and accepted Edmondson's explanation that this had looked like a good opportunity to show that he was evenhanded in dealing with Mayall and Miller. Thereupon, Menzel met separately with Mayall and Miller and gave each one a "Dutch Uncle" talk. A few days later, Mayall and Miller drove to Kitt Peak where they had a long discussion without any interruptions.[19] After that their relationship changed back to what it had been before the space program controversy. This had an immediate beneficial impact on the planning and execution of the November trip to Chile.

Hiltner, Menzel, Harrell, and Carson had agreed to be on the site selection team, but were forced to withdraw to meet more pressing obligations. This left Lee as the only administrator in the group, so Edmondson invited Wiggins. His response[20] was a strong negative: "It seems to me that the Board has a responsible working organization in the Observatory Director and his staff, and that the Board should look to that group for detailed reports and analyses of the project." He was bothered by the thought that a large delegation of the Board would make the trip. "I would view this as an unjustifiable waste of the taxpayers' money, taking on the aspects of a 'pleasure junket'." Edmondson's two and a half page response[21] was equally strong: "I believe the Board would be abdicating its responsibilities if a substantial number of its members were not fully informed by personal experience regarding the nature and complexity of the problems confronting us in this Chile Project. I did not realize the importance of this until I made my trip to Chile last April." He expressed regret that Wiggins had not been present at the June and September meetings of the Executive Committee, when the visit was discussed and unanimously approved. Wiggins did go to Chile twice in later years.

The final list was:[22] Edmondson, Mrs. Edmondson, Miller, A.G. Smith, Mayall, Mulders, Whitford, Lee, Linnell, and Lorenz, the new administrator from Wisconsin.

The dedication of the McMath Solar Telescope took place on November 2, 1962.[23] Edmondson and Mrs. Edmondson began their trip on November 8, and stopped in Rio de Janeiro on November 9–10 to touch base with the NSF representatives in the Latin American Regional Science Office. Mayall and Mulders stopped in Rio for this purpose after Tololo was chosen.

The Edmondsons joined Mayall and Mulders in Buenos Aires, and visited the La Plata Observatory on November 12. Jorge Sahade was the gracious host. They flew to

Mendoza, Argentina on November 13 by commercial aircraft, and from there to El Barreal in three small aeroclub planes in order to visit the Yale–Columbia site at nearby El Leoncito. The night was spent in San Juan, where Carlos Cesco and his wife hosted a fine dinner in their home starting at 11:00 p.m. The trip across the Andes from Mendoza to Santiago was made the next day, November 14.

The morning of November 15 was spent at Cerro Calan, where the group met the four Russian astronomers, headed by M.S. Zverev, Deputy Director of the Pulkovo Observatory near Leningrad (its name in 1962). Prior to lunch, Edmondson, Mayall and Miller returned to Santiago to meet with Dean Mori. Edmondson told Mori that we were dismayed by the news that the customs legislation had not been introduced in the regular session of the Chilean Congress.[24] He told Dean Mori that we were going to choose a site, but could not lift a shovelful of dirt until the customs legislation was passed. Dean Mori shrugged his shoulders and said: "President Alessandri is a former student of mine. I'll get you an appointment with him."[25] After lunch, some of the group visited Dr. Smith's radio astronomy installation at Maipu, while Edmondson and Miller conducted some business at the embassy. In the evening, Ambassador Cole entertained the AURA group and about one hundred local dignitaries at a reception in his new residence.

The group flew to Copiapo on November 16. AURA hosted a dinner for the Intendente and other officials that evening. On the next day Edmondson, Whitford, Linnell, Mayall, and Mulders went to La Peineta, and spent the night. The others were taken by Hugo Moreno and Carlos Torres to inspect the port facilities at nearby Caldera. On the 18th the Caldera group and two geologists, Dr. Kenneth Segerstrom (US Geological Survey) and his assistant, travelled to La Peineta, arriving about noon.

After refreshments at the residence of the Intendente on the morning of the 19th, the group travelled to La Serena. Mayall and Whitford drove with Stock and the others flew. Lee, Lorenz, and Miller inspected the port facilities in Coquimbo in the afternoon. Again, AURA hosted a dinner for the Intendente and other local officials in the evening. Whitford, Linnell, Smith, Mulders, Mayall, Stock, Lorenz, and geologist Dr. Eger departed the next morning for Tololo on horseback. The Edmondsons, Lee, Miller, Hugo Moreno, and Carlos Torres went to Vicuna by taxi to participate in a luncheon attended by all the leading citizens of the community. About 4 p.m. the group departed for Los Nichos, "a one hour drive up the Elqui Valley." It took three hours including a stop at the Tres Cruces vineyard for refreshments. The group arrived back in Vicuna about midnight. The Tololo group returned to Vicuna in time for lunch on November 21. The entire group returned to La Serena in time for dinner.

During the trip, Edmondson had been composing the talk he planned to give on Thanksgiving evening, November 22, in Santiago at a dinner hosted by the Rector of the University of Chile. He planned to start with a few words in English, and then have Stock read a Spanish translation. Hugo Moreno worked closely with Edmondson to achieve a translation, not necessarily literal, that would convey the meaning intended by the English words. This was important, because Edmondson planned to say that AURA would send Stock to Argentina to test the seeing with his double-beam telescope while we were waiting for the customs legislation. Edmondson phoned Ambassador Cole from La Serena and read the talk to him. The Ambassador approved the entire talk, including the "threat" of sending Stock to Argentina to make some test observations.

On the morning of the 22nd Smith, Mayall, Stock, and Miller went to Coquimbo to attend a big turn-out of officials and businessmen connected with the shipping industry. The others paid a formal call on the Commandante of the Army Regiment in La Serena. In the afternoon the group flew to Santiago.

When the group arrived at the Union Club for dinner, Dean Mori ran across the room and said: "You have an appointment to see President Alessandri at 11:30 tomorrow morning." Edmondson thanked him, and then went into a huddle with Stock to remove the paragraph about the site testing in Argentina. The talk was well received, and the food was excellent.

The Chile site survey team convened at 9:15 a.m. on Friday, November 23, 1962 at the Hotel Carrera in Santiago.[26] Those present were Edmondson, Lee, Linnell, Lorenz, Mulders, Rutllant, Whitford, A.G. Smith, Mayall, and Miller. Stock presented his report on the site survey, and explained how all sites were eliminated except the two still under consideration. La Peineta had more hours of clear sky than Tololo, but it also had higher wind velocities. He pointed out that the higher altitude was needed at the northern site to get above the dust and haze. Whitford said this was the best site survey he had ever seen, and he requested that the record show that according to the data available, the Chile sites were well ahead of any sites available in the United States. All present agreed that the differences between the sites were so small that other considerations would dictate the site selection.

Counsel Puga joined the group at this time. He recommended Tololo because it was private land, and he felt it would be important for AURA to own the land. Government land was available only on a 20-year renewal basis, but subject to negotiation and change, which could be detrimental to AURA's long-term interests.

Edmondson recessed the meeting at 11:30 for the audience with President Alessandri. The group walked across the Plaza to the Palace of the Moneda, where they were joined by Dean Mori, who served as the interpreter for the conversation between Edmondson and the President. When the problem with the legislation was described to President Alessandri, he agreed to introduce the bill in the Extraordinary Session of the Chilean Congress that very day. A photograph of the group with the President, and an interview with Dean Mori were published in El Mercurio the next day.[27]

The meeting reconvened at 12:15 p.m. to discuss selection of the site. Stock said he was reluctant to recommend one site over the other, but he did have a personal preference for Tololo. Edmondson asked if anyone had a strong preference for La Peineta, and no one responded. He ruled that the vote would be limited to the seven AURA Board members who were present. Rutllant abstained, and six voted for Tololo. A list of thirteen names for the observatory that had been proposed was then presented by Miller. Mayall recalled the frequent practice of identifying an observatory with its location, and the group unanimously chose the name "Cerro Tololo Inter-American Observatory."

Other actions included appointment of Cole as architect with Sr. Marchetti as associate architect; approval of an administrative structure with Stock reporting to Mayall, and an administrator reporting to Stock; and authorizing Miller to purchase Cerro Tololo as soon as possible. Cerro Tololo is located on a large estancia (ranch) called

EL MERCURIO

Santiago de Chile, Sábado 24 de Noviembre de 1962

EL TIEMPO.— Hoy: Bueno. TEMPERATURA DE AYER: Máxima: 25.6 grados a las 16; mínima: 10 grados a las 6.30 hrs.

P A G I N A   2 5

Científicos Informaron a
S. E. sobre Instalación de
Observatorio Astrofísico

Grupo de investigadores norteamericanos indicó que se proyecta su construcción en un punto de la zona norte.— Será instalado con fondos de la Fundación Nacional de Ciencias de EE. UU.— Gobierno acordará franquicias aduaneras

El Jefe del Estado recibió ayer a un grupo de científicos norteamericanos que le informaron sobre la próxima instalación en Chile de un observatorio astrofísico que será el segundo del continente, luego del de Palomar, en Estados Unidos.

Dicho observatorio, que constará de dos telescopios reflectores, uno de 90 centímetros y otro de 150, comenzará a ser construido el próximo año con fondos proporcionados por la Fundación Nacional de Ciencias de los Estados Unidos, como parte de sus programas para centros de investigación.

El grupo que se entrevistó con S. E. fue encabezado por el doctor Frank Edmondson, presidente de la AURA, Association of Universities for Research in Astronomy, que estará a cargo de ese observatorio. Lo acompañaban el Decano de la Facultad de Ciencias Físicas y Matemáticas de la Universidad de Chile, señor Carlos Mori, y el director del Observatorio Astronómico Nacional de esa Universidad, señor Federico Rutllant, miembro del directorio de AURA.

FACILIDADES DEL GOBIERNO

La delegación de científicos informó al Jefe del Estado que se realizan estudios para decidir el lugar en que estará situado el observatorio. Le manifestaron que los sitios más propicios son los cerros Tololo, cerca de Vi-

cuña, y La Peineta, en la zona de Copiapó. Se eligió Chile para este programa de investigación, en razón de que ofrece inmejorables condiciones para los estudios astronómicos en el hemisferio sur.

El Primer Mandatario les manifestó su interés por esta iniciativa, y ofreció el apoyo para liberar de derechos aduaneros la internación de los equipos respectivos.

CIENTIFICOS NORTEAMERICANOS CON S. E.— Una delegación de científicos de Universidades norteamericanas, acompañada de catedráticos chilenos, se entrevistó ayer con el Jefe del Estado. Le informaron sobre la proyectada instalación de un observatorio astronómico en zonas cercanas a Copiapó o Vicuña y que será el más grande en el continente, luego del de Palomar. En el grabado se observa a S. E. junto al Decano de la Facultad de Ciencias Físicas y Matemáticas de la Universidad de Chile, don Carlos Mori (extremo derecho); señor Frank Edmondson, quien preside el grupo de científicos, y el director del Observatorio Astronómico, señor Federico Rutllant, a la izquierda.

"El Totoral." The owner, Juan Orrego, had been reported as saying he would be willing to sell this land for 30,000 Escudos ($13,000). Counsel Puga recommended that the purchase of El Totoral should be negotiated before the owner saw news reports saying that he owned the best observatory site in the world, because he might raise the price. AURA was host for a dinner that evening that included the Chilean astronomers and their families, and the four Russian astronomers. Zverev, an accomplished pianist, brought the evening to a memorable conclusion by playing several numbers on a piano that was in the private dining room at the restaurant.

Most of the members of the site survey team departed for home on Saturday, November 24.[28] On Miller's instructions, Counsel Puga had called the Intendente in La Serena on the evening of the 23rd and asked him to arrange to have Sr. Orrego present in Vicuna on Sunday, November 25.[29] Miller, architects Cole and Marchetti, and Luis Pascal (a lawyer associated with Puga) flew to Vicuna in two small chartered planes on the morning of the 25th. Sr. Orrego was in custody at the Governor's office in Vicuna at 11:00 a.m. when they arrived, and he had not been allowed to read any newspapers. Captain Munoz had sent two policemen on horseback to find Sr. Orrego, and they had located him halfway between two of his ranches, an eight hour ride from Vicuna.

The negotiations for the sale took two hours. 30,000 Escudos ($13,000) was to be paid by AURA for 120 square miles of arid land that included Cerro Tololo and four other mountains within its boundaries. There followed a Fiesta Grande in the evening with 45 in attendance, including the Intendente who came from La Serena accompanied by two officials from Coquimbo. Captain Munoz presented an 8-foot Chilean flag to Miller for installation on Cerro Tololo along with a US flag.

On Monday, November 26, Miller, Stock and Pascal closed the purchase in the office of Sr. Orrego's attorney, where Sra. Orrego also signed the agreement. Title clearance was expected to take between two weeks and two months. As it turned out, El Totoral became AURA property at noon on January 19, 1963.[30] Miller, Stock, and Pascal called on the Intendente while Cole and Marchetti toured La Serena and found a possible site for the city headquarters, should La Serena be chosen over Vicuna by the Executive Committee. The two final days were spent on business in Santiago before Miller and Cole departed for the United States on November 28.

The AURA Executive Committee approved the selection of Cerro Tololo and all of the other recommendations of the site survey team at a meeting on December 1.[31] Stock's appointment as Associate Astronomer without tenure, with the additional title of Director of CTIO (reporting to Mayall), would be recommended for action by the AURA Board. It was also decided to locate the city headquarters in La Serena.

Stock thought the two acre lot found by Cole and Marchetti on November 26 would be too small,[32] and he sought help from the Governor, Hernan Carrasco, to find a better site. They were not successful until they went to see a potential site the day before Christmas.[33] This land was owned by the Caja de Colonizacion, a Government institution, and a 20 acre lot next to the La Serena branch of the University of Chile was still available for approximately 12,000 Escudos ($5,000). Stock sent a favorable report to Mayall.[34] The Caja was supposed to sell to small land owners, but the local Caja official felt there would not be any difficulties in selling to a scientific institution. The Executive Committee authorized purchase of this land for an amount not to exceed $10,000.[35] There were some bureaucratic delays and the paperwork took much longer than had been expected. The purchase was finally completed on October 5, 1963 for 15,000 Escudos ($4,500).[36]

Construction of a road to the summit of Cerro Tololo became the top priority as soon as AURA became the owner of El Totoral. Stock had already engaged a local engineer, Sr. Zoltan Timkovic, to make a detailed study of the road and make an estimate of its cost. To provide the needed information, several sample sections of the road had been constructed in December 1962. After AURA took possession of El Totoral, Stock took over Timkovic's crew of 25 experienced workers and personally supervised the construction work by initiating the blasting operations on the most difficult section.[37] In February 1963 Richard Kroecker, the Tucson contractor who had constructed the first Kitt Peak road and a neighbor of Miller, spent three days with Stock looking at the road project. Kroecker felt that a new Caterpillar D-8 tractor could do much of the work that Stock was doing by drilling and blasting, and he told Stock it would take 60 days to complete the tractor work. They discussed this with Miller in Santiago, and Miller telephoned the AURA office in Tucson and told them to order the D-8 and several other pieces of equipment that Kroecker recommended. Stock continued the blasting that was needed to make it possible for the D-8 tractor to go through once in order to open up the road for heavier rock drilling equipment. In April 1963 Stock, Miller, Mayall and Edmondson decided to enter into a contract with Kroecker to do the tractor work for $20,000 and all expenses paid, and this was signed at the end of June. The road construction equipment arrived at the Coquimbo docks on July 6, and Kroecker and his assistant, Winston Shumaker, arrived in La Serena on July 11. At this time Kroecker stated several times that he planned to finish his part of the work in 30 days.

A personality conflict between Stock and Kroecker developed soon after the work started. Kroecker was giving orders to Stock's employees without Stock's knowledge or consent, and there were frequent disagreements about the routing of different sections of the road, Stock's blasting techniques, and Kroecker's practice of burying vegetation, including giant cactus plants which later punctured tires on some of the construction vehicles. Kroecker left the site on September 8, leaving several short sections near the summit unfinished. On September 25 he asked for an adjustment in his contract with AURA.[38] Kroecker said the work could have been completed in five weeks, including the bad weather, and that AURA (i.e. Stock) was responsible for the delays that caused him to spend eight weeks on the job. He said his Tucson business lost $4,000 per week while he was away, and claimed that AURA owed him an additional $12,000 for the extra three weeks he was in Chile.[39] AURA rejected the claim as being without merit, and Kroecker did not pursue the matter in court.

Korp's previous experience in heavy construction[40] paid off at this point, and Stock put him in charge of finishing the road.[41] He put the most experienced men on the job to finish the work near the summit, and Stock and Korp drove to the new house at the end of the road on the afternoon of September 10,[42] marking the end of horse- and mule-back travel to Tololo. In a letter to Stock, Edmondson had proposed bringing a small AURA group to Chile for a ceremony to inaugurate the opening of the road:[43] "Last month you mentioned that the opening of the road should be an occasion for ceremony. This would also be a good time for a visit of a group representing AURA. ... Opening a road in the United States usually consists of some speechmaking

Waiting to cut ribbon to open road to Cerro Tololo, 14 December 1963.
NOAO Photo Archives.

followed by cutting a ribbon which has been stretched across the road. If we have such a ceremony in Chile, it should be followed by a trip to the top of Tololo for the AURA group and invited guests."

Eugenio Gonzalez Rojas was elected in mid-1963 to succeed Juan Gomez Millas as Rector of the University of Chile, and a friendly exchange of letters was initiated by AURA.[44] Two months later Edmondson received another letter from the Rector stating:[45] "I regret to inform you that Mr. Federico Rutllant, Director of the Astronomical Observatory, is very ill. For that reason he has left the Observatory." He suggested that Dean Enrique d'Etigny would be the appropriate person to replace Rutllant on the AURA Board. At the November 22, 1963 meeting of the Executive Committee, the Nominating Committee recommended that d'Etigny be appointed as a consultant now, and that he should be nominated for a full three year term as a director-at-large at the 1964 Annual Meeting of the Board. This was approved unanimously.[46] Just before the break for lunch Miller announced that President Kennedy had been shot in Dallas, and news of the President's death was received during lunch. The remaining business, including the plans for the road opening ceremony, was finished as quickly as possible after lunch.

The road opening ceremony took place on December 14, 1963.[47] AURA was represented by President Edmondson, Vice-President Harrell, and Chairman of the Scientific Committee Code. About 30 special guests and 20 others attended. The ceremony began when the Archbishop of La Serena, Monsignor Alfredo Cifuentes, sprinkled holy water on the road. Short speeches were then given by the US Ambassador, Charles W. Cole, the Intendente of the Coquimbo Province, the former Dean of the Faculty of Physical Sciences and Mathematics of the University of Chile, Carlos Mori, the new

Frank K. Edmondson (AURA President) cutting ribbon to open road to Cerro Tololo. William B. Harrell (AURA Vice-President), and Dean Enrique d'Etigny (University of Chile) are holding the ribbon. NOAO Photo Archives.

Dean, Enrique d'Etigny, the Mayor of Vicuna, and the President of AURA. Finally, the red, white and blue ribbon that was stretched across the road was cut by Edmondson, assisted by Harrell and d'Etigny. Waiters immediately appeared bearing trays loaded with glasses of champagne, and the ceremony concluded with a series of toasts. Twenty cars carried the participants to the summit of Tololo where lunch was served in a tent on loan from the Army Regiment in La Serena. There were more speeches and toasts after the lunch, and this was followed by an inspection of the mountain facilities before the participants returned to La Serena.

Construction of the city headquarters building, two residences, and a caretaker's house in La Serena became the top priority following the purchase of the city land on October 5 and the formal opening of the Tololo road on December 14, 1963. Seven Chilean construction firms were invited to submit bids for the city buildings, but only two of them showed up at the bid opening meeting in Santiago on December 10. The low bid was $40,000 more than the $133,000 budget figure,[48] which caused the AURA officers to decide that AURA should hire its own crew to construct these buildings, with Korp in charge. If the results were satisfactory, this procedure would also be used for the mountain construction. Thus was born what soon came to be called the "AURA Construction Company." Sr. Jose Guarini, the representative of the low bidder on

December 10, was hired by Korp to be the construction supervisor. By the end of January 1964 the foundations for all buildings were laid, and parts of their superstructures were rising. It was expected that these buildings would be ready for occupancy in July.

On Tololo, Stock began in January to level the summit to a level 42 feet below the highest point. This would produce a plateau about 550 feet long and 200 feet wide.[49] Mayall had warned Stock that "the levelling of the highest point should not be overdone, but only enough to provide for domes for the 60-inch, and eventually a 150-inch telescope in such a manner that there is no mutual interference."[50] He suggested a different location for the proposed 36-inch telescope. Stock proceeded to do it his way, and the levelling was completed on June 5 to a level slightly lower than the original plan, with a flat area of 560 × 330 feet.[51] Six telescopes were eventually erected on this area: two 16-inch, a 36-inch, a 60-inch, a 150-inch and the Michigan Curtis Schmidt. Permanent construction on the mountain would begin after completion of the construction in La Serena.

Alan T. Waterman retired as Director of the National Science Foundation on June 30, 1963, and was succeeded by Leland J. Haworth,[52] a member of the Atomic Energy Commission and former Director of the Brookhaven National Laboratory. He became familiar with AURA problems by attending the September 20, 1963 meeting of the Executive Committee in Washington,[53] and a year later he arranged to have the National Science Board meet in Tucson.[54] Edmondson wanted the NSB to see the quality of the people AURA was working with in Chile, and he persuaded d'Etigny to come to Tucson for the concurrent meeting of the AURA Executive Committee.[55] d'Etigny was seated at the head table at a dinner hosted by AURA on the first evening of the NSB meeting, November 19, 1964, and his after-dinner remarks made the intended good impression on Dr. Haworth, the NSB members and the NSF staff members.

Even so, CTIO almost went down the drain during the NSB meeting earlier in the day. The NSB was on the point of dropping CTIO because of the inflation in Chile, and Board member Father Theodore Hesburgh talked them out of it.[56] He wrote to the author: "All I can say about Cerro Tololo is that I greatly favored this project and really had to fight for it during the Board meeting which was going to drop it. Their reason was inflation and I had to remind them that one day we, too, might face that problem." The NSB had approved a third $1,000,000 for CTIO at its July 22, 1964 meeting, and four months later at this meeting in Tucson Keller requested a $300,000 supplement due to the serious inflation in Chile. After listening to Father Hesburgh, the NSB gave Haworth the authority to "review, approve, and take final action" for an amount not to exceed $300,000 for fiscal year 1965. Five members abstained from this vote.[57]

Edmondson had informed Haworth prior to the NSF meeting that a visit to Chile by several AURA Board members was being planned for shortly after Thanksgiving, and he asked Haworth to designate one or two NSF staff members to participate.[58] Keller was designated, and the AURA group consisted of Edmondson, Wildt, Harrell, Wiggins, Franklin, Mayall, and Miller. Edmondson and Wildt arrived in Santiago on Thanksgiving day, and the others arrived by Sunday. d'Etigny had arranged for the group to meet with the new President of Chile, Eduardo Frei, on Monday, November 30, at 5:00 p.m. Frei's English was excellent, and it was a friendly and useful meeting.[59]

Viewing the image of the Sun at the prime focus of the solar telescope, 20 November 1964. From left to right: William B. Harrell (AURA Vice-President), Leland J. Haworth (Director of the National Science Foundation), Eugene Johnsor. (Chairman, Papago Tribal Council), Frank K. Edmondson (AURA President), and Nicholas U. Mayall (KPNO Director). NOAO Photo Archives.

A visit to the Yale–Columbia site in Argentina had been arranged to give the AURA–NSF group an opportunity for a first hand comparison with CTIO. The group flew from Santiago to Mendoza, Argentina on December 1st.[60] Three CTIO vehicles, two carryalls and a pickup truck, had been driven across the Andes by Stock, Korp, and Carlos Rodriguez, one of the CTIO drivers. These vehicles were used to transport the group to the Yale–Columbia site at El Leoncito, and for the return to Chile. The observatory buildings had been constructed since the November 1962 visit by Mayall, Mulders, and the Edmondsons. The group arrived in time for lunch on December 2nd and spent the afternoon inspecting the facilities: the 20-inch double astrograph, the hydroelectric power plant (which was in operation in November 1962), the library–office building, and the house for the resident astronomer. Individual rooms were provided for the group in the dormitory–dining hall building, and a fine dinner was served there. The hosts were Dr. Arnold Klemola (Yale), Mr. Cyril Jackson (resident astronomer), Dr. Carlos Cesco (Cuyo University), and Sr. Victoria (project engineer).

The group began the long day's trip to Chile before 5:00 a.m. on December 3rd, and had breakfast at a restaurant near the border. The mountain pass, with the famous Christ of the Andes statue, is at an elevation of 13,000 feet, and the group made a one

Group assembled for cornerstone ceremony, 6 December 1964. NOAO Photo Archives.

hour rest stop there. After lunch in the town of Los Andes, one vehicle went to Santiago and the other two went to La Serena.

The La Serena headquarters building and the residences were finished and partly occupied by the time of this visit. Guarini had done such a good job as construction superintendent in La Serena that he was given the responsibility for the construction on Tololo. Construction sites for the two 16-inch telescopes had been prepared in November by the recently hired mountain superintendent, Sr. Harold Schaeffer, and a corner-stone laying ceremony for one of these buildings took place on Sunday, December 6, 1964.[61] Stock introduced the AURA–NSF visitors, and Mayall brought greetings from KPNO. The ceremony was initiated by the new Archbishop of La Serena, Monsignor Arturo Mery, who sprinkled holy water on the site. A visit to the Coquimbo Province by the new Minister of Mining prevented several invited guests from attending, and the new Intendente, Eduardo Sepulveda, was represented by his predecessor, Tulio Valenzuela. Following several formal speeches, a sodawater bottle containing a piece of paper signed by the AURA and NSF members was cemented in the corner-stone.

The ceremony was followed by lunch, again under a tent loaned by the Army Regiment in La Serena. There were many toasts and more speeches, and entertainment was provided by the Grupo Folklorico de Profesores de La Serena, who sang Chilean songs and danced the Cueca and other Chilean national dances. One of the ladies took the President of AURA firmly by the hand and taught him how to dance the Cueca. The Secretary of AURA also danced, somewhat more willingly.

Group talking after cornerstone ceremony: Frank K. Edmondson, Intendente
Don Tulio Valenzuela, L.Gard Wiggins, Rolf Korp, and the Archbishop of
La Serena, Monsignor Arturo Mery. NOAO Photo Archives.

The group flew to Santiago on Tuesday, December 8, and all except Wildt and
Edmondson continued on to Miami on the evening flight. At noon on Friday,
December 11, Edmondson received the Chilean "Al Merito" decoration, established
by Bernardo O'Higgins in 1817, from the Minister of Education, Juan Gomez Millas,[62]
who was Rector of the University of Chile when Cerro Tololo was chosen. The citation
said the decoration was given in recognition of his services to Astronomy in Chile. The
award ceremony was attended by Wildt, Stock, Sanduleak (new CTIO staff member),
d'Etigny, Adelina Gutierriez de Moreno, Hugo Moreno, Claudio Anguita (Ruttlant's
successor as Director), and Dr. Dmitri Polojentsev (head of the USSR group). This
was Edmondson's last trip to Chile during his three-year term as President of AURA.

Wildt and Lee were elected for one-year terms as President and Vice-President of
AURA at the annual meeting of the Board of Directors on March 23, 1965. Hiltner was
elected Chairman of the Scientific Committee, and Miller was re-elected Secretary.[63]
Edmondson's final report to the AURA Board said that he had made 55 trips, flown
190,000 miles, and had been away from home 323 days on AURA business during his
three-year term as President. Also, he had made or received 637 AURA long distance
telephone calls and written 351 AURA letters. He also said: "The By-laws now provide
that the President and Vice-President are elected annually up to a limit of three years
consecutively. I feel that one year is too short a time for a President to be effective,
but I also feel that three years are plenty." The report also mentioned the progress that
had been made in purchasing land adjacent to the KPNO headquarters for parking and

Edmondson wearing Chilean "huaso" hat and serape.
NOAO Photo Archives.

possible other future needs, a program that he had initiated at the end of his first year.

Wildt had planned to have an official AURA Board visitation to Chile in February 1966,[64] but this was superseded by an unanticipated development. Carson visited Chile in early October to take a close look at the organization and operation of the observatory, and Wildt, Mayall and Miller followed in mid-October. The Chile Committee met in Washington to discuss their findings on October 25, and followed this with a report at the November 11 meeting of the Executive Committee.[65] Miller was back in Chile November 13–26, primarily to initiate termination of the contract with architect Marchetti. He returned again in December accompanied by Richard M. Bilby, AURA's Tucson counsel. Bilby presented a report to the AURA Chile Committee on December 21[66] and Stock resigned in Tucson on December 31.[67] It is unfortunate that the attributes that made both Stock and Meinel so effective in conducting the field work of a site survey did not carry over into the day-to-day administration of an observatory.

Wildt felt that Edmondson's "Al Merito" decoration from the Government of Chile

Frank K. Edmondson wearing Al Merito decoration and rosette.
NOAO Photo Archives.

made him a known quantity to the authorities in Chile and, therefore, the logical AURA person to step in and be the first Acting Director of CTIO. Furthermore, he had been deeply involved in the development of AURA relations with the European Southern Observatory (Chapter 18) and in the current negotiations with the Carnegie Southern Observatory (Chapter 19). This logic was hard to resist, and Edmondson was the first of the five (Edmondson, Alex G. Smith, Hiltner, Mohler, Hoag, and Hiltner again) who served as Acting Director of CTIO for short terms. W.W. Morgan called them the "Director of the Month." Edmondson was appointed by the Executive Committee for a six week period beginning February 2, 1966, and Mayall was authorized to appoint any future Acting Directors.[68]

Korp had obtained a 90-day option to buy a house and $2\frac{1}{2}$ acres adjacent to the AURA land on the hillside to the west. AURA had been renting this property and the Edmondsons would be living in the house. The Executive Committee agreed that the property should be purchased using corporate funds, subject to a favorable

Frank K. Edmondson handing the AURA President's gavel to his successor, Rupert Wildt. NOAO Photo Archives.

recommendation from the Edmondsons. The house was called "Casa Tres" after the purchase, and was occupied in turn by all of the Acting Directors. The Edmondsons arrived in Santiago on February 3, 1966, and slept in Casa Tres for the first time on Saturday, February 5. They went to Tololo the next day, and returned for an overnight stay on Wednesday–Thursday, February 9–10.

Miller and Len Gardner, Robertson's administrative assistant, arrived on February 8. Warren J. Hynes, NSF audit manager, conducted a detailed audit of NSF Grant 6-16371 and NSF Contracts C270 and C300 between February 7 and March 4, and was in La Serena during a large part of this time. Edmondson and Miller felt their sessions with Hynes were professional and productive, but they had mixed feelings about Gardner's attempts at micromanagement. Edmondson reacted by writing a report[69] that persuaded Robertson to overrule Gardner and approve enlarging the dormitory-dining hall from six bedrooms to twelve bedrooms.[70]

Stock arrived in La Serena on February 16, and he and Edmondson were on Tololo for the next two days preparing to remove Stock's personal property from the house that he and his family had lived in. A notary from Vicuna arrived on the afternoon of the second day to notarize the signatures of Stock and Edmondson on the document that listed what was being taken. The truck was finally loaded and departed at 4:00 p.m.[71]

The Observatory was host to the US Ambassador, Ralph A. Dungan, and his wife shortly after the mid-point of Edmondson's term as Acting Director.[72] The embassy group arrived in La Serena on February 25 in a US Air Force DC-3 assigned to the embassy, and they spent the night on Tololo. Dinner was served in the construction workers' dining hall, the only place on the mountain large enough to hold the embassy group and the large number of invited guests from the local area. The early evening was spent viewing a number of objects through one of the 16-inch telescopes. The following day the group had lunch at the headquarters building in La Serena, prior to an early afternoon departure for Santiago. Edmondson and Miller were given a ride to Santiago, where they connected with the evening flight to Miami and on to Chicago for a very important Special Meeting of the AURA Board to discuss the Ford Foundation–Carnegie Southern Observatory problem (Chapter 16). Edmondson returned to Chile on March 2, and the Edmondsons left Chile on March 10 to fly directly to Tucson for the annual meeting of the AURA Board. They returned home to Bloomington on March 16.

The Kitt Peak mountain superintendent, Stuart R. Hurdle, was sent to Chile in November 1965 to assist in finishing the construction of the buildings for the 36-inch and 60-inch telescopes.[73] He returned to Tucson for Christmas, and was back in Chile when the Edmondsons arrived. Mrs. Hurdle joined him in early April, and they lived on Cerro Tololo until June in the house that had been occupied by the Stocks. Hurdle returned in February 1967 to supervise installation of the telescopes and rotating domes. He left Kitt Peak for the final time in the spring of 1968 to take on the big job of construction superintendent for the building and dome for the CTIO 150-inch telescope, and after this was finished he retired in December 1970.

During Hiltner's third period as Acting Director, January–May 1967, he had the 36-inch telescope mounted in the open on a concrete pad in order to carry out observations while the dome was being installed. There was little risk in doing this because it never rained at this time of year. Ludden and Graham set up the telescope, and it was used by Hiltner in the latter part of February and most of March. Erection of the dome was completed on March 24, and the telescope was in operation on April 9.

Plans had been made for Mohler to return as Acting Director following Hiltner, but this was abandoned after Dr. Victor M. Blanco accepted the position of Director of CTIO, effective July 15, 1967.[74] He visited CTIO for a week in May and spent several weeks in Tucson during July for orientation before taking up residence in Chile with his wife with their two sons in early August. In addition to his normal duties as Director, Blanco also became responsible for the final stages of planning for the dedication of CTIO on November 6, 1967.

An "Inter-American Conference on Astrophysical Photometry" was organized by Anguita and Blanco, and held in Santiago on November 3. The formal dedication ceremony was held on the grounds in front of the La Serena headquarters building at 10:00 a.m. on November 6, with about 100 invited guests.[75] Music was provided by the Orquesta Sinfonica de Ninos de la Serena before the speeches. Blanco welcomed the guests and participants, and there were six speakers: Eduardo Sepulveda, Intendente of the Coquimbo Province; Edward Korry, US Ambassador; Philip Handler, Chairman,

President Eduardo Frei at the 60-inch CTIO telescope.
NOAO Photo Archives.

National Science Board; Rupert Wildt, President of AURA; Enrique d'Etigny, Dean of the Faculty of Physical Sciences and Mathematics, University of Chile; and Otto Heckmann, President of the International Astronomical Union and Director General of the European Southern Observatory. The benediction was given by the new Archbishop of the Archdiocese of La Serena, Monsignor J.F. Fresno, who became a Cardinal on May 25, 1985. Lunch was served, and Chilean singers and dancers entertained until mid-afternoon.

The President of the Republic of Chile, Eduardo Frei, arrived on Tololo in a Chilean Air Force helicopter about 4:30 p.m., and was greeted by the AURA and US Government group and by Chilean officials who had driven to Tololo after the lunch in La Serena. Dinner was served in the recently opened dining hall, after which most of the participants returned to La Serena. The President and his party and the AURA officers spent the night on Tololo, and the President enjoyed looking at various objects through the 60-inch telescope until almost 2:00 a.m.

Those who had slept in La Serena returned to Tololo for lunch the next day. A few lucky people, including the author and his wife, flew to Tololo in the President's helicopter. Frei flew to Los Nichos to place a wreath on the grave of Gabriela Mistral, who won the Nobel Prize for poetry in 1923, and returned to Cerro Tololo in time for lunch.

President Eduardo Frei looking through the 60-inch CTIO telescope.
NOAO Photo Archives.

AURA Board members and guests returned to Santiago on November 8, and were received that evening, with other guests from Chile, by Ambassador and Mrs. Korry at the Ambassador's residence. His speech in Spanish at the dedication was an act of bravery; it was his first public speech in Spanish, a language he was still learning.

Stock's site survey reports had attracted the attention of the European Southern Observatory (ESO) and the Carnegie Southern Observatory (CARSO). ESO sent two observers to Chile shortly after Cerro Tololo had been chosen, and they worked on La Peineta December 6–19, 1962 and on Cerro Tololo December 30, 1962–January 13, 1963. CARSO representatives initiated informal discussions with AURA in April 1963, and CARSO observers began test observations on Cerro Morado in mid-1964. The complicated story of AURA's interaction with ESO and CARSO will be told in Chapters 18 and 19.

15

A project manager is appointed for the 150-inch telescope, and the bid for the fused quartz mirror blank by the General Electric Company is $500,000 lower than the bid by the Corning Glass Works. A $5.0 million grant from the Ford Foundation in April 1967 provides funds for half the cost of a 150-inch telescope for CTIO. Congress adds the other half to the National Science Foundation appropriation. Bids for two 150-inch telescope mountings are opened on October 18, 1967. Installation on Kitt Peak is completed in March 1973, and on Cerro Tololo in October 1974.

As reported in Chapter 12,[1] McMath proposed at the time of the 1961 annual meeting that AURA should initiate a program of design and location studies for a 150-inch telescope. A 150-inch Telescope Review Committee consisting of Mayall (chairman), Shane, Bowen, Code, Hiltner, Mohler, Rule, and Stromgren was appointed, and held its first meeting on September 23, 1961.[2] The discussion included optical parameters, mechanical design, automation, instrumentation, dome and building, seeing tests and site selection, and an estimate of the cost. $200,000 was requested in the FY 1963 budget for the preliminary design and location studies,[3] and on January 16, 1962 the Executive Committee approved a recommendation from the Scientific Committee that funds for the mirror blank be included in the FY 1964 budget.[4]

Three Corning Glass Works officials came to Tucson on February 6, 1963 for a meeting with AURA and KPNO representatives.[5] They said a new structure would be needed at Corning to make a 150-inch Pyrex blank, and between two and four years would be required from date of order to date of delivery. The Corning method for fabricating large fused silica blanks was described, and they said the time schedule for a 150-inch would be one and a half to two years from the date of order. Corning would need a "letter of intent" from AURA before Corning management would be willing to spend any money on the research and development and plant expansion needed for such a large project. Miller asked if Corning were the only firm that could do this job, and the answer was in the affirmative. Edmondson asked for an estimate of the costs, and Phillips gave them as approximately $1 million for Pyrex and $1.5 million for fused silica.

Mayall gave a detailed report of this meeting to the Executive Committee three days later.[6] Prices were received from Corning on February 25, 1963, and these were reported to the Scientific Committee on April 15:[7] Pyrex solid, $1.2 million; Pyrex ribbed, $1.3 million; fused silica solid, $1.5 million. Mayall recommended that fused silica be chosen, and the Scientific Committee approved unanimously. Mayall also recommended that Skidmore, Owings, and Merrill be appointed architect for the initial design studies for the dome and base structure; this was also approved unanimously. These two recommendations were approved by the Board of Directors the next day.[8]

Several attempts had been made to recruit an associate director for the Stellar Division whose duties would include responsibility for the 150-inch telescope. It was finally decided that these two functions should be separated. After careful review by the KPNO staff, Mayall recommended to the Executive Committee on June 28 that Dr. David L. Crawford should be appointed Project Manager for the 150-inch telescope. This was approved unanimously.[9] The position of Associate Director, Stellar Division was finally filled more than a year later when Arthur A. Hoag accepted a second offer from Mayall.[10] He assumed his duties in January 1965.

The 150-inch Telescope Advisory Committee held its second meeting in Tucson on September 14[11] and the Executive Committee met on September 20.[12] Both groups were brought up to date on the discussions with Corning, the proposed signing of a contract with Skidmore, Owings, and Merrill for preliminary design studies for the rotating dome and building, and the contract that had been signed with Westinghouse Electric Corporation of Sunnyvale, California for the preliminary design of the telescope mounting. Westinghouse was the low bidder when bids for construction of the CTIO 60-inch telescope mounting were opened on November 21, 1963.[13] Westinghouse had also been the low bidder for the heliostat and the No. 2 and No. 3 mirror mountings for the solar telescope in 1960, and was the high bidder for the 80-inch telescope in 1959.

The February 6, 1963 meeting with Corning officials in Tucson was followed by a visit to Corning headquarters and the fused silica facility in Bradford, Pennsylvania on September 25–27, 1963.[14] Shortly after this Corning quoted a price of $1,510,000 for a 150-inch fused silica blank, valid until December 31, 1963. Miller felt this price was too high, and began to investigate other firms to see if Corning was truly a sole source. The New York offices of General Electric had no competition to offer, but the Amersil Quartz Division of Englehard Industries, Inc. offered an alternative: a blank of industrial quartz, topped with a cap of optical quartz made by Hereaus in Germany.[15] After giving this report Miller discovered that the Lamp Glass Division of General Electric in Richmond Heights, Ohio would be interested in competing, and he reported to the Executive Committee on January 31, 1964 that GE would be invited to send representatives to Tucson for discussion.[16] He also reported that representatives of AURA, the European Southern Observatory (ESO), and the Carnegie Southern Observatory (CARSO) had met with representatives of Hereaus in Frankfort and Hanau, West Germany on January 23 and 24, 1964 following AURA–ESO meetings in Paris (Chapter 17). Hereaus was dropped from further consideration when they gave a cost estimate of $2.3 million for a single purchase or $1.7 million for a multiple purchase with ESO.[17]

The 150-inch Telescope Advisory Committee held its third meeting on October 12,

1964.[18] The negotiations for the mirror blank were summarized; final bids were expected in time for action at the November meeting of the Executive Committee. The GE bid was $1,151,750 (FOB plant); the original Corning bid was $1,510,000 and it was increased to $1,600,000 on January 1, 1964.[19] Corning told Miller they had a copy of the GE bid, and sent an unsolicited bid, $1,148,000, a week before the Executive Committee meeting. Miller told GE about this, and they reduced their price to $1,025,000. The Executive Committee engaged in a lengthy discussion of the ethical and legal aspects of the situation, and Counsel Bilby was present to give any advice that might be needed. Miller felt that AURA would be in a better position to stay with GE on the basis of the bona fide quotations from the two vendors. The Executive Committee agreed, and unanimously approved that the 150-inch fused silica blank be purchased from the General Electric Company at the original bid price of $1,151,750, subject to the approval of the National Science Foundation. The contract was signed on December 31, 1964 by Edmondson and Miller for AURA, and by R.A. Popp and Robert E. Holmes for GE.[20]

The need for a 150-inch telescope at CTIO was discussed when the KPNO 150-inch was proposed.[21] It was the major item included in the 1966–70 budget projection for CTIO,[22] but was dropped from the approved budget for FY 1967.[23] The Scientific Committee debated the choice between a large Schmidt Telescope and a 150-inch, and preferred continuing to push for a 150-inch for CTIO.[24] The Carnegie Institution of Washington and the European Southern Observatory were also planning for large telescopes in the southern hemisphere (Chapters 18 and 19), and so was the University of California.[25]

The Trustees of the Carnegie Institution of Washington (CIW) approved a Southern Hemisphere Project[26] on May 10, 1963; this was two years after the AURA Board had first discussed the need for a 150-inch telescope at CTIO. The CIW proposal for $19 million to fund construction of a 200-inch telescope in the southern hemisphere (CARSO) was presented to the Ford Foundation on July 1, 1964.[27] It was tabled by the Ford Foundation Trustees on March 15, 1966,[28] but a formal declination letter was not written until April 25, 1967.[29] On January 23, 1967 the Ford Foundation invited AURA to submit a proposal for $5.0 million to cover half the cost of the CTIO 150-inch,[30] which would be granted if the National Science Foundation could obtain a matching $5.0 million from Congress. The details of the CARSO–Ford Foundation–AURA episode which led to this invitation will be given in Chapter 16.

The unsolicited invitation from the Ford Foundation was presented to the Executive Committee in Executive Session on January 26, 1967,[31] and KPNO staff members were informed at the start of the open session that followed. The AURA proposal was mailed to the Ford Foundation on February 11[32] and the grant letter was dated April 19.[33] Bids had not yet been solicited for the mounting, and this meant that AURA would now have the unprecedented opportunity to write a single purchase order for two very large telescopes.

Crawford gave a summary of the fourth meeting of the 150-inch Telescope Advisory Commttee, held on May 8, 1967, at the meeting of the Scientific Committee on June 12, 1967.[34] He reported that the Observatory was planning to go to bid on June 19 for two domes and the Kitt Peak building, and on July 1 for two mountings. The Scientific Committee approved Mayall's recommendation that Dr. Victor M. Blanco be appointed

Arrival of 158-inch (4.0 meter) fused quartz telescope mirror blank in Tucson on 30 October 1967. NOAO Photo Archives.

Director of the Cerro Tololo Inter-American Observatory, and the Executive Committee made the appointment on June 16.[35] Blanco entered the meeting following the unanimous vote. The Executive Committee also added the duties of project manager for the CTIO 150-inch to Crawford's responsibilities for the KPNO 150-inch.

There was an *ad hoc* meeting in Tucson on July 17–19 for a complete review of the 150-inch telescope project and long-range plans for CTIO, both on the mountain and in La Serena.[36] Bids were opened on September 13 for the dome and building for Kitt Peak and the dome for Cerro Tololo.[37] Ten general contractors had submitted bids and the low bidder was the M.M. Sundt Construction Company of Tucson. NSF approved the contract for the full amount, even though it was $600,000 over budget. The contract for construction of the Kitt Peak dome and building was signed on October 14, and the add-on contract for fabrication of the Cerro Tololo dome was signed in November. An AURA crew would construct the CTIO building and assemble the dome.

The GE mirror blank was inspected and accepted on September 28; it was shipped from Cleveland by rail on October 25, and arrived in Tucson on October 30, 1967.[38] The bids for the CTIO mirror blank were solicited on a multiple purchase basis in cooperation with the Anglo-Australian Observatory and French National Observatory large telescope projects, and were opened on October 13, 1967. Corning Glass Works,

158-inch (4.0 meter) telescope building under construction – Kitt Peak.
NOAO Photo Archives.

General Electric Company, and Owens–Illinois submitted bids, and Owens–Illinois was
the low bidder with a new material, CER-VIT (from ceramic-vitreous), which could
be made with a zero coefficient of expansion. Their bid of $663,266 each for three
blanks was about half the cost of fused silica, and the purchase was authorized by the
Executive Committee.[39] The blank was poured on June 25, 1969; it cracked following
a power outage during the cooling process and a second blank was poured on August
7.[40] The blank was accepted on January 7, 1970,[41] and was shipped to Tucson, where
it was put in storage until the KPNO mirror was finished.

The bids for the two mountings were opened on October 18, 1967. One was to be
erected by the contractor on Kitt Peak, and the other was to be packed for overseas
shipment for erection by AURA personnel on Cerro Tololo. Bids were submitted by
Boller and Chivens, Westinghouse Electric Corporation, Western Gear Corporation,

158-inch (4.0 meter) telescope building under construction – Cerro Tololo.
Note: The KPNO building was constructed using a welded steel framework.
The CTIO building was constructed using a bolted steel framework mounted
on reinforced concrete pillars because of frequent earthquakes in Chile.
NOAO Photo Archives.

and C.W. Jones. Western Gear was the low bidder for the two mountings and control
systems: $1,695,000 for Kitt Peak, and $1,408,400 for Cerro Tololo. Erection costs
were included in the Kitt Peak bid. An analysis of the bidding firms by the KPNO
Engineering Department ranked Western Gear highest, based on shop facilities and
engineering and management backup, and the Executive Committee authorized Miller
to execute a contract with Western Gear in an amount not to exceed $3,103,400.[42] This
was $1,600,000 under budget, which more than compensated for the over budget cost
of the two domes and the Kitt Peak building.

Observatory workers started site preparation in April 1967 on Kitt Peak and in
December 1967 on Cerro Tololo.[43] Construction of the building on Kitt Peak by the
contractor started in March 1968, and construction by observatory workers began on
Cerro Tololo a month earlier.[44] S.R. Hurdle left the position of Kitt Peak mountain
superintendent in early 1968 to take charge of construction of the CTIO 150-inch
telescope building.[45] A small amount of interior finishing was all that remained to be
done when he retired in December 1970, and observatory workers finished this in
mid-1971. Construction of the building and dome on Kitt Peak was finished in

Final inspection of the 158-inch (4.0 meter) CERVIT telescope mirror blank at Owens-Illinois Plant in Toledo, Ohio on 7 January 1970. Seated in the cassegrain hole are W.A. Hiltner and N.U. Mayall with a representative of Owens-Illinois. A. Keith Pierce and Mrs. Edmondson are talking with a representative of Owens-Illinois, and Frank K. Edmondson and Orren C. Mohler are talking with two Owens-Ilinois representatives. NOAO Photo Archives.

September 1970,[46] and the "AURA Construction Company" was dissolved at the end of the year.[47]

Several other KPNO employees were sent to Chile for short periods to assist with the construction. Those who stayed longer than a month were W. Johnson, carpenter foreman; C.R. Lelo, chief instrument maker; N. LePore, Assistant Mountain Superintendent; and W. Baustian, KPNO Chief Engineer. D.J. Ludden, KPNO Assistant Chief Engineer, had been sent to Chile to assist with the installation of the 36-inch and 60-inch telescopes. He was transferred to the CTIO budget in early 1968 with the title Senior Telescope Engineer-CTIO. His title was changed to Chief Engineer-CTIO effective April 1, 1970, and he continued to live in Chile after he retired.[48]

The grinding and polishing machine was delivered in August 1967, and was used for smaller mirrors for several months. The 150-inch fused quartz blank was placed on the machine on August 20, 1968, and grinding and polishing began on October 24.[49] The mirror was finished in November 1971, and stored pending installation of the mounting.[50] The CER–VIT blank was then placed on the machine, and grinding and polishing began in early 1972.[51] The finished mirror was shipped to Chile, and arrived in the port of Coquimbo on September 2, 1974.[52]

The Kitt Peak mounting was installed by the contractor between February 1971 and December 1972, and the mirror was placed in the telescope in February 1973. The first images were seen visually on February 27 by Mayall, Crawford, and Hoag, and the first photographs were taken during the last week of March 1973.[53] The CTIO mounting was shipped to Chile in February 1973; installation began in June and was finished before the mirror arrived in September 1974. The mirror was placed in the telescope on September 28, 1974, and the first photographs were taken on October 18, 1974.[54]

16 The complicated story of the Ford Foundation grant

The search for the funds needed to build and operate the Carnegie Southern Observatory[1] began when Ackerman made an approach to the Ford Foundation on February 28, 1964.[2] Haskins and Ackerman met with Borgmann and Heald on July 1, 1964[3] and presented them with a "small portfolio" containing a revised edition of the Prospectus[4] and other material. Borgmann wrote a Discussion Paper[5] for presentation at the December 10–11, 1964 meeting of the Ford Foundation Trustees: "Thus, the issue is presented to the Board: should the staff proceed toward a grant which, though it is within existing Science and Engineering granting policy, is likely to be in the range of $10 to $30 million and therefore to require some form of special budgetary designation?" Borgmann's memorandum to the files said:[6] "The Trustees' reaction to that paper can best be described as a lukewarm approval for the staff to inquire further into the reasons why the Foundation might support an observatory in the Southern Hemisphere equipped with a very large optical telescope." His letter to Haskins[7] was equally pessimistic: "While this consideration did not result in a decision not to support such facilities, it is only honest to say their enthusiasm was not high."

Borgmann scheduled a meeting with Haskins and Ackermann in early January 1965 to discuss the questions that had to be addressed before the CARSO proposal could be submitted to the Ford Foundation Trustees.[8] Borgmann felt their response to his questions "was for the most part general rather than detailed. They had not yet developed the factual bases for their answers to a degree which would satisfy our staff needs in shaping a recommendation."

Borgmann's next step was to visit the Mount Wilson–Palomar Observatories in Pasadena, the Lick Observatory and the Kitt Peak National Observatory.[9] He had several conversations with Ackerman and Haskins before and after this trip, and "discussed the proposal with three key members of the Cal Tech administration, President DuBridge, Provost Bacher, and Professor Anderson, Chairman of the Division of Physics, Mathematics and Astronomy. From each I received a statement of the general need for this facility." He looked into the use of the Mount Wilson–Palomar facilities by astronomers from other institutions and found "that use by outsiders does not include

Cerro Tololo from the air: 158-inch (4.0 meter) telescope building under construction. NOAO Photo Archives.

any appreciable time on the Hale 200-inch telescope. . . . For example, last year the staff booked the telescope for between 97 and 98 percent of the time, whereas use of the 100-inch telescope by Observatories staff accounted for only about 25 percent of the time." He wrote to President Heald:[10] "I have now reached a point where I must either do the kind of detailed staff work that could be considered negotiation and which inevitably raises the lively expectations of the potential grantee, or I must shelve or deny the request. Because I believe the Board's reservations have been answered, and because I am favorably inclined toward recommending a major grant on the order of $20 million, I now seek your assent to enter serious negotiations." Presumably Heald agreed, because Borgmann scheduled a trip to Chile for October.[11]

Borgmann and his wife, Ackerman, and Horace Babcock, Director of the Mt. Wilson and Palomar Observatories, arrived in Santiago on October 10, 1965, and flew to La Serena the next day.[12] They visited Morado and spent the night of the 12th on Tololo. They returned to La Serena on the 13th, visited the Coquimbo docks, and attended the AURA open house at the new headquarters which was hosted by Mayall and Wildt, the new President of AURA. They drove from La Serena to Santiago on the 14th, and joined Haskins and his wife, who had arrived the day before. During the next few days meetings were held with US Embassy officials, Rector Gonzalez[13] and Dean d'Etigny of the University of Chile, and government officials. Borgmann's memorandum to the files[14] and a letter from Babcock[15] indicate the trip provided useful information for the Ford Foundation.

Borgmann continued to be concerned about the "availability of the unique major Mount Wilson–Palomar and Carnegie Southern Observatory (CARSO) telescopes to competent astronomers from other observatories and other astronomy departments."[16] He summarized his understanding of the "plans of the Carnegie Institution (Washington) and the Mount Wilson–Palomar Observatories" in a memorandum (five major topics and seven specific questions) for his own use.[17]

It was Ford Foundation policy, prior to starting final staff work and serious negotiations, to ask if a prospective grantee would be willing to accept the grant in common stock of the Ford Motor Company with no present intention of selling, assigning or otherwise disposing of the shares for a period of five years from the date of transfer. If the answer was "no" that ended all discussion, but if the answer was "yes" this did not guarantee that the grant would be made.

The Ford Foundation's form letter relating to acceptance of stock was received by the Carnegie Institution on December 3, 1965,[18] more than three months before the staff recommendation concerning the grant would be brought to the Annual Meeting of the Ford Foundation Board of Trustees. The CIW Executive Committee promptly gave the needed approval, and Borgmann continued his discussions with Babcock and Ackerman.

A CIW Board of Trustees document dated March 9, 1966[19] includes the following statements:

"B. The Ford Foundation has notified the Institution of its intent to make the grant in the form of common Stock of the Ford Motor Company, etc.

D. The Board of Trustees of the Institution, upon recommendation of the Executive Committee, has approved acceptance of the grant, etc."

The date of this statement was five days before the Ford Foundation Trustees voted to table the CARSO proposal (see Reference 54). It seems clear from these words that the CIW Trustees had been led to believe the grant had already been approved by the Ford Foundation.[20]

Dr. Julius A. Stratton, President of MIT, was a member of both the National Science Board and the Ford Foundation Board of Trustees, and would become Chairman of the Ford Foundation Board on January 1, 1966. His knowledge of AURA's plans for a 150-inch telescope for CTIO and the CARSO proposal for a 200-inch telescope in Chile had an impact on the discussion at the November 1965 meeting of the National Science Board. Geoffrey Keller, now the Division Director of Mathematical and Physical Sciences (MPS), wrote to NSF Director Haworth:[21]

> You will recall that at the recent Board meeting, during the discussion of funding for CTIO, the question was raised as to whether adequate consideration had been given to economies which might be obtained by sharing support facilities for CTIO and the proposed CARSO (Carnegie Southern Observatory). You will recall that the establishment of the latter observatory, being proposed by Mt. Wilson and Palomar Observatories in conjunction with the Carnegie Institution of Washington, would hopefully be financed by a grant from the Ford Foundation. In response to questions

raised by the Board I tried to indicate that, so far as CTIO, AURA and the Foundation are concerned, we have made every effort to offer our help and facilities to the representatives of the CARSO group. On the other hand we have tried not to overwhelm them with offers of cooperation for fear that they might feel that we were trying to take over their project. Dr. Stratton remarked at this point that he hoped that we would have further conversations with him about cooperation in Chile, implying that, in connection with his forthcoming responsibilities as Chairman of the Board of the Ford Foundation, he had an interest in maintaining cooperation and reducing costs thereby.

He added that he and Mulders (Program Director for Astronomy) had lunch with Ackerman on November 26 as part of the exploratory process, and asked permission to send the following note to Stratton:[22]

> Dear Dr. Stratton:
>   You will recall that at the meeting of the National Science Board on November 18–19 we discussed the possibility of reducing expenses of building and supporting a 200-inch CARSO telescope in Chile by exploring the possibility of further cooperation between CARSO and CTIO.
>   The purpose of this note is to say that we are actively looking into costs that might be saved as well as other mutual benefits to be derived from working together on this, and hope to make some concrete suggestions to you shortly.

This could not have come at a worse time. Stock's resignation was imminent, and Edmondson had agreed to go to Chile for a short term in February and March as Acting Director of CTIO.[23]

Keller reached Borgmann by telephone at his hotel room in Washington on the morning of November 30 and told him about the discussion at the NSB meeting and his talk with Ackerman.[24] Borgmann reported this to Ackerman and Haskins[25] at lunch, and "Ackerman then reported on his end of a previous conversation with Keller. The location of the proposed 200-inch telescope was apparently the only matter discussed." Borgmann urged Ackerman to widen the areas of possible collaboration by looking at how other facilities and services might economically be shared by the two organizations (AURA and the Mount Wilson–Palomar Observatories). Ackerman called later to say that he expected to discuss this with Wildt (the President of AURA), Mayall, Babcock and Keller. Borgmann thought this "shotgun marriage" might result not only in useful collaboration, but also in closer working relationships between the two observatories.

Ackerman called on Wildt at Yale University a few days later and invited him to attend a conference to discuss telescope locations and possible sharing of some support facilities. Wildt then called Borgmann, and arrangements were made for him to visit the Ford Foundation on Wednesday, December 15.[26] During a discussion that lasted an hour and a half, Borgmann learned that Wildt's concern was with increasing micro-management by NSF, not the relationship with CARSO. Wildt invited Borgmann to attend the meeting, which would be in Tucson, and Borgmann said he did not feel he should be there if NSF was not represented. Wildt later withdrew his opposition to

NSF participation in the meeting, and Mayall told Babcock it seemed likely that either Keller or Robertson would attend.[27] Mayall learned that Mulders was to be in Tucson on January 14 and said he would also be invited.[28] A notice of the meeting that was sent to AURA Officers and Committee Chairman included a draft of agenda items that Babcock had prepared.

The CARSO–AURA meeting was held at the KPNO headquarters in Tucson on January 14, 1966.[29] Except for the introductory statements by Babcock, Wildt, and Mayall, the rest of the meeting bore no relation to Babcock's agenda. Babcock said CARSO would like to have all of Morado, but Wildt and Mayall said that AURA also had an interest in future use of Morado. The meeting recessed at 10:10 a.m., and the AURA group met in Mayall's office to hear and discuss a presentation by Keller. He began with a discussion of the current tight budget situation caused by President Johnson's decision not to fund the war in Vietnam with a tax increase, and continued with a description of the Daddario report and proposed reorganization of the National Science Foundation. He concluded this part of his presentation by pointing out that NSF had been designated as the "principal Federal agent for ground-based astronomy in the Office of Science and Technology, emphasizing long-range planning not only for Federal funds, but private sources as well. He then set forth seven questions to be presented to CARSO and AURA:

1. If the CARSO project goes ahead, and institutes arrangements for the appointment of visiting fellows and observing time for other visiting U.S. astronomers, would the Government be justified in saying that this constitutes a major step toward meeting the needs of U.S. Astronomy and astronomers for ground-based optical astronomy in the southern hemisphere?

2. To what extent does CARSO look forward to U.S. Federal fiscal and diplomatic support for its Chile observatory–both now and in the long range? That is, should NSF, on behalf of the government, include CARSO's needs in its long-range planning?

3. If the CARSO project goes forward, and initiation of the construction of other major U.S. telescopes in Chile is deferred for perhaps ten years, is it felt that AURA can staff, manage and justify the cost of its CTIO operation with the presently authorized facilities? The total dollars committed (including probable commitments for FY 1966) are of the order of $4.5 million.

4. If Federal support were planned for both endeavors, what provisions need we make to avoid unnecessary duplication of offices, shops, scientific and management staff in Chile?

5. Similarly, what plans are needed to avoid unnecessary further duplication of offices, shops and staff (specifically for support of the Chile observatories only) in the parent institutions in the U.S.?

6. How can NSF avoid being placed unnecessarily in the position of arbiter of the competing requests for support for two similar institutions in Chile?

7. In view of the remoteness and alien milieu of Chile, which may be expected to continue to generate political, social and economic problems for the observatory managements, what steps should be taken to join hands to provide

(a) A resident community of viable size–one which is scientifically and socially stable;

(b) A situation which will attract resident directors or superintendents of the highest caliber;

(c) Strong, flexible and financially well-backed U.S. support?

Keller made the same presentation to the CARSO group in the conference room at 11:15 a.m.

The meeting reconvened after the lunch break. Keller said he had made the same presentation to the two groups, but AURA's discussion had centered on possibilities of cooperation, while the CARSO group had discussed particularly question number 2. Borgmann asked what was AURA's response to question number 3, and Wildt said that the AURA group was dismayed that support for its operation in Chile might be withdrawn. Ackerman felt the two observatories were serving different needs, and were not competitive, and Babcock said the needs of astronomy in the southern hemisphere were so great that the CTIO facilities would be very much in demand. After a brief recess for separate discussions by the two groups, it was agreed to meet again on Wednesday, January 26. Borgmann said he would not plan to attend.

The official minutes, as written by Miss Elliott, did not reflect the strong emotional undercurrent that developed during this meeting. It was shown in Borgmann's memorandum to W. McNeil Lowry written a week after the meeting.[30] He was anxious to bring the CARSO proposal to the Board of Trustees in March 1966. After a perceptive account of the Tucson meeting he wrote: "Obviously, I cannot make a serious proposal until not only the land question is answered, but also agreements are reached concerning future collaboration."

The agenda of the AURA Chile Subcommittee meeting on January 19, 1966 included routine business and a report on the AURA–CARSO meeting.[31] Wildt said that shortly before that meeting Ackerman had informed him that CARSO wanted to use all of Morado, and Babcock confirmed this in his introductory remarks. There had been no meeting of minds at that time on seven questions that Keller had presented. The consensus of the Chile Subcommittee was that the answer to question number 1 was "no", on the basis that CARSO was primarily for use by Mount Wilson–Palomar staff and Observatory Associates they would appoint. Edmondson said CARSO's answer to question number 2 was "no", but he felt this would change. The Chile Subcommittee felt the answer to question number 3 had to be "yes", and that AURA could justify its operation in Chile. Lee felt that the answer to question number 6 was for both sponsors to insist on cooperation in one way or another. Questions 4 and 5 were covered by a resolution[32] presented by Edmondson and adopted by the Chile Subcommittee. Edmondson presented four administrative alternatives, and the Chile Subcommittee adopted the first: "That the Carnegie Institution of Washington be invited to accept membership in the Association of Universities for Research in Astronomy, that the CARSO proposal be modified by mutual agreement, and that the Ford Foundation be requested to make the grant to AURA, Inc." Wildt reported this to Borgmann two days later.[33]

Ackerman and Edmondson had developed good rapport at the November 22, 1963

Executive Committee meeting, the January 1964 AURA–ESO meeting in Paris, and on several later occasions in Washington. This background made it natural for Ackerman to wish to talk to Edmondson after the Chile Subcommittee meeting and before the January 26, 1966 meeting in Tucson. They agreed to meet in Bloomington, Indiana on Saturday and Sunday, January 22 and 23, and Edmondson reported this to the Chile Subcommittee. They met at the Indianapolis Airport on January 22, with Edmondson arriving from Tucson and Ackerman from Washington at nearly the same time.

Travel to Bloomington was in a car rented by Ackerman, with ice and snow on the road.[34] He said the CIW Board would be meeting in Washington on Monday, and he would give them a report of these discussions. Edmondson described the suggestion to build two 150-inch telescopes for the price of one 200-inch, which would serve twice as many astronomers. Dinner was at the Edmondson residence, with Ackerman getting his first taste of persimmon pudding.

Discussion was resumed after dinner with Edmondson presenting the Chile Committee Resolution and the suggested implementation. Ackerman proposed setting up a joint AURA–CARSO Advisory Committee now. At the end of an hour and a half they agreed that neither AURA nor CARSO could afford not to cooperate to make this operation go. Ackerman said we had done everything he had hoped to accomplish and in less time, so there was no need to meet on Sunday. He would present the AURA ideas to Haskins, Babcock and others.

Wildt, Edmondson, Hiltner and Mayall participated in a conference call on Sunday, January 23.[35] Edmondson summarized his meeting with Ackerman, a phone call to Keller, and a "thank you" call from Ackerman.

The second CARSO–AURA meeting was held at the KPNO Headquarters in Tucson on January 26, 1966.[36] Ackerman, Babcock and Boise represented CARSO, and the NSF and AURA representation was the same as at the first meeting. Edmondson read the Resolution of the Chile Subcommittee, and Ackerman said that Carnegie would prefer "an observatory complex" or "a group of observatories" to "a single major observatory in Chile." Keller said the important thing from the NSF point of view was achievment of a good program, meeting the needs of U.S. astronomers, and avoidance of unnecessary duplication of facilities. Next, Edmondson proposed that two 150-inch telescopes be built in Chile, instead of one 200-inch telescope. The meeting then recessed for a 40 minute tour of the KPNO optical shop and engineering addition.[37]

Edmondson read the proposed implementation of the resolution, and Ackerman responded that the Carnegie Institution of Washington "has stressed independence and the power to make its decisions free from outside pressure." Operations in all departments of the Institution have been based on the use of endowment income and limited acceptance of Federal grants, and further, the Carnegie Institution of Washington has engaged in cooperative ventures, but none were ever integrated into another organization. "By policy and tradition, therefore, the Carnegie Institution is barred from accepting this invitation to become a member of AURA."

Ackerman presented two quite different organizations for discussion. The first proposal was that the Carnegie Institution, with a substantial operating organization contemplated in Chile, would operate the facilities under contract with AURA, with the

Cerro Tololo Inter-American Observatory instruments to be managed as directed by AURA. As for Carnegie instruments, both northern and southern hemisphere 200-inch telescopes would be available one-third of the time to outside astronomers, including CARSO's Observatory Associates. The second proposal was to have a joint administrator (a "major domo") appointed by both Directors responsible for southern hemisphere operations. Separate staffs would be responsible to each Observatory Director. A joint committee, advisory in nature, was contemplated, to which common problems could be addressed. Its first task would be lease or purchase of a site for CARSO, the second, pooling of certain facilities. This plan, he felt, would meet the Carnegie Institution's need for preserving its identity and independence. Ackerman said the advisory committee would not allocate observing time; this would be done by each group, the mechanism for CARSO being the Program Committee.

There was a caucus of the AURA members after the lunch break. When the meeting reconvened, Wildt requested Edmondson to present the sense of the AURA caucus, which was: (a) the AURA group was still interested in the possibility of two 150-inch telescopes, and hoped for more discussion of this proposal; (b) two such telescopes would make possible a better organization than that presently contemplated; (c) as an alternate suggestion, AURA would propose that observing time on the CARSO 200-inch telescope be shared equally by AURA and CARSO, each group to administer its share of time. Mr. Lee added that AURA representatives felt that the advisory committee should be equivalent in authority to a board or executive committee. Dr. Keller and Mr. Moore thought that some legal entity was needed to enter into contracts and open bank accounts.

The CARSO and AURA groups met separately between 3:30 and 4:30 p.m. Ackerman reported that he had been unable to reach Dr. Haskins, President of the Carnegie Institution during the CARSO caucus. Regarding the proposal of constructing two 150-inch telescopes instead of one 200-inch telescope, he said that the Carnegie Institution had received financial commitments on the basis of a 200-inch telescope which might be lost if the size was reduced, and Babcock felt that a 200-inch telescope would arouse the interest of the best observers, wherever they might be found. Also, the CARSO group favored the administrative arrangement wherein administrative responsibility lay with the two directors.

The meeting adjourned at 5 p.m. and it was agreed that the group would meet again the following morning. Ackerman said he would continue to try to reach Dr. Haskins by telephone.

The meeting reconvened on January 27, 1966, at 8:35 a.m. in the Library of the Tucson Headquarters, with all present as on January 26, plus Mr. Richard Bilby, AURA Counsel. Ackerman said that since adjournment on January 26, he had talked with both Dr. Haskins and Dr. Borgmann of the Ford Foundation. He then presented the following proposals: (a) that one-third of the observing time on only the CARSO 200-inch telescope be placed at AURA's total disposal, CARSO's two-thirds time to include the Observatory Associates, staff members and Guest Investigators; (b) AURA's share of operating expenses of the CARSO 200-inch telescope would be 33 percent. Ackerman assumed that the site on Morado would be included in this understanding, and Carnegie representatives felt there must be some legal arrangement whereby the land on which CARSO telescopes are located is under Carnegie control, either by ownership or long-term lease. Regarding organization, Ackerman said Dr. Haskins did not favor a joint committee endowed with executive responsibility.

Babcock, in response to an inquiry of Mayall, said that in regard to the so-called "major domo," he contemplated seeking an individual capable in management, either an engineer or an astronomer, to be responsible for observatory operations, with scientific direction coming from Tucson and Pasadena.

Wildt said these proposals would be presented to the Executive Committee at its meeting to follow, and representatives of both groups agreed to meet in Washington, DC, the following week. Lee also suggested that the advisory committee be appointed at an early date. The meeting adjourned at 9:10 a.m.

The AURA Executive Committee met the next day.[38] Wildt first reported on the January 14 meeting with CARSO representatives. At Wildt's request, Keller summarized the remarks he had made at that time, which had changed the direction of the meeting, and he also read the seven questions he had presented. Edmondson read the resolution and suggested implementation adopted by the Chile Subcommittee on January 19th, and mentioned his discussion of this with Ackerman in Bloomington on January 22nd. The proposed implementation was not acceptable to the Carnegie Institution on the basis of its history and policies. Wildt also reported on the discussion at the second CARSO–AURA meeting, before the recess for lunch.

After lunch, the Executive Committee approved the purchase of three acres adjacent to the La Serena Headquarters, the purchase of additional land in Tucson, and the appointment of Edmondson to be Acting Director of CTIO for six weeks beginning February 2, 1966. The present status of negotiations with Stock was described.

The discussion about CARSO then continued, with emphasis on the organization and joint operation of the observatory. The areas of agreement, including shared observing time on the 200-inch telescope, were discussed and summarized in three recommendations to be accepted depending on the adoption of "mutually agreeable organizational arrangements." The Committee favored an operating board or committee with line authority, and this seemed to be the remaining major point to be negotiated.

Ackerman called Borgmann on the morning of the 27th to report what happened on the first day of the CARSO–AURA meeting, and later in the morning Haskins called him to tell him what Ackerman had reported about the early morning meeting.[39] Borgmann viewed this with optimism, and wrote: "A further meeting is planned at which these oral agreements will be put in writing. ... Hence today the picture is almost reversed from what it was yesterday. I will look into the details of the agreements when written, to be sure that our interests are preserved. If satisfied, I will be ready to move toward a recommendation for the March meeting."

The Edmondsons left for Chile on February 2, 1966 and this was also the date of the CARSO–AURA meeting in Washington at the CIW Headquarters. Dirk Brouwer died unexpectedly on January 31, and Wildt's already heavy burden increased when he was asked to take on the duties of Chairman of the Yale Astronomy Department.

The position of Chief Administrative Officer (CAO, or "major domo") that Babcock wanted proved to be the only serious obstacle to final acceptance of the CARSO–AURA agreement. The job description called for this person to have two bosses, the Directors of Mount Wilson–Palomar and the Director of KPNO, and to report separately to them. The CARSO–AURA meetings on February 2 and 4 produced five drafts of the agreement. Ackerman mailed copies of a sixth draft, dated February 8, 1966, to Edmondson and Miller in Chile,[40] and this was revised on February 12.[41] None of these drafts resolved the questions about the CAO that had been raised by AURA, and AURA counter proposals were rejected out-of-hand by CARSO.

Time was growing short, and Borgmann told Ackerman that he "would need to have the final agreement in the form approved by the responsible bodies of both agencies before [he] would be prepared to bring the proposal to the attention of the Trustees."[42] This presented a serious problem because the annual meeting of the AURA Board was scheduled for March 15, and the Ford Foundation Trustees were scheduled to meet on March 14. Borgmann's memorandum continued: "I learned only yesterday that my insistence has resulted in a meeting of the AURA Board in Chicago for March 1." Elliott mailed the formal call for a special meeting on February 11,[43] but only after AURA had been told that "Borgmann thinks the CIW proposal has to be considered now, for chance of approval."[44] This was because McGeorge Bundy had replaced Henry Heald as President of the Ford Foundation, and big changes were in the offing.[45]

Haworth told the National Science Board[46] that the AURA Board of Directors was "planning a Special Meeting on March 1 for the purpose of discussing and ratifying (if it so agrees) the attached Agreement. Dr. Haskins, President of CIW, has already been authorized by a majority of the Executive committee of the Governing Board of CIW (which has power to act in this case) to sign this Agreement. The Ford Foundation has stated that these ratifications must be achieved by the time of a meeting of its Board on March 14 as a pre-condition to its consideration of the proposed grant at that time."

The radio link between Tucson and La Serena was very helpful during this period. Edmondson and Miller were kept informed about new developments, and were able to make their views known without delay. For example, Edmondson said:[47] "If we develop this joint organization, we should not jeopardize the positions of the present staff in La Serena and on Cerro Tololo. If CARSO and AURA go into joint operation, we should make use of the existing staff as a nucleus. . . . He may go to Chicago to present these views himself." In a later radio conversation,[48] Edmondson said he had decided to attend the Chicago meeting, and this had necessitated a change in the schedule of Ambassador Ralph Dungan's visit to Tololo.[49]

Babcock had some second thoughts about the February 8 draft of the "Memorandum of Agreement" between AURA and CIW.[50] He said the Agreement "had to be formulated under certain pressures arising from changes within the Ford Foundation and within the National Science Foundation, changes which would have been quite unpredictable a few months ago." The final draft, dated February 21, 1966, was signed by Haskins, and 21 copies were sent to Mayall for consideration at the special meeting of the AURA Board.[51]

The special meeting convened at 9:00 a.m. on March 1, 1966.[52] All members of the Board were present, except Spitzer and d'Etigny. Wildt asked Edmondson to report on CTIO activities. Next, Miller read the draft of the AURA–CIW Agreement, and the Board discussed it and made suggestions for changes in several paragraphs. Harrell asked if this Agreement was good for astronomy. Whitford said "yes", because this seemed to be the only way that construction of a large southern hemisphere telescope could begin at the present time, within boundary conditions that AURA was powerless to change. Code disagreed, and said he could not conclude that the proposed joint arrangement was in the best interests of astronomy. He was not happy about the arrangement and had talked with no one who thought it was a good idea. He also objected to AURA's having to settle all the details in three months, after the years spent in its site selection and telescope design. Miller asked if AURA could afford to refuse the request of two sponsors, and possibly place CIW in a position where its application to the Ford Foundation might be turned down. Even if the cooperative arrangement was later broken up, funding would have been obtained for CARSO's observatory, and a major development in the southern hemisphere would have been accomplished. In response to an inquiry of Miller, Keller said the National Science Foundation was interested in having an effective observatory in the southern hemisphere doing as much good astronomy as possible, and the National Science Foundation attitude toward an "annulment" would depend on the facts at the time.

Wildt then asked if the Board would delegate to the Officers authority to proceed along these lines, and the Board agreed:

"*ACTION:* Upon motion of Mr. Wiggins, seconded by Mr. Woodrow, the Board of Directors unanimously agreed to authorize the Officers of AURA, Inc. to continue negotiations along the lines of the present draft of the Memorandum of Agreement between AURA and CIW as discussed above and, on the agreement of both parties, to execute it."

Keller read a resolution approved by the National Science Board the preceding Saturday, as follows: "Resolved, that the Board regards with favor the proposal that the 200-inch optical telescope and associated facilities to be constructed in Chile with the aid of such funds as may be obtained by them from private sources be operated jointly with the facilities of AURA, and authorizes the Director to take such steps as he deems necessary to implement the plan. The Board feels that the implementation of this proposal would constitute a major step toward the advancement of research by U.S. astronomers." Keller added that Dr. Haworth did not see any possibility of obtaining money from government sources for a large telescope in the southern hemisphere.

Wildt next suggested a discussion of the proposed Chief Administrative Officer and his functions as drafted by CIW. This discussion continued until the lunch break at 12:30 p.m., and resumed at 1:30 p.m.

Miller and Lee reached Ackerman by telephone during the lunch break. They returned to the meeting at 2:00 p.m. and reported that Ackerman had agreed in principle to the changes suggested by AURA, except for an item he would have to check with Babcock.

When the discussion of the CAO ended, the Board turned to the composition of the Advisory Council that was referred to in the draft. There was some more general discussion, and the meeting ended with a vote of confidence, and a round of applause, for the officers who had negotiated the present draft of the agreement. The meeting adjourned at 3:00 p.m.

Elliott mailed the "original and duplicate original copies" of the final revision of the Memorandum of Agreement to Wildt[53] on March 4, and she sent unsigned copies to Ackerman and Borgmann.[54] Wildt signed for AURA and mailed the original and duplicate to Ackerman on March 7; Haskins signed these for CIW on March 9.[55] Ackerman sent a facsimile of the signed document to Borgmann on March 10.[56]

On March 14 the Ford Foundation Board of Trustees received and discussed the staff recommendation to grant $19 million to the Carnegie Institution of Washington to fund the construction of a 200-inch telescope in Chile. The Trustees voted to table this for further study and exploration of alternate means of support of astronomy. "The amount recommended was considered to be more than the Foundation should devote to astronomy at that time, and the proposed provision of access for astronomers other than those on the Mount Wilson–Palomar staff was considered inadequate."[57]

The Annual Meeting of the AURA Board[58] took place on March 15, the day after the meeting of the Ford Foundation Trustees. Wildt said that following the Special Board Meeting in Chicago on March 1, he had received reactions from several Board members, and had decided to re-open discussion of the AURA–CIW agreement for about an hour. Most of the comments were critical of the role that NSF had played. Menzel said it was obvious that AURA must defer to the wishes of its sponsor, but merely setting up a 200-inch telescope in Chile, to which AURA has only limited access, does not begin to fulfill the needs of astronomy. He also felt that AURA should place its reservations and disappointment at the present course of events on record with NSF. Carson felt that nothing further could be done until word was received from CIW about the decision of the Ford Foundation Trustees. It took a week before this became known outside the Ford Foundation,[59] and Miller then informed the AURA Board.[60]

McGeorge Bundy became President of the Ford Foundation on March 1, 1966, and there were some who were quick to blame him for the decision to table the CARSO proposal.[61] Stratton, Menzel, Goldberg, and even AURA[62] were soon added to the list of villains. The last was particularly unfair, because AURA reluctantly signed the agreement with CIW, including the objectionable CAO position, in an effort to help CARSO get the grant from the Ford Foundation. Furthermore, the National Science Foundation and the Ford Foundation conducted their later negotiations in secret, and did not inform AURA of the outcome until December 23, 1966.

Bundy phoned Goldberg, whom he had known in his days as Dean at Harvard, on March 1st or 2nd to ask for his opinion of the CARSO proposal.[63] He called Haworth on March 3rd,[64] presumably at Goldberg's suggestion, for the purpose of informing himself about the status of the NSF plans for a large telescope in Chile. Between March 3 and 8, Bundy phoned Whitford and several other astronomers who had written letters to Babcock supporting the CARSO proposal;[65] he phoned Haworth on March 9, probably to tell him about these calls. Following the action of the Ford Foundation Trustees, Bundy asked Haworth to suggest names for a committee to advise the Ford Foundation with regard to CARSO.[66]

Babcock wrote to eleven prominent American astronomers soliciting new letters of support to submit to the Ford Foundation.[67] Goldberg's reply[68] said: "The only real doubts I have about your proposal concern the provisions for the use of the facilities by visitors from outside institutions. I consider it absolutely essential that a 200-inch telescope in the Southern Hemisphere be operated as a national or international facility." Goldberg sent a "blind copy" to Bundy, and wrote to Bundy[69] again after he took Menzel's place on the AURA Board. The final paragraph says: "I should make it plain that I would consider it a disaster if the end result of the present controversy were to be no large telescope at all in the Southern Hemisphere. At the same time, I hope that the Ford Foundation will give careful consideration to a possible alternative to the CARSO proposal, namely, that the funds be granted to AURA with the understanding that the National Science Foundation would undertake to provide a long-range support for the operation of the telescope. In effect, this was my advice when you asked what I would do if I had the $20,000,000 and I suppose I could have made this letter a whole lot shorter by simply saying that I had not changed my mind."

Bundy's reply,[70] written after he returned from two weeks in Japan, said: "This is just a line to thank you for your letter of May 5 about the Southern telescope. I am very glad to have this careful account of your own thinking, and I can assure you that I share your fundamental conclusion that it would be very bad if the Ford Foundation's initial interest in a 200-inch telescope were to have the net result of preventing or even delaying some large-scale installation in the Southern Hemisphere. ... I am making this an early order of business, and I can promise you that your view of a solution will be very carefully considered as we go about our work on the problem."

Bundy also wrote to Stratton:[71] "I was grateful to you for raising the telescope issue the other day, because I have been slow. I entirely agree with you that we must not let this one simply fade out. I am talking to Carl Borgmann about it today. ... It remains highly important that the Ford Foundation's activities in this field should not have the net result of retarding optical astronomy in the Southern Hemisphere." He also said he had read the file of letters which Babcock had collected, and had talked to some of those who had written them, and "what they write to Babcock and what they say to me are not identical, to put it mildly."

Bundy looked at several other possibilities before he gave serious consideration to Goldberg's suggestion of a grant to AURA. He discussed the situation with Borgmann on June 13,[72] and Borgmann met with Haskins and Ackerman on June 29.[73] He told them Bundy's primary concern was to avoid perpetuating or increasing the monopoly of a single academic department (Cal Tech), and the Ford Foundation support would be $10 million or less. They were not receptive to the suggestion of building and operating a southern hemisphere observatory in a second, separate partnership with the astronomy department of a university other than Cal Tech.

Borgmann requested a meeting with Haskins and Ackerman on August 24[74] to discuss a three-way partnership, Carnegie, Cal Tech, and another university. They said they would be willing to consider it, provided the observing time was divided equally among the three institutions. Borgmann said he would explore the question with Harvard, Princeton, Chicago, and possibly others. He telephoned Ackerman on September 27 and said his visits with the presidents of these universities "revealed little interest" in joining his proposed three-way partnership. All of them were "satisfied with their membership in AURA, and considered it desirable to continue participation in that organization." He also told Ackerman that if the Foundation were to make any grant for a southern hemisphere telescope it would be made to AURA or to the National Science Foundation for AURA.

On September 12, Borgmann reported to Bundy:[75]

Following your suggestion after my memorandum of July 1, 1966, I have conferred with Goheen, Beadle, Pusey, Robertson (Associate Director-Research, NSF), Purcell (once by phone and more recently, in person), and Haskins. None of the university presidents reacted favorably either to the idea of a select consortium or to that of a tripartite operation as a full partner of Carnegie Institution and Caltech.

It would appear as a result of these discussions that the most practical way in which the Foundation can contribute within its fiscal and educational guidelines towards a 200" telescope in the Southern Hemisphere is to collaborate with NSF in making the necessary funds available to AURA. Therefore, unless you have other suggestions for me to investigate, I would propose to meet with Lee Haworth, John Wilson and Randal Robertson.

Borgmann met with Haworth, Robertson, and Keller on September 26, 1966.[76] He told them that if the Ford Foundation made any grant for a large telescope it would be to AURA and would cover only a part of the cost. He said he wanted to explore the interest of NSF, for it appeared to be the most likely collaborator in such a project. Haworth was interested, but said timing and ability to fund would be a problem. The first opportunity to include such an item in a budget request would be for Fiscal '69. Both Keller and Robertson questioned whether time-and-money-wise a 150-inch wouldn't be better. Haworth agreed with Borgmann's position that the seeing conditions and the sky areas available really called for a 200-inch telescope. NSF was quite satisfied with the AURA organization and the Kitt Peak staff, and felt they are well qualified to design and engineer the job.

The meeting ended with an agreement that Haworth would explore the NSF potential with appropriate members of the National Science Board, of the President's Science Advisory Committee (PSAC), and of the Bureau of the Budget. Borgmann would make it clear to the Carnegie Institution that any decision to make a grant to AURA was on his judgment and not on the initiative of NSF. "Also, no approach to AURA would be made by either party" until Haworth could make a judgment following his conversations with the NSB, *et al.*

Keller told Borgmann after the meeting that NASA was seriously considering a grant to the University of California, to be matched by Australia, for a 150-inch telescope in Australia.[77] If this grant were made, Congress probably would not be willing to provide funds to match the Ford Foundation offer, so Haworth asked if NASA was planning to make the grant. Homer Newell, the NASA Associate Administrator for Space Science and Applications, wrote to Haworth:[78] "I understand that you have inquired about NASA's current interest in participating with the Lick Observatory of the University of California to create a 150-inch telescope in Australia jointly with the Commonwealth government. ... The availability of NASA funds to support the Lick Observatory – University of California proposal is highly questionable for Fiscal Year 1967 or 1968. Therefore I encourage you to pursue urgently the Chilean telescope plan, which will be the cornerstone of American astronomy's southern hemisphere program."

During this period of uncertainty, Sir Richard Woolley, the Astronomer Royal, proposed cooperation between the Royal Greenwich Observatory and the Mount Wilson and Palomar Observatories in the establishment of a 200-inch Southern Telescope for

Astronomical Research (STAR).[79] This muddied the waters, but did not slow down Haworth's efforts. He told the National Science Board in Executive Session about the discussions with the Ford Foundation at the November 17–18, 1966 meeting,[80] and the Board encouraged him to continue the negotiations.

On November 28, Haworth wrote to Charles L. Schultze, Director of the Bureau of the Budget, and Donald F. Hornig, Special Assistant to the President for Science and Technology, requesting their approval of NSF funds to match the Ford Foundation's $5 million offer.[81] He proposed to reprogram $650,000 in FY 1967 funds, and to allocate $1.2 million from the FY 1968 budget. This, plus the Ford Foundation matching funds, would make $3.7 million available, and the balance of the Federal share would be budgeted in FY 1969 and 1970.

Haworth told Hornig: "Members of the NSF staff feel that there is much to be said for constructing a 150-inch rather than a 200-inch instrument." Scientific, technical, scheduling and cost considerations were discussed in a separate document.[82]

> Our cost estimates at the present time are admittedly rather rough, primarily because the Ford Foundation has asked us rather particularly not to let the word of the current discussions get out to the astronomical community. For this reason the NSF staff has not felt free to call on the expertise which would normally be available to us at KPNO and elsewhere. However, it does seem reasonable to suppose that the cost of a second 150-inch telescope would be comparable to that of the one now under way at KPNO, which with housing and auxiliaries, should cost in the neighborhood of $10 million. Savings of the order of ten to twenty percent might be effected through use of the same designs and parallel production. The telescope building for Chile might be less elaborate because of the intrinsically better observing conditions. On the other hand, construction costs in Chile are very high and difficult to predict, so that it should be recognized that an allowance for contingencies of as much as 20% over the estimated $10 million would not be unreasonable.

At their December 8–9, 1966 meeting the Ford Foundation Trustees authorized continuation of the negotiations with NSF, and authorized the officers to commit up to $5.0 million toward the cost of a 150-inch telescope in the southern hemisphere.[83] It was agreed that NSF would provide any funds needed in addition to $10.0 million.

Two Bureau of the Budget officials initially opposed the use of Federal funds to match funds provided by a private foundation. William D. Carey, later the Executive Officer of the American Association for the Advancement of Science (AAAS), and Hugh Loweth didn't want to set a precedent.[84] Loweth didn't want "any goddamn foundation putting money in the federal budget."[85] Haworth, T.O. Jones, Randal Robertson and Aaron Rosenthal met with Carey and Loweth and persuaded them to reconsider. After this, personal approval of President Johnson was required for any "late add-on" to the budget. Haworth and Jones went to see the President about the CTIO 150-inch telescope and an additional Antarctic Research Station. Approval was given for both,[86] and the NSF budget narrative for CTIO was rewritten "in the period around Christmas."[87]

AURA finally learned about the Ford Foundation–NSF negotiations when Haworth talked to Mayall on December 23rd.[88] Mayall was strongly warned that it was vital that word about this should not become known until after the budget had been "sent to the Hill." President Johnson had a history of canceling actions that were "leaked" in advance. Jones recalled:[89] "Joe Califano got me over in his office and asked me the details on this, and he says, 'By God, the President will kill you if anybody knows about this.'" Mayall called Wildt, who informed Edmondson, Hiltner and Lee. No one else in AURA or KPNO knew about this for another three weeks.

The officers and Edmondson had met with Haworth the first week in May.[90] They told him how the AURA Board felt about the arrangements with CARSO, and its concern about future relations with NSF. During this meeting, Haworth had asked if there were any scientific merits in placing the 150-inch telescope in Chile rather than on Kitt Peak, and this was discussed by the Scientific Committee on August 30, 1966.[91] There was more discussion the next day, and the Committee tried, without success, to interpret the meaning of Haworth's suggestion. The Committee endorsed a statement, prepared by Hardie, that described the scientific advantages of having a 150-inch telescope on Cerro Tololo; it did not state a position on where the first 150-inch should be located.

Mayall wrote to Mohler[92] on September 13 that Robertson had called to inform him of another large budget cut. Then, on his own initiative he had said that AURA should completely soft-pedal at this time any thought of proposing to put the first 150-inch telescope in Chile. Mayall continued: "It seems that the Ford Foundation has asked Carl Borgmann to try to come up with some new formula, or rationale, for them to consider seriously their going in to support astronomy on the scale proposed. Randy thinks we should not do anything to interfere with, or influence in any way, this new move by the Ford Foundation. He asked me whether Borgmann had talked with me again, and I replied "No". He ended the conversation by saying, "Do not do anything that would let them off the hook!" Mohler's reply[93] was uncharacteristically gloomy: "The last paragraph of your letter is especially disturbing to me since I feel it contains an amplification of the signs pointing to decreased support not only for Cerro Tololo but also for KPNO. ... Indeed, I feel that my enthusiasm for the transfer of the Michigan Schmidt to Chile is reaching a very low point. If the future of the Tololo venture is as dim as your report of the new budget cuts and the tacit support of the CARSO–Ford consortium by the NSF makes it appear, I certainly do not want to ship the telescope to Tololo and have it stranded there for lack of funds."

These letters were written two weeks before Borgmann's meeting with Haworth, Robertson, and Keller.[94] Hindsight tells us that Robertson was trying to tell Mayall that something good was going to happen, and that AURA should be patient. Mohler's gloom described the way AURA saw things for another three months.

Borgmann met with Wildt in New Haven on January 12, 1967.[95] He described the arrangement with NSF for shared funding of a 150-inch telescope for CTIO. The discussion included the possibility of the Carnegie Institution joining in if they could find the additional funds needed for a 200-inch telescope. He phoned Ackerman on January 16[96] to inquire about the likelihood of Carnegie being able to raise the additional $7.0 to $8.0 million that would be needed to build a 200-inch. Ackerman asked for a day or two in order to consult with Haskins, and he called back on January 17[97] to say "they would like to make an all out effort to raise the funds." Borgmann phoned

Haworth at 2:25 p.m. on January 17,[98] presumably to tell him about Ackerman's call.

Borgmann wrote to Wildt on January 23, 1967 to confirm the substance of their January 12 conversation.[99] He said the Ford Foundation "would be interested in receiving a proposal from AURA for the construction of such a telescope."

Wildt reported this welcome development to the Chile Subcommittee on January 24, 1967.[100] Borgmann had read his letter to Wildt on the telephone, and it was now in the mail. The group agreed that the formulation of a response should await actual receipt of the letter.

Ackerman called Borgmann on January 25[101] to report: (a) They had not taken the problem of participation in the new plan for a 200-inch telescope to the CIW Trustees or pursued the question of availability of funds in any official way, and (b) he and Babcock were to see Haworth and Robertson at 4:00 p.m. Borgmann called Robertson on January 26[102] to learn the nature of the NSF discussion with Ackerman and Babcock. Babcock had said that the conditions for constructing and operating a 200-inch telescope would be (1) construction would be under the control of Carnegie, and (2) they would hope to reach an operating agreement with AURA that would provide Mount Wilson–Palomar with a share of the time proportional to the funds invested. Borgmann wrote: "I suspect the terms expressed at the above conference would not be acceptable to AURA and consequently Carnegie may never need to seek funds."

The two hour difference between Eastern Time and Mountain Time made it possible for Borgmann to talk to Wildt before the start of a meeting of the AURA Executive Committee.[103] He told him there was nothing further to report on the likelihood of Carnegie having funds with which to participate, and that the Executive Committee should proceed today "under the assumption that the 150-inch telescope was the one of immediate interest." This was also good news for NSF because of "the possibility of an alternative messing up their request for an appropriation for a 150-inch telescope."[104]

Borgmann's letter to Wildt was the first item of business for the opening Executive Session of the Executive Committee.[105] Wildt read it and then reported the good news in Borgmann's phone call. The draft response to the letter was modified by deletion of the second paragraph, which expressed concern about the uncertainty and resulting delay that would be caused by the proposed Carnegie participation.

At the beginning of the open session, "Wildt informed KPNO staff members present that AURA had been invited to submit a proposal for joint funding by the Ford Foundation and the National Science Foundation of 150-inch telescope for Chile, and expressed gratitude for this vote of confidence in AURA."

The next step was preparation of a proposal to the Ford Foundation by February 15, to be acted on at the March Meeting of the Ford Foundation Trustees. The proposal was mailed to Borgmann on February 11, 1967.[106] His reply said:[107] "Thank you for accepting our invitation to submit a proposal. It is heartening to know that the AURA Executive Committee has given their enthusiastic and unanimous approval to the project."

Wildt's "Report of the President" to the AURA Board on March 14[108] summarized the history of the Ford Foundation–NSF negotiations and said a proposal for $5.0 million had been sent to the Ford Foundation at their request. The Ford Foundation Trustees approved the grant at their March meeting, and the Grant Letter[109] was dated April 19, 1967. Bundy phoned Haworth at 11:40 a.m.,[110] probably to inform him that

Cerro Tololo from the air. The long building is the dormitory and dining hall for observers, the circular building is the scientific office building, and the buildings at the upper right are the maintenance facility area. The domes include two 16-inch (0.4 meter) telescopes from the US site survey, a 36-inch (0.9 meter) telescope, the Michigan Curtis–Schmidt, the 40-inch (1.0 meter) Yale telescope, a 60-inch (1.5 meter) telescope, and the 158-inch (4.0 meter) that was jointly funded by the Ford Foundation and NSF. The 24-inch (0.6 meter) NASA–Lowell telescope is not in this picture. NOAO Photo Archives.

the Grant Letter was in the mail. The Ford Foundation wrote to Haskins on April 25:[111] "Final considerations have been given to your proposal for support to the proposed Carnegie Southern Observatory dated June 25, 1964, as modified by addenda dated June 1, 1965 and February 10, 1966. I am sure that you will know from the discussions during the past year that the Ford Foundation is unable to accede to your request. This letter is simply to complete our formal records."

The Ford Foundation funds could not be spent, or even obligated, unless and until Congress passed the National Science Foundation FY 1968 appropriation with the 150-inch line item included. Seven months elapsed before it was passed and then signed by President Johnson on November 3, 1967. Nevertheless, President Johnson and President Frei of Chile announced at the Punta del Este (Uruguay) Conference on April 13, 1967 that a 150-inch telescope would be installed as a joint US–Chilean Project at the Cerro Tololo Inter-American Observatory.[112] An internal Ford Foundation memorandum comments:[113] "Where does the FF grant stand at this time? I had understood that we were waiting until NSF had funds in hand so as not to appear to be pressuring

Congress in any way. (Johnson may have made his announcement for the opposite reason.)"

The Ford Foundation grant specified that cash would be made available to match funds provided by NSF as bills came due and were paid.[114] This created a problem because major funds had to be obligated when contracts were signed, long before the cash was needed. As soon as this was explained,[115] the Ford Foundation gave AURA permission to obligate the entire $5.0 million in FY 1968.[116] Installation of the 150-inch telescope on Cerro Tololo was completed in October 1974. The final report for the Ford Foundation files[117] mentioned the earlier grant to ESO and summarized the history and purpose of the grant to AURA. The conclusion was: "The Foundation can well be proud of this grant."

Haworth had worked hard to persuade the Bureau of the Budget, President Johnson, and the House and Senate Appropriations Committees to approve the funds for the NSF share of the cost of the 150-inch telescope in Chile. In 1979 he wrote to the author[118] that the 150-inch telescope "is one of the few things during my tenure at NSF of which I am most proud."

Part IV
Cerro Tololo's neighbors

17 The USSR (Pulkovo Observatory) astrometric expedition to Chile.

Four astronomers from the Soviet Union were among those who greeted the AURA site selection team at Cerro Calan on November 15, 1962, and they were among the guests at the AURA dinner on the evening of November 23. They had arrived in October to initiate and carry out a program of measuring the accurate positions of 11,376 stars from 25° south of the celestial equator to the celestial south pole. Ten other observatories were also participating in this Southern Reference Star Program, sponsored by Commission 8 (Meridian Astronomy) of the International Astronomical Union.[1]

The leader of the Soviet group was Mitrofan S. Zverev, Deputy Director of the Pulkovo Observatory. His colleagues were Bronislav Bagildinsky, Valentina Shishkina (Mrs. Bagildinsky), and Vladimir Bedin.[2] The Pulkovo Observatory, founded in 1839 and located near St. Petersburg (it was called Leningrad in 1962), was the Palomar of the nineteenth century. It was there that F.G.W. Struve was one of the first to measure the distance of a star by the trigonometric parallax method. Observations with the Pulkovo meridian circle have made major contributions to the catalogs of star positions from the early years to the present time. Zverev and all of the Soviet astronomers who worked at Cerro Calan from 1962–1973 were among the world leaders in fundamental positional astronomy.

As early as 1932 the Pulkovo astronomers became interested in sending one of their instruments to the southern hemisphere. They discussed this with their colleagues during the meeting of Commission 8 at the 1932 General Assembly of the International Astronomical Union in Cambridge, Massachusetts. Commission 8 submitted a resolution endorsing the proposal to the General Assembly, where it was approved.[3]

Whatever progress had been made in implementing this project was brought to a halt by World War II. It reappeared at the Tenth General Assembly of the IAU, the fourth after the War, in Moscow in 1958.[4] This was followed by a conference at the University of Cincinnati on May 17–21, 1959, attended by 33 participants from eleven countries.[5] Zverev read a paper entitled: "Reference Stars in the Southern Hemisphere," and A.A. Nemiro and Zverev read a paper entitled "A Plan of U.S.S.R.

Participation in Astrometric Observations in the Southern Hemisphere." Rutllant's paper summarized the history of the 111-year-old observatory in Santiago, Chile, which was transferred to the University of Chile in 1927. He described the plans for the new location at Cerro Calan, where construction was already under way. In response to questions, he said the pivots and circles of the meridian circle were in good condition. It was the same type as the meridian circles in Cordoba and La Plata, and had been delivered in 1909.

The conference adopted ten resolutions, of which six stressed the importance of southern hemisphere astrometric observations. The second resolution included a recommendation: "That the Academy of Sciences of the U.S.S.R. proceed with its plan of organizing in the near future an expedition to the southern hemisphere to engage in an intensive program of astrometric observations." It also urged the five South American countries with astrometric equipment to participate in the program.

The Cincinnati Conference was followed six months later by an Inter-American Conference in Argentina.[6] The Chairman, Gerald Clemence, opened the business session by saying that this conference was a continuation of the Cincinnati Conference. Eleven resolutions were adopted, the first being a blanket endorsement of the Cincinnati resolutions. It was reported that Argentina and Chile had received letters from the U.S.S.R. Academy of Sciences, in accord with the language of the Cincinnati resolution. The replies would be dealt with by the governments of Argentina and Chile.

Rutllant described the present status of the U.S.S.R. proposal on August 10, 1960 during the Conference on the Chile Observatory in Tucson.[7] The University of Chile had agreed to the proposed cooperation, and two Pulkovo astronomers had visited Santiago. Recently the Rector of the University of Chile had visited the Pulkovo Observatory, but nothing had been heard since that time. The decision to locate the Pulkovo Observatory Southern Station at Cerro Calan was announced by the special committee of Commission 8 at a conference held at the La Plata Observatory on November 7–8, 1960.[8] Zverev spent several days in Santiago after this meeting discussing arrangements, and two years later four Pulkovo astronomers were at Cerro Calan starting their work, thirty years after the initial proposal at the 1932 General Assembly of the International Astronomical Union.

The USSR expedition was initially equipped with a new photographic vertical circle, designed by Zverev, and a Zeiss transit from Pulkovo. The Cerro Calan Repsold meridian circle was modernized with microscopes and cameras, manufactured in Leningrad, to photograph the circles. A large transit instrument and a 40-inch meniscus astrograph (Maksutov telescope) were under construction;[9] Stock told the AURA Scientific Committee the 40-inch would be erected on Cerro Robles, one of the first sites he had investigated.[10] These instruments were installed in 1967-68.[11]

Zverev got the observing program started and stayed in Chile for two years. Dimitri Polojentsev was head of the Pulkovo expedition for the next two years, and attended the ceremony when Edmondson received the "Al Merito" decoration.[12] Zverev returned after two years, and he and his wife were living in the residence at Cerro Calan when Edmondson was Acting Director of CTIO in February and March, 1966.[13]

The AURA Executive Committee met in Chile on November 1970, and Hugo Moreno and Carlos Torres took the Edmondsons to Cerro Robles to see the 40-inch Maksutov telescope. Nemiro had just arrived to take over as head of the Pulkovo group, and he arranged to make his first inspection of the Cerro Robles site on this day. It was an enjoyable reunion. The Maksutov telescope was a well made, impressive looking instrument, and the photographs taken with it were of excellent quality. The control panel had the function of each switch, etc. engraved on metal tabs in Cyrillic characters. The Chilean astronomers had used a Dymo label maker to produce the Spanish equivalent; it was truly a bi-lingual control panel. The telescope had a field of 25 square degrees, and a start had been made on photographing the first epoch plates of 160 areas with galaxies in the declination zone −90° to 0°.

The progress of the observing program can be followed by reading the reports and minutes of Commission 8 in the Transactions of the International Astronomical Union.[14] The report at the Prague General Assembly said that 81 percent of the observations had been completed by July 1, 1966, and 20 percent had been reduced. Six years later it was reported at the Sydney General Assembly that the observations had been completed for the 11,496 stars on the program by June 1, 1973, and that reductions were 75 percent finished for the stars within 43° of the celestial south pole and 60 percent complete for the others.

Improvement in the accuracy of fundamental star positions from pole to pole was a major international effort, of which the Pulkovo–Chile cooperation was an important part. The FK4 (Fundamental Katalog No. 4), published by the Astronomisches Rechen-Institut (ARI) at Heidelberg, was the best available in 1962. The goal of the international program was to provide data that would enable the ARI to produce an improved FK5. The observations made by the Pulkovo and Chilean astronomers at Cerro Calan revealed a large systematic error in the FK4 right ascensions (the east–west coordinate).[15] The corrections to the FK4 ranged from +0.01 to −0.03 arc seconds as a function of declination (the north–south coordinate), and were confirmed by comparison with work at other observatories.

A difference of opinion developed between Pulkovo and the US Naval Observatory (USNO) about which system should be used for publication of the Southern Reference Star Catalogue (SRS). Commission 8 at the Patras General Assembly (1982) reluctantly agreed to the publication of two catalogs, one from USNO referred to FK4 and another from Pulkovo referred to an improved SRS system. This was changed at the Dehli General Assembly (1985) in two recommendations from the SRS Committee: "1) That the Pulkovo Observatory and the US Naval Observatory should produce a single catalog of SRS positions which should be both completed and published jointly within a reasonable time in the FK5 system; 2) That the preliminary catalog of SRS positions should be referred to the system of the FK4 catalog and later transferred to the FK5 system." The SRS Committee reported at the Baltimore General Assembly (1988) that actions in accord with the second recommendation had led to the production of a preliminary catalog of 20,488 positions. The activities of the SRS Committee ended at this meeting, with thanks from Commission 8. Completion of the Basic FK5 by the ARI was reported at the Buenos Aires General Assembly (1991).

Jorge Alessandri, a conservative, was President of Chile when the Pulkovo astronomers arrived in October 1962. He was succeeded by Eduardo Frei, a Christian Democrat, who served as President from 1964 to 1970. Salvadore Allende, an avowed Marxist, was elected in 1970, and was overthrown by a military Junta in September 1973. The Pulkovo astronomers were immediately ordered to return to the Soviet Union, and this was the end of the Pulkovo work in Chile. Fortunately, the observations of the SRS stars had been completed, and the reductions of the observations, which were well advanced, could be finished in Pulkovo.

The momentum of the SRS program has continued, with new programs of meridian observations at Cerro Calan initiated by Claudio Anguita and his staff. A Danjon astrolabe has been operated at Cerro Calan in cooperation with ESO, and the Maksutov telescope has been kept busy by Carlos Torres and his associates. It is now more than 20 years since the first epoch plates of the 160 areas with galaxies were taken, and the time has come to start the second epoch.[16]

The installation of the Maksutov telescope on Cerro Robles was reported to Commission 9 (Instruments and Techniques) at the Brighton General Assembly of the IAU in 1970. At this time the USSR Academy of Sciences became interested in having a larger telescope in Chile, a duplicate of the 104-inch telescope of the Crimean Astrophysical Observatory.[17] Stock was invited to spend two months in the Soviet Union observing with this telescope and assisting in the planning for a site survey in Chile. "The two largest double-beam telescopes that have ever been built so far" had been used in the site survey for the 6-meter telescope and another site; and were available. After the equipment for the site survey arrived in Chile, Stock and Polojentsev visited a few sites in the north, including one north of Antofagasta. Stock helped choose several for testing, but had moved to Venezuela by this time and did not participate in the work of the survey. The list of new activities reported to Commission 9 at the Sydney General Assembly of the IAU 12 August 1973 includes: "Pulkovo and Crimean Observatories organized a joint site testing expedition to Chile." The political upheaval in Chile a month after this meeting led to the termination of the USSR site survey.

It should be noted in conclusion that the astrometric observing program took place during the cold war, and several members of the AURA Board expressed concern about the Soviet presence at Cerro Calan. It was necessary to inform the astrophysicists and administrators on the Board that this was international cooperation in real science, not politics or idealogy. The letter from Scott[18] was written at Edmondson's request, read by him at the September 22, 1962 meeting of the AURA Executive Committee, and included in the minutes as Attachment I.

18 The European Southern Observatory (ESO).

The definitive official history of the European Southern Observatory has been written by its second Director General (1970–1974), Adriaan Blaauw.[1] A French perspective is given in Chapters 20–26 (100 pages) of Charles Fehrenbach's history of French astronomy.[2]

On January 26, 1954 twelve leading astronomers from the Federal Republic of Germany (Heckmann and Unsold), Belgium (Bourgeois), France (Couder and Danjon), Great Britain (Redman), The Netherlands (Oort, Oosterhoff, and van Rhijn), and Sweden (Lindblad, Lundmark, and Malmquist) met in Leiden to discuss Walter Baade's suggestion, made seven months earlier during a visit to Leiden, that a joint western European observatory should be established in South Africa. They signed a statement that explained the need for more telescopes in the southern hemisphere, and that expressed their wish that the representative scientific organizations in their home countries would recommend to the qualified authorities the construction in South Africa of a joint observatory with a telescope of 3 meters aperture and a Schmidt telescope of 1.2 meters.[3] An Executive Committee (later called the ESO Committee) with headquarters at Leiden was formed consisting of one representative from each of the six countries: Heckmann, Bourgeois, Danjon, Sir Harold Spencer Jones, Oort, and Lindblad.

In the first exchange of letters between McMath and Struve concerning the agenda of the proposed NSF Advisory Panel for a National Astronomical Observatory, Struve told McMath[4] he had received a letter from Oort, dated February 1, 1954, that "brings the news that several European countries are now definitely interested in the creation of a South African observing station equipped with a large telescope, perhaps of the order of 120 inches, such as the one at Mount Hamilton." Oort's letter[5] asked for Struve's opinion about the possibility of assistance from private funds and mentioned the Ford Foundation. Struve replied on February 16: "I had recently a long talk with Dr. Hutchins of the Ford Foundation and he told me emphatically that his organization, according to its statutes, cannot support work in the physical sciences. I assume that it would even be difficult, and perhaps impossible, to induce them to support this project in the light of its international aspects." He also told Oort: "In the meantime I would like to mention confidentially that there is now under way in the United States (in the discussion stage) a project for the creation of a large national observatory under the auspices of the National Science Foundation." A month later Struve wrote to Oort about conversations he had with Waterman and Bowen, and the two letters were acknowledged by Oort on March 30, 1954. It is worth pointing out that the McMath NAO Panel was appointed effective April 1, 1954.[6]

The first meetings of the ESO Committee were held in Paris on November 8–9, 1954, and in Bergedorf on April 20–21, 1956. At this time it was decided to make an approach to the Ford Foundation emphasizing the international cooperation aspect of the proposed western European observatory. Oort wrote to Shepard Stone, the Ford Foundation's Director for European Affairs, on August 2, 1956:[7] "I am writing to inquire whether the Ford Foundation might be interested to assist in instigating this enterprise of Eurpoean cooperation. There is little doubt that a substantial financial support from your Foundation could be a deciding factor in setting the project on foot." Twelve days later the Secretary of the Ford Foundation replied:[8] "I regret to inform you that the Foundation is not in a position to respond favorably to your letter of August 2. . . . The Foundation does not have a program in the natural sciences and therefore we have been compelled to turn down many important requests in this area." This confirmed what Struve had written to Oort in February 1954.[9]

It turned out that Oort's inquiry was just a bit too early. Henry Heald, an engineer, became President of the Ford Foundation in 1956, and in 1958 he hired Carl Borgmann, President of the University of Vermont and also an engineer, to start a program in science and engineering. The emphasis was planned to be on engineering, and a hundred million dollars was the anticipated funding level. Astronomy became a beneficiary of this program, partly because Borgmann "had a personal interest in Astronomy", starting with his early association with Walter Roberts in Boulder, and "when it came to chances, I jumped at them; they [the Astronomy proposals] looked good."[10] Two months before Borgmann arrived, an informal inquiry by Lindblad in March 1958 received another negative response.[11]

British interest in ESO diminished as soon as Richard Woolley succeeded Sir Harold Spencer Jones as Astronomer Royal in 1956.[12] This was reflected in Oort's July 18, 1958 letter to Henry Heald[13] which said: "Discussions with Great Britain have thus far led to a negative result. Moreover, the list of ESO Committee members in this letter did not include a British representative. Oort had been encouraged to write directly to Heald by C.D. Shane and Otto Struve during a visit to the Lick Observatory in May. The three of them visited the Chancellor of the University of California, Clark Kerr, and he agreed to put in a good word for the project with Heald. Heald's reply to Oort[14] confirmed that Kerr had indeed communicated with him, and he agreed to meet with Oort and Lindblad in October following their attendance at a meeting of the International Council of Scientific Unions (ICSU) in Washington. Heald did not give Oort much encouragement, saying "In fairness, I think I should say in advance that it seems to me quite unlikely that the Ford Foundation will be able to participate in the establishment of the new observatory, but we shall be glad to visit with you anyway."

The Ford Foundation lost no time in starting to investigate the merits of the project that Oort had described. Ten days after Heald wrote to Oort, Borgmann asked his colleague, Paul Pearson, to make some inquiries in Washington, and warned "Make your inquiries discreet and do not indicate at all that we might have any interest."[15] Pearson's report was quite favorable.[16] Skapski sent Borgmann a five-page report covering: Purpose of the ESO Project, History of the ESO Project, Present Situation of the ESO Project, Evaluation of the ESO Project, and Conclusions.[17] He wrote: "The realization of such study by building the ESO Observatory is considered by both astronomers and physicists whom I interviewed, as one of the most worthy astronomical projects that could be devised in our time." The report included high praise from Niels Bohr.

An exchange of letters,[18] established 3:00 pm on October 9, 1958 as the time for the meeting of Oort and Lindblad with Heald, who wrote that he would ask Borgmann, Director of the Science and Engineering Program, and Stone, Director of the International Affairs Program, to join them in the meeting. Borgmann promptly wrote to Lindblad inviting Lindblad and Oort to have lunch with him before the meeting.[19] The letter began: "I feel that I have known you for many years, largely because Dr. Adam Skapski has told me so much of you and your work." He also invited Stone to join them for lunch. The meeting with Heald was cordial, but no promises were made.

A new problem was created when the French astronomers decided not to attend the October 31–November 1, 1958 meeting of the ESO Committee because their government had not contributed their share of the expenses of the ESO Committee in 1958 and would not be able to contribute financially in 1959.[20] The seven-year civil war in Algeria, which began in 1955, was very costly for the economy of France, and the Ministry of Finance was opposed to funding the expensive new projects, such as ESO. Oort felt that a grant from the Ford Foundation would be helpful in persuading the French Government to decide to participate in the creation of ESO.[21] Stone was told by Gaston Berger, Director of Higher Education in the French Ministry of Education and a close personal friend:[22] "There is no doubt that the support of the Ford Foundation would be most valuable in conquering the last hesitations (financial) and in accelerating the carrying out of the project."

The year 1959 started with a letter from Borgmann to Oort[23] requesting information about recent developments: "We are quite confused as to the real need of Foundation support and particularly as to the present status of official governmental participation in the cooperating countries. Are you in a position to give us a detailed statement covering as exactly as possible the present status of the Observatory? Such background information is seriously needed so that we can give proper consideration to your request." Oort's three-page reply[24] covered the situation in each country: Germany is "ready to go, even if France does not immediately take part in an agreement"; in Belgium "the way is clear to enter upon a preliminary agreement"; in Sweden and the Netherlands "the project has been officially approved and accepted by the research councils" but government approval has not yet been received owing to the financial situation; in France "It has, indeed, been the hesitation and the difficulties of France which, for several years, have prevented the realization of the project."

Borgmann's reply[25] expressed concern about conditions in South Africa, and asked if "consideration has been given to possibilities of locating the observatory in Australia or Argentina." He asked if there had been any preliminary contact with the South African government, and what is its attitude. Will ESO be associated with a local university or scientific group in the country where the Observatory is located? The letter concluded: "I believe we have the background on all other factors but need the above information in order to have the total picture on which to make a judgement."

Oort's three-page reply to Borgmann's letter[26] addressed the questions of choice of country and contact with official authorities. He said the political and social problems were recognized, but they were outweighed by the astronomical advantages. The four South African observatories "where work requiring good seeing was being done" had conditions "as good as those found in the large observatories in California." The conditions at Mount Stromlo in Australia and La Plata in Argentina "are much poorer than in South Africa, and certainly not good enough for the establishment of a large new observatory." The shorter distance from Europe also favored South Africa. The South African government had been approached "only in an informal way, through the Netherlands ambassador." There had also been frequent contact with the Director of the South African Council for Scientific and Industrial Research. The Ford Foundation staff continued to be concerned about South Africa. Pearson noticed an article in *The New York Times* about Gerard Kuiper's plans for an observatory in Chile,[27] and Borgmann called this to Oort's attention a month later.[28]

The recommendation for "Approval of an appropriation of $1 million for partial support of the construction and equipment of a major European optical observatory in the southern hemisphere" was approved by the Board of Trustees of the Ford Foundation on September 25, 1959.[29] The staff summary includes a statement that had been made several times by others: "An appropriation of $1 million by the Ford Foundation would, it is felt, be the catalytic agent for acceptance by four or more nations of a convention in which they would agree to provide the balance of the capital funds and all of the operating funds needed."

Borgmann telephoned Oort on October 1, and wrote a confirming letter the next day.[30] This was an important first step. $1,000,000 had been appropriated (i.e. set aside), and would be granted after the following three conditions were met:

(1) A convention must be authorized by the governments of at least four of the following countries: Belgium, France, German Federal Republic, the Netherlands, and Sweden

(2) The terms of the convention must include assurance of the balance of the capital funds; assurance of adequate operating funds; appropriate means of including other countries; a minimum of seven years notice of withdrawal by member nations; and prorating of national contributions on a basis related to national income.

(3) An agency must be established with authority to accept grants of funds for the purpose of constructing and operating an observatory in the southern hemisphere.

The promise of $1,000,000 by the Ford Foundation had the desired effect, and nine months later the French Government approved the first installment of funding for ESO.[31] After this problem was solved it took another two years and three months before agreement was reached on the terms of the Convention. It was signed in Paris on October 5, 1962 by the ambassadors of the foreign countries and by the Secretary-General of the Ministry of Foreign Affairs for France.[32] No financial commitments could be made until the Convention was ratified, and this took more than a year. The first meeting of the ESO Council took place in Paris on February 5–6, 1964, with Oort serving as President. The Ford Foundation was informed following the meeting that the three conditions for the grant had been satisfied.[33] There were some ESO delays in completing the formal paper work, and the $1,000,000 check was finally mailed to the bank on September 15, 1964.[34]

The Ford Foundation had accepted the ESO Committee's reasons for choosing South Africa at the time the $1,000,000 appropriation was approved. Nevertheless, there was continued concern as shown in a memorandum written by Pearson on March 24, 1960.[35] The final paragraph says: "Considering the reactions of the English, French and Belgians and recent statement of the State Department on the situation in South Africa, it may be appropriate for the Foundation to review its position relative to locating the observatory in South Africa. Pertinent to this is the report of the National Science Foundation that there is a location in Chile which affords equally as good 'seeing conditions' as does any known location in South Africa."

Oort wrote a long memorandum to the members of the ESO Committee on April 22, 1960[36] two months before the French Government agreed to participate in ESO. It included a statement about South Africa:

> The recent difficulties in South-Africa have of course, also come into discussion. Not so much in connection with the momentary troubles – which are likely to be transient – but because they enhanced once more the general risk involved in establishing in South-Africa an institution intended to be operated through at least several decades. Personally I do not yet believe this risk to be so large that we should abandon the plan of going to South-Africa. The same was in general felt by our German colleagues. We felt that we would be justified in continuing our test observations and provisional astronomical programmes in South-Africa.
>
> At the same time it would appear to be wise to keep an open eye for other possibilities. In Australia no evidence has yet been given that a really first-class observing site can be found fulfilling all requirements. On the other hand, recent seeing expeditions to Chile, organized, first by the Yerkes, McDonald and Santiago Observatories, and now transferred to AURA and the Chilean National Observatory, seem to indicate that first-rate astronomical conditions may exist in that country, and at not too great distance from Santiago.
>
> It would seem that we could well start with the designs and even construction of the large telescopes without having made a binding choice of the country, and that it would be wise to await further tests in South-Africa as well as in Chile during the coming two years before making such a decision.

Oort wrote a confidential letter to Shane a week later:[37]

> Dear Donald,
> As a consequence of the recent serious unrest in South-Africa, I (and several others connected with the project of the European Southern

Observatory) have anew considered the problem – which we had often considered before – whether we can assume the responsibility for establishing a large observatory in the Union of South-Africa. You and I have talked so much about this that there is no need to enumerate again the advantage there would be to establish the observatory in South-Africa, nor to discuss the risks involved. At present, on the one hand, the rapidity with which the risk has increased, and, on the other hand, the fact that favorable preliminary reports have reached me about Chile, have once more suggested the possibility of a change of our plans.

Oort asked Shane to send him the information then available about cloudiness and seeing at the sites being investigated in Chile. He concluded: "I should point out that I am asking this informally, and that the ESO Committee has not even discussed the possibility of considering Chile as a place for the observatory. For this reason I do not at this moment want to write to Rutllant or to any of the authorities in Chile."

Shane gave Oort a prompt reply.[38] He talked to Kuiper, and passed the information on to Oort. He also added: "Negotiations are underway to transfer the Chile Project from the University of Chicago, Texas and Chile to AURA." He welcomed the possibility that ESO might come to Chile, and said: "I believe if the ESO Committee is interested, we could cooperate in the matter of the site survey to the advantage of both groups." Shane wrote to Oort again on June 3:[39] "When I wrote to you May 6, I failed to mention the fact that I regarded your information about a possible interest of ESO in Chile as confidential and mentioned it only to Mayall, who will beyond question be the new director of KPNO. If I have authorization from you, I will tell Kuiper what the situation is and ask for copies of Stock's very complete and informative reports to be sent to you." Oort replied:[40] "there is no objection to your discussing the ESO with Kuiper or Mayall, as long as you ask them not to spread any rumor that the ESO Observatory might possibly come to Chile. At present that is no more than a vague possibility. I know by experience how quickly such rumors will spread and how soon they will then be considered as a practical certainty."

Correspondence during the next two years led to the formulation of plans for ESO observers to join the AURA site testing in Chile. In June 1962 Blaauw, the Secretary of the ESO Committee, wrote to Stock and Mayall:[41] "The ESO Committee now plans to send two observers to Chile to join in the AURA site tests for a period of about two months, probably from about October 15 to December 15, 1962. We do not intend to conduct an independent survey for suitable sites. The ESO Committee hopes that in due time the results of the AURA tests will become available to ESO. We therefore at the moment are rather interested in establishing a basis for the comparison between the scales of observations collected by AURA and those obtained in South Africa. We expect in this way to be able eventually to judge the relative merits of the Chilean and the South African sites."

Blaauw also wrote to Edmondson, the new President of AURA, about ESO's plans to send observers to Chile.[42] Two experienced members of the ESO site testing team, A.B. Muller and P. McSharry arrived in Chile in late November 1962.[43] They worked on La Peineta from December 6 to 19, and on Cerro Tololo from December 30 to January 13, 1963. Their general conclusions included: image quality was better than

AURA–ESO group on horseback at the saddle between Tololo and Morado.
On horseback, from left to right: Ch. Fehrenbach, O. Heckmann, Sr.
Marchetti, J.H. Oort, N.U. Mayall, F.K. Edmondson and A.B. Muller, 10
June 1963. Edmondson photo.

in South Africa, photometric quality was good, and long spells of clear weather
appeared to be a common feature of the climate in the Andes whereas they were rare
in South Africa. Muller's report[44] was submitted at the February 1963 meeting of the
ESO Committee, and it was decided that several members should visit Chile for further
investigation.

Five members of the ESO Committee visited Chile in June 1963.[45] Edmondson and
Mayall made a special trip to Chile to meet with this delegation: Oort (Chairman of
the ESO Committee), Heckmann (recently appointed to be the Director of ESO during
the construction period[46]), Fehrenbach (Chairman of the Instrumentation Committee),
Siedentopf (Chairman of the Site Selection Committee), and Muller. The dates, June
5–15, were chosen by Edmondson and Heckmann to avoid conflicts with already scheduled AURA and ESO activities.[47] Edmondson and Mayall arrived in Santiago on June
5; members of the ESO group had arrived a few days before this. The group visited
Cerro Calan on June 6, drove to La Serena on the 7th, and rode horseback to Tololo
on the 8th. A fine dinner was served by Senora Ramos when the group arrived at Los
Placeres. Oort, Heckmann, architect Marchetti, and Edmondson slept at Los Placeres,
while Stock and the others rode to the top of Tololo in the moonlight. The four rode
up on the morning of the 9th. A helicopter of the Chile Air Force, arranged by Heckmann through the German Embassy, arrived on the 10th and took the group first to
Cerro Morado, and then back to the place where the automobiles were parked. A draft
agreement between ESO and AURA was drawn up on June 6 at Cerro Calan and

revised on June 9 on Cerro Tololo.[48] The group spent the night of the 10th in Vicuna, the night of the 11th in La Serena, and drove back to Santiago on the 12th. The 13th was a Chilean holiday, a time for rest and relaxation after the previous five strenuous days. The concluding formal event was a reception given by the German Ambassador at noon on the 14th for the AURA–ESO visitors and their respective ambassadors.

Edmondson described the trip to Chile and other AURA activities in a letter to Shane.[49] He said: "These five men are firmly convinced that ESO should build in Chile. . . . Nick and I hope they can persuade the ESO Council in October that Chile is preferable to Africa." The ESO Committee met on July 23–24, 1963 and instructed Heckmann to start discussions with the government of Chile, and to pursue the negotiations with AURA. The final meeting of the ESO Council prior to the ratification of the Convention on January 17, 1964 took place in Bonn on November 15, 1963, and Oort wrote to Edmondson:[50]

> Dear Frank,
> Yesterday the provisional council of ESO has met in Bonn. As you know, the ratification of the ESO convention has not yet been completed in all of the countries concerned. We are, therefore, still working with a provisional council. This is not authorized to take the decision concerning the country where the European Southern Observatory will be established. Officially, therefore, no decision on this point has been taken. The provisional Council has, however, decided unanimously to recommend to the definitive Council that in the first meeting of this Council Chile be chosen as site for the Observatory. In practice we shall proceed with our activities on the assumption that ESO will be established in Chile. But this should not be publicized at the present moment.

A full report was given to the AURA Executive Committee on June 28th.[51] The draft AURA–ESO agreement was discussed, and the President appointed a committee to formulate a resolution welcoming the interest of ESO in establishing an observatory in Chile, and in particular on AURA land. The Executive Committee approved the resolution and directed the President to forward it to the National Science Foundation. Edmondson's letter to the new Director of the National Science Foundation[52] summarized the history of ESO, described the June 5–15 trip to Chile, and asked approval to communicate the resolution to Oort. Haworth gave his approval "with the express understanding, of course, that before final arrangements are entered into between AURA and ESO they shall first be submitted to the National Science Foundation for approval."[53] The resolution was sent to Oort on August 20 and was acknowledged on August 23.[54]

The AURA Executive Committee met in Washington on September 20.[55] Edmondson reported that a meeting had been held with NSF legal staff on September 18 to discuss how AURA could legally deal with a treaty organization like ESO. The questions that were asked in this meeting were sent to Heckmann with a request of a copy of the ESO Convention.[56] Copies of the Convention and related material were sent to Mayall and Edmondson,[57] and Heckmann answered the questions.[58]

Mayall told the Executive Committee that he had been informed by Dr. Horace Babcock just before the Annual Meeting of the AURA Board in April 1963 that the

Carnegie Institution of Washington was planning to seek private funds to finance a large telescope in the southern hemisphere, to be called the Carnegie Southern Observatory (CARSO). AURA's relations with CARSO will be described in Chapter 19.

Heckmann came to Tucson in October for an in-depth discussion of AURA–ESO relationships with Edmondson, Wildt, Mayall, and Stock. Minutes were taken by Julia Elliott, Assistant Secretary of AURA.[59] Several major points emerged from the discussions during the two-day meeting: (1) ESO would like to have the entire Morado area; (2) ESO strongly preferred purchase to lease; (3) ESO was negotiating a Convention with the Government of Chile, and expected to have diplomatic status and extraterritorial privileges similar to UNESCO's; (4) water supply would be a serious problem, and Heckmann suggested that the possibility of a pipeline from the Elqui River should be investigated; (5) CARSO's potential interest in Morado presented a new problem.

After dinner on the second day, Edmondson summarized the attitude of the AURA group in two points: (1) AURA would be prepared to reserve a temporary site on Tololo to enable ESO to start scientific work at the earliest possible date; (2) the group would recommend to the AURA Executive Committee the lease, but not the sale, of a part of Morado. Concern about the negative impact that ESO's extraterritorial privileges might have on AURA if the land were sold came up repeatedly in the discussions, and was a factor in this decision. Heckmann asked that these points be presented in a formal letter.

Finally, Heckmann asked if it would be possible for someone representing AURA to go to Europe for further discussions. Edmondson and Wildt said a time after January 1, 1964 would be convenient for them, and Heckmann felt a date before January 15 would be preferable.

Heckmann and Wildt continued informal discussions on Sunday, October 13, and Wildt told Edmondson:[60] "There was a marked difference between Heckmann's opinion and mine as to what had been accomplished . . ., and by the formal statement you made in the last session." Heckmann hoped that Edmondson's formal letter would bypass the issue of lease versus purchase. He also surprised Wildt with a promise that the draft of the ESO–Chile Convention would be submitted to AURA for comment and advice before it was signed by either party.

Heckmann summarized his impression of the Tucson meeting in a letter to Edmondson,[61] and repeated the unexpected promise he had made to Wildt about allowing AURA to comment on the ESO–Chile Convention before it was signed. The formal letter from Edmondson that Heckmann had requested[62] confirmed that AURA preferred lease to purchase of Morado, and included a statement that "any decisions made by AURA must be approved by the National Science Foundation before they become official."

Oort's letter of November 17[63] said the ESO Committee had reacted negatively to a long-term lease of Morado. Oort was also disappointed that AURA now wished to reserve part of Morado as a possible site for CARSO. He wrote: "Because of these various circumstances it was not possible for our Council to come to a decision regarding the collaboration with AURA. It was felt that further discussions would be required to obtain a sound basis for such a cooperation."

The meeting with Heckmann, Edmondson's formal letter to Heckmann, and the letters from Heckmann and Oort were discussed by the AURA Executive Committee on November 23, 1963.[64] Moreover, Oort had telephoned Mayall on November 21 to discuss the Minutes of the meeting with Heckmann, and Wildt and Edmondson were in on the conversation. Oort also promised to send Mayall a copy of the "Agreement between the Government of Chile and ESO" that Heckmann had signed in Santiago on November 6, 1963.[65]

The question of lease versus purchase had been considered by the Scientific Committee, and a general policy resolution was presented to the Executive Committee,[66] which unanimously voted to recommend it to the Board of Directors. It was agreed that Edmondson should send a copy of this resolution to Oort. It was also agreed that AURA should send representatives to Europe in January to discuss an agreement for cooperation for presentation to the AURA Board in March. At this time Miller announced that President Kennedy had been shot in Dallas. As reported in Chapter 14, the remaining business was finished as quickly as possible after lunch.

Edmondson and Mayall wrote to Oort following the meeting.[67] Oort's reaction to the AURA policy resolution was:[68] "This clarifies the position considerably. I think that ESO will now start to investigate other sites, if possible close to the AURA domain." Actually, the day after the ESO–Chile Convention was signed Heckmann viewed the mountains in the AURA region from a Chilean Air Force plane.[69] He invited Babcock and Stock to accompany him.

The AURA–ESO meeting was held on January 21–22, 1964 in the Conference Room at l'Observatoire de Paris.[70] Oort presided. Some of the ESO representatives felt that the Observatory should be on AURA land, and others felt ESO should look elsewhere for a site. The possible conflicting interests of CARSO and ESO were also discussed. Two documents were the final result of the meeting: a draft Cooperation Agreement, and four recommendations. These would be brought to the AURA Executive Committee on January 31 prior to the first official meeting of the ESO Council on February 5–6, 1964.

The AURA group visited Frankfort and Hanau, West Germany, on January 23 and 24 to discuss the production of a 150-inch quartz mirror blank with representatives of Hereaus (see Chapter 15).

The AURA Executive Committee approved the Paris recommendations in principle, subject to approval by NSF,[71] and this was communicated to Oort.[72] Oort's reply[73] said the ESO Council on February 4–5, 1964 had decided to build the Observatory in Chile. He added: "Naturally we were not able to take a decision on the site, as this problem will have to be studied in the country itself. For this purpose Drs. Heckmann, Fehrenbach and Rösch will go to Chile, where they will look at Cinchado and other mountains within the AURA domain and its general vicinity. Upon their return they will prepare a report for the Council containing a proposal for the definite site. We believe that it will be expedient to await the outcome of their expedition, before working out the agreement with AURA."

Heckmann left for Chile on March 18 and returned to Europe at the end of April.[74] Rösch had to leave Chile on April 4, and Fehrenbach arrived on April 6.[75] Mayall and Hiltner, in place of Wildt, arrived on April 8. The ESO and AURA representatives met on Tololo on April 12 and discussed an agreement for AURA to sell the entire summit of Cerro Cinchado to ESO.[76] Heckmann had some objections to the draft report of the discussion that Counsel Puga sent to him, and these were answered by Puga and Mayall,[77] Mayall also reminded Heckmann that NSF had the final approval on the terms of sale.

Rösch and Heckmann had explored the area to the north of the AURA domain before April 12. After the meeting Fehrenbach, Heckmann and Muller visited the five or six sites they had judged as most promising, travelling by auto, on horseback, and by helicopter.[78] Finally, Heckmann and Fehrenbach visited the summits of the two most promising mountains, La Silla north of Tololo and Guatulame to the south. Without the need for any discussion, they agreed that La Silla was the best place for the Observatory. This was in an area owned by the Government of Chile, and this would simplify acquisition. Indeed, as Fehrenbach recalls: "Heckmann is very enterprising; he pursued his contacts with the Chilean government and effectively obtained placing at our disposal the terrain of La Silla as well as certain advantages, notably exemption from certain taxes." The purchase of 627 square kilometers, including La Silla was consummated on October 30, 1964.[79]

Heckmann and Fehrenbach presented their report to the ESO Council on May 26–27, 1964, and Oort wrote to Edmondson:[80]

> Dear Frank,
>
> At the meeting of the ESO Council at the Observatoire de Haute Provence in the last week of May, it has provisionally been decided to start observations for the European Southern Observatory on the mountain La Silla, some 100 km to the North of the AURA domain. This is the mountain which, during the recent visit of ESO astronomers to Chile, had erroneously been called Cinchado North.
>
> Professors Heckmann and Fehrenbach had prepared a detailed report on the various possibilities they had investigated, amongst which the "real" Cinchado, in the AURA domain, had particular significance because of the closer connections with your group, which this site would offer. The possibility of disturbance by the nearby mountains to the North and rather rugged top part were, however, considered to present a disadvantage compared to some other mountains studied. La Silla, in particular, should be quite free from disturbing other mountains and has a surface which lends itself very well for the construction of an Observatory.
>
> Though it may reasonably be expected that seeing conditions on La Silla will be similar to those found on Tololo as well as near Copiapo, we naturally want to confirm this by observations on the site itself. Also the water problem requires further investigation. Therefore, the choice of La Silla is still provisional.

The replies from Mayall and Edmondson expressed their hope for continuing AURA–ESO contacts.[81] Seventeen years later, Oort recalled:[82] "But there were other considerations. On the part of some of the European governments it was felt that it would be wiser to have this observatory entirely independent. It would be independent financially in any case, but to have it also in appearance quite independent of the Observatory on Tololo. And for this reason, although we had initially thought that perhaps a site on Morado, a mountaintop near Tololo, which looked as though it might have equally good seeing conditions as Tololo itself, would be a possibility, it was later decided that it would be better to look entirely independently for a site that the European Observatory could own itself. And that is how finally the choice of La Silla came to be made."

The ESO Council met in Chile in March 1966 for the dedication of the road to the summit of La Silla.[83] In March 1969, the Council came to Chile for the second time for the dedication and celebration of the completion of the first stage of construction on La Silla and the headquarters building in Santiago. They visited Tucson and Kitt Peak on March 17, 1969 on their way to Chile. The dedication on La Silla took place on March 25. The President of Chile, Eduardo Frei M., participated in the ceremony, and more than 300 people attended. Mayall, AURA President Hiltner, AURA former President Edmondson and Mrs. Edmondson left the US on the last day of winter and arrived in Chile on the first day of autumn.

AURA had never considered a location other than La Serena for the CTIO headquarters. ESO made a different choice. The headquarters would be in Santiago and an ESO supply office would be in La Serena. The Government of Chile donated land next to the United Nations building, and construction began in early 1967. The impressive building was ready for use at the time of the 1969 dedication. A mansion had been purchased in March 1967 for use a Guesthouse. It was also used for administrative offices until the headquarters building was finished.

A major restructuring of ESO was started in 1974 when L. Woltjer became the third Director General of ESO. The headquarters was moved to Garching, German Federal Republic, and the building in Santiago was sold. Major new telescopes have been planned and funded in recent years, and ESO seems to be on the way to become the premier observatory in the southern hemisphere.

19 The Carnegie Southern Observatory (CARSO).

The Carnegie Institution of Washington was founded on January 28, 1902, and one of the first actions of the Trustees was the appointment of an Advisory Committee on Astronomy: E.C. Pickering (Chairman), Lewis Boss, G.E. Hale (Secretary), S.P. Langley, and Simon Newcomb. The Committee submitted its first report before the end of the year.[1] In one section the report pointed out that there were ten times as many working observatories in the northern hemisphere as there were in the southern hemisphere, and proposed a long-range plan: "The scheme for an observatory in the southern hemisphere presented in Appendix A may not be realized in full for many years to come; but we have thought that it may be possible to make the preliminary studies for its scope and location now."

Hale renewed the effort in 1925.[2] He wrote: "I tried several years ago to raise funds for this purpose, but I did not succeed." He said what was really needed was a 100-inch telescope, but if this or a new 60-inch could not be funded by CIW the Mount Wilson 60-inch should be moved to the southern hemisphere. Walter Adams prepared a budget for a new 60-inch the following year,[3] and continued to push for a Carnegie southern hemisphere observatory after he became Director of the Mount Wilson Observatory.[4] Other astronomers supported the project and urged inclusion of a very large telescope.[5] Adams sent a proposal for a 120-inch telescope to CIW President Merriam in September 1934, but the reply was not optimistic.[6] The economic depression of the 1930s meant that the top priority for the Carnegie Institution was to obtain $5 million from the Carnegie Corporation to add to its endowment.[7]

Mayall had worked with Hubble and Humason for two years, 1929–31, before returning to the Lick Observatory to finish his Ph.D. Hubble hoped to move the 60-inch to the southern hemisphere to extend his work on galaxies, and he told Mayall that he would recommend that he be employed after he got his degree to initiate the program, to take the telescope to the chosen site, set it up, and start observing.[8] Mayall recalled: "But everybody knows what happened at that time. This was nearly a couple of years after Black Friday on the stock exchange. ... And so they were unable to consider the southern hemisphere telescope project."

215

A third of a century later, the Trustees of the Carnegie Institution of Washington approved funds for a site survey for a southern hemisphere observatory at their May 1963 meeting.[9] This was a direct consequence of a visit to Pasadena by Merle Tuve in mid-December 1962. He had been urging all of the CIW research centers to make long-range plans for future growth, and his discussions with Bowen, Horace Babcock and Robert Leighton produced a list of six major projects for the Mount Wilson Observatory, including a "southern observatory." Bowen's follow-up letter to the President of CIW,[10] Caryl P. Haskins, described the ESO, Anglo-Australian and AURA activities and said: "I do not believe we should plan a station with a 60-inch or even a 100-inch as its major instrument but should at least match the 140 or 150-inch aperture of these observatories now planned." Three months later Olin Wilson wrote to Tuve:[11] "Bill Baum informs me that you would appreciate receiving support from the staff here in your discussions with the trustees." He made a strong argument for a CIW 200-inch telescope in the southern hemisphere, and wrote: "I have always admired George Ellery Hale's advice to his colleagues, never to think in small terms." He raised the question of how this might affect the future of the Mount Wilson Observatory, and said: "Whether the Institution would wish, or be able, to give the necessary support is something that would have to be decided on the basis of the development of the entire picture."

The American Astronomical Society met in Tucson on April 17–20, 1963, immediately after the Annual Meeting of the AURA Board, and this gave Babcock an opportunity to talk to Mayall and Stock about the plans for a Carnegie Southern Observatory (CARSO)[12] and to meet Rutllant.[13] He sent a detailed report of what he had learned about AURA and ESO to Haskins.[14] A short time later Mayall sent Babcock copies of the AURA–University of Chile Cooperative Agreement, a translation of the customs legislation, and Stock's Technical Report No. 2,[15] and Babcock sent copies to Haskins.[16]

Mayall gave Babcock a full report[17] on the June 1963 AURA–ESO meeting in Chile (Chapter 17). Babcock exchanged several letters about the plans for a CARSO site survey in Chile with Mayall, Stock and Miller, and there were further discussions when he came to Tucson for a meeting of the Advisory Committee for the Kitt Peak 150-inch telescope on September 14. The CARSO plans for a site survey in Chile were described to the Executive Committee on September 20.[18] Mayall brought Babcock up-to-date in a phone call on October 21, 1963 when he volunteered a large amount of information about: (a) recent changes at the University of Chile, (b) ESO and the October 11–12 meeting with Heckmann in Tucson, and (c) the Corning bid for the 150-inch quartz mirror blank and Miller's search for a less expensive source. Babcock's summary of this information for Haskins[19] filled three single-spaced typed pages. Mayall also had sent Babcock a copy of his notes on the meeting with Heckmann; he had second thoughts about the propriety of this and asked to have the notes returned.[20]

Mayall's October 21 phone call and October 24 letter did have a positive effect, in that Babcock finally became more forthcoming about the CARSO project in a two-page letter to Mayall.[21] A copy of their 58-page proposal[22] was enclosed with the letter, which happened to be written on the sixth anniversary of the incorporation of AURA. Babcock also sent a carbon copy of the letter to AURA President Edmondson. Mayall wrote to Edmondson:[23] "I must say that until I received this letter from Horace, and the lengthy written proposal, I had no clear or definite idea about their plans. If he had taken me into his confidence last April when I first learned about CARSO, I might not have advocated consideration of Morado so strongly for the ESO group." Mayall reported this new development to the Executive Committee on November 22, 1963.[24] The Minutes include: "Dr. Mayall then said that if the AURA group had known the scale and definiteness of CARSO's plans at the time of the discussions with Dr. Heckmann, these discussions would have taken a different tack." Mayall had been told just before this meeting that the reluctance to inform anyone of Carnegie's plans was due to the fact that the southern hemisphere project was not approved until the Carnegie trustees met on May 10. This explanation, of course, did nothing to solve the messy problem that had been created by CARSO's secrecy.

Bowen and Edward Ackerman (CIW Executive Officer) attended the meeting of the Executive Committee by invitation. Babcock was already in Chile starting the site testing, and Haskins (CIW President) had a schedule conflict. Bowen and Ackerman in a 20-minute presentation described the CARSO project, and said the observatory would have a duplicate of the 200-inch Hale telescope, a 48-inch Schmidt telescope, and one or two smaller instruments. They also said that $30 million dollars would be sought from non-government sources to construct and operate the observatory. After they left the meeting the Executive Committee continued its discussion of AURA's relations with ESO and CARSO. It was agreed that AURA would send representatives to Europe in January for discussions with ESO. It was also suggested that CARSO should be invited to send representatives to this meeting; Babcock and Ackerman attended as reported in Chapter 18. Word that President Kennedy had been shot in Dallas was received just before the lunch break, as reported in Chapters 14 and 18.

The January 21–22, 1964 meeting in Paris provided a forum for AURA, ESO, and CARSO to tell each other about their plans and answer questions.[25] It became clear that neither ESO nor CARSO wanted to share Morado with the other, or even with AURA at some future time. In retrospect, it is also clear that ESO had decided prior to this meeting to choose a site outside the AURA property.[26] The formal choice of La Silla by the ESO Council in May[27] opened the way for uncomplicated bi-lateral negotiations between AURA and CARSO.

Babcock's November 4–December 21, 1963 trip to Chile is documented in a 21-page report.[28] He arrived the day before Heckmann signed the controversial agreement between ESO and the Government of Chile,[29] and left a week after the Tololo road opening ceremony.[30] James N. Hanson, staff associate and field astronomer for the site survey, and his wife had arrived in early October.

After two days in Santiago and two days in La Serena, Babcock spent six nights on Tololo, November 9–14. On the 10th he walked with Stock from La Mollaca Pass (5000 feet), between Tololo and Morado, to the summit of Morado (7100 feet), and reported: "We actually smelled lunch cooking on Tololo, some miles away, and correctly identified the dish!" He walked with Stock, Hanson and Arturo Garrote to the summit of Pachon (8900 feet) on the 12th, and on the 14th rode horseback with Stock, Don Rogelio Ramos and Arturo Garrote to Collinda Pass (5900 feet) and walked the rest of the way to the summit of Cinchado (7050 feet). After the weekend in La Serena, he flew to Santiago on the 18th, and took possession of the GMC truck on the 19th at the dock in Valparaiso. Babcock and Hanson drove the truck to La Serena on the

20th. They learned of the Kennedy assassination two days later while in the office of the import agent, Sr. Jorge Wilson, in Coquimbo. A concert by the La Serena Philharmonic that evening "began with a rendition of the Star Spangled Banner, the entire audience standing."

Babcock went to Coquimbo on the 23rd to take possession of the air shipment of material for the site testing (13 pieces), and drove to Santiago on Sunday, November 24, to get the sea shipment, which had bypassed Coquimbo. He took possession of the big box on November 26 after it was cleared by customs officials at Cerro Calan, and Hanson drove it to La Serena on the 27th. Babcock flew to La Serena with Miller and five others on the 29th.

Babcock and Hanson spent three nights on Tololo adjusting and working with one of the CARSO ASMs (automatic seeing monitors). They made a trip to Copiapo and La Peineta on December 4–6 to dismantle the abandoned observing shed for re-erection on Tololo to house one of the ASMs. Babcock returned to Tololo for eight more nights, the last one following the road opening ceremony. He drove to Santiago on Sunday, December 15, and left for home the next day.

Babcock returned to Chile, accompanied by Ackerman, on February 10, 1964.[31] Their primary interest was to look at Pachon as a possible site for the 200-inch CARSO telescope, and for this purpose they made an overnight pack trip from Tololo on February 15–16.[32] They drove from Vicuna to the town of Hurtado, in the next valley to the south of the Elqui valley, on Monday, February 17. There they managed to find Sr. Manuel Penafiel, the owner of the land on the south slope of Pachon abutting the AURA property. They were accompanied by Sr. Igor Stanciç of the Puga law firm, who investigated land title records, and the title held by the present owner appeared to be remarkably clear. They visited Sr. and Sra. Penafiel on February 18 and made an offer for the entire estancia. This was declined, but it was understood that Sr. Penafiel would get in touch with the Puga law firm if he wished to consider the offer further.

Babcock left for Australia and New Zealand on March 17,[33] taking along an ASM. He found there was too much cloud cover in New Zealand, and two sites in Australia, Mount Stromlo and Siding Spring, did not have really good seeing. Tests of other sites in Australia were made by Daniel Crotty.[34] This work was terminated as soon as the superiority of the sites in Chile was demonstrated.

Hanson's work proved to be unsatisfactory[35] when he was not directly supervised, and he was recalled from Chile after less than six months on the job.[36] Fortunately for CARSO, a senior astronomer with strong feelings about the astronomical importance of the southern hemisphere was available. John B. Irwin replaced Hanson on June 8, 1964,[37] and was in charge of the site testing for the next three years.[38] Irwin and his wife Ruth set up housekeeping on Morado in a new house with modern conveniences. Access was on a spur road from La Mollaca Pass, built by AURA personnel and equipment at CARSO's expense. There were favorable reports very soon about the vigorous way Irwin had taken charge of the CARSO site survey.[39]

The AURA Board of Directors approved formal "Regulations for Temporary Occupancy of El Totoral" on March 10, 1964.[40] In accordance with this policy, Babcock applied for permission "for temporary occupancy of a site or sites on El Totoral for astronomical testing by the Carnegie Institution of Washington (CARSO). This request is for the period ending June 30, 1965, although we shall probably be interested in an extension." The paperwork was finished and permission was given a few days before Irwin arrived in Chile.[41]

The CARSO–Ford Foundation–AURA imbroglio during the period from December 1965 to April 1967 had little impact on the CARSO site survey. Babcock wrote to Mayall on April 11, 1967:[42] "The proposal of the Carnegie Institution for a large southern reflector is still quite an active one and is receiving our best efforts." He requested permission to continue the CARSO work on Morado for another year. Mayall consulted Wildt before sending a reply that gave formal AURA approval.[43] He also told Babcock: "Yesterday Randal Robertson informed me that the Chairmen of the House and Senate Appropriations Committees had each sent letters to Dr. Haworth giving their approval of the proposed schedule of funding for NSF's share." Matching the $5.0 million from the Ford Foundation was now assured.

There was some apprehension on the part of AURA and NSF about AURA's obligations to CARSO under the terms of the March 1966 agreement, and Miller asked Counsel Bilby for a legal opinion.[44] Bilby's reply[45] discussed three separate situations that could arise: (1) If CARSO failed to obtain firm funding for a 200-inch telescope by March 9, 1968 the agreement would be terminated. (2) If CARSO had a firm commitment for the funding, but failed to commence construction of the 200-inch by March 9, 1968 the agreement would be terminated. (3) If construction started on or before March 9, 1968, AURA would have a legal obligation to allow CARSO one-third of the observing time on the 150-inch telescope during the five-year period from March 9, 1968 to March 9, 1973 even though the 200-inch was not in operation. Babcock was worried for a different reason.[46] He was afraid that AURA might claim observing time on a CARSO 200-inch telescope in Chile that was funded without support from the Ford Foundation. He wrote: "Dr. Haskins might wish to consider whether or not a definite abrogation of the agreement should be arranged with the officers of AURA."

The Carnegie Institution of Washington continued to seek funding for a 200-inch telescope, and a slightly revised version of their proposal was sent to NASA.[47] The project was deemed to be worthy of support, but there was no way NASA could provide funds in the amount of $20 million. The proposal was also submitted to the Kresge Fund,[48] and an unsuccessful effort was made to organize a joint Canadian–Carnegie project for a 200-inch telescope in Chile.[49] A 40-inch telescope at Las Campanas was dedicated in 1971, and named for astronomer Henrietta Swope whose substantial 1967 gift had been used in site development and in the purchase of the telescope. An agreement with the University of Toronto allowed the latter to erect a 24-inch reflector nearby. Although funds for the 200-inch were not obtained, the Institution's astronomers urged the construction of a major instrument of lesser aperture, if only as an interim measure. A gift of $1.5 million by trustee Crawford H. Greenewalt and Mrs. Greenewalt made possible the completion in 1976 of the 100-inch Irénée du Pont telescope, named in honor of Mrs. Greenewalt's father. Designed by the staff of the Mount Wilson and Palomar Observatories, principally Bowen, Babcock, and chief engineer Bruce Rule, the du Pont telescope has proven a world-class instrument in all respects, especially in its wide-angle features.[50]

A meeting was held in Chicago on March 18, 1968 to discuss a plan for the long-term development of Morado that had been prepared by CARSO.[51] The question of lease versus purchase of Morado had been a problem for ESO, and this was still true for CARSO. AURA and NSF preferred a lease or a long-term easement, while Ackerman said the Carnegie Trustees would prefer purchase. He described Carnegie's historical lease problems with Stanford, Cold Spring Harbor and Mount Wilson, and said these were the reason for preferring purchase. After the meeting Babcock wrote to Mayall:[52] "My personal hope is that the various parties concerned will agree to an outright purchase of the specified property by the Carnegie Institution." Mayall's reply[53] included some new information: "In Chile, it now appears that there may be explicit legal restrictions or unfortunate consequences of the sale of any part of El Totoral." This was based on information that Blanco had reported to Mayall in a radio conversation, after he had consulted Counsel Puga, the Chilean Minister for Agriculture, and the Agency for Agricultural Reform (CORA).

Actually, Irwin had been instructed to look at sites outside of El Totoral shortly after the Ford Foundation tabled the CARSO proposal on March 15, 1966.[54] He drove north on July 2, 1966 to look at a mountain they called Campanita,[55] and he returned in September to climb to the top, a five-hour round trip.[56] Babcock climbed this mountain (correctly called Las Campanas) with Irwin when he visited Chile in October.[57] Shortly after this Irwin decided to return to the United States to continue his professional career. The Irwins left Morado on August 26, 1967,[58] and Donald L. Buck was appointed Project Supervisor.[59]

Representatives of AURA (Mayall, Blanco, Miller, Hiltner) and CIW (Babcock, Rule, Buck) met on Morado on May 23–24, 1968, and drafted a memorandum of understanding for the terms of a ninety-nine year lease.[60] Bilby polished it up for discussion by the Executive Committee[61] and transmittal to Babcock.[62] Babcock forwarded the draft to Washington to be reviewed by CIW officers and legal counsel.[63] He told Miller he had been in touch with Hiltner by telephone, mainly to discuss the question of sale versus lease. Hiltner wrote to Babcock[64] emphasizing that lease, not sale, was the policy of the AURA Board by a unanimous vote at the recent Annual Meeting.[65] Mayall sent him a copy of the "Open Door Policy" that the Board had approved.[66] Babcock's quick reply to Mayall[67] said CIW legal counsel had some questions about the lease agreement. His reply to Hiltner[68] said the CIW negotiations with the Canadian Government were still in progress, and he suggested postponing a little longer the detailed discussion of the lease agreement. Hiltner agreed to the postponement.[69]

The AURA Executive Committee met on Cerro Tololo on November 21, 1968, following a few days in Santiago. Queen Elizabeth was in Santiago at this time,[70] and her presence complicated life in the Hotel Carrera. The Queen and Prince Philip occupied rooms on the top floor, directly above the room occupied by the Edmondsons. Armed guards were stationed at the stairways leading to the top floor, and the elevators were blocked off from general use when she was leaving or returning to the hotel. On the afternoon of November 19 several members of the AURA group were standing near the ground level elevator door waiting to see the Queen leave the hotel. Edmondson happened to notice Babcock standing nearby, and this led to an invitation for Babcock and his two associates to join the AURA group for dinner. Babcock sat next to Mayall for almost two hours at the dinner, but did not mention that he had met with President Frei earlier in the day and had arranged for the Carnegie Institution to buy Cerro Las Campanas, a mountain outside the AURA property and a short distance north of La Silla. The members of the AURA group learned about this the next morning when they saw a front page story in *El Mercurio*.[71] At Hiltner's request, Blanco translated the article orally during the meeting of the Executive Committee.[72]

Mayall was especially hurt by Babcock's failure to confide in him during the dinner, because their long friendship began when Babcock was his first Ph.D. student at the Lick Observatory.[73] The full story finally became known to Edmondson and Mayall seventeen and a half years later when they read the historical essay in the "Carnegie Evening 1986" booklet.[74]

> One afternoon in late 1968 Horace Babcock was in Santiago, en route to the United States after a visit to Las Campanas. He had made up his mind to recommend to Haskins that Carnegie take the decision to go ahead at that site. Babcock was surprised to learn that an earlier effort to arrange an interview with Chilean president Eduardo Frei had now borne fruit, and that Babcock was on Frei's calendar for the next morning, November 19.
>
> Cordially received by the president, Babcock outlined Carnegie's five-year testing effort in Chile and said that the Institution hoped to enlist the president's interest and support. Babcock mentioned that the Institution might decide to seek property rights for the land around Las Campanas – nearly 100 square miles then owned by the Chilean government.
>
> To Babcock's surprise, Frei picked up the telephone and said to his minister of Land that he wanted an agreement to be rapidly completed to Carnegie's desires. Putting down the telephone, Frei told Babcock that "the land is yours. You can telephone to the United States to start construction of the telescope immediately," he said, and he jokingly volunteered to serve as the project lawyer.
>
> The unexpected events made Babcock uneasy, for he had been given no specific authority to commit the Institution to this direction. He was further startled when photographers and reporters appeared. Indeed, the Institution was embarrassed when AURA officials first learned of the matter from the Santiago newspapers.
>
> But there was no question of turning back from the remarkable opportunity. The purchase contract was smoothly negotiated, and in July 1969 the Institution found itself owner of a large piece of the Norte Chico. An agreement followed with the University of Chile for entry of Carnegie personnel and equipment and for future cooperation in using the telescopes; a Carnegie fellowship was established for the education of young Chilean astronomers in the United States. The Institution purchased a plot of land in La Serena for use as administrative headquarters. Site development began at Las Campanas: roads and buildings were built, the water supply was developed, and detailed atmospheric and geological testing were carried out – a prerequisite to choosing the precise sites for future telescopes.

This account makes it clear that Babcock's failure to tell Mayall about his meeting with President Frei was caused by his short-term panic for committing the Carnegie

Institution to the land purchase without specific authority from CIW headquarters in Washington.

Hiltner reviewed the November trip and the surprising CARSO development at a meeting of the Organization Committee on December 13, 1968.[75] At this time Miller read aloud Babcock's letter to Hiltner,[76] dated December 4, officially advising the President of AURA of CARSO's decision. After extensive discussion of the reaction to CARSO's decision, it was agreed "that although feelings initially were mixed, from an administrative point of view there was a sense of relief."

The CARSO decision was also reported to the AURA Board of Directors on January 23, 1969.[77] Hiltner reported that he had received a letter from Babcock, dated January 14, 1969, stating that the Carnegie Institution of Washington would like to give to AURA the access road and the larger frame house on Cerro Morado, and to remove the steel towers and prefabricated buildings. The AURA Board voted unanimously to authorize the requested removal. Buck wrote to Blanco on April 29:[78] "We have removed all CARSO equipment and buildings from the summit of Cerro Morado. . . . I thus advise you that CARSO has definitely abandoned the Cerro Morado summit site and consider the property, the mud house, the main house, one steel interferometer support (in place), a blasting cap storage, and the summit access road as delivered to you, as of the date of this letter." The letter concluded: "I wish to thank you for the fine assistance and cooperation that the AURA staff has given us during our stay at Morado." Thus ended the "shotgun marriage."

Ray Bowers' essay[79] concluded:

> Today, the Las Campanas observatory is in full operation. Astronomers routinely travel to Chile for periods of observation of a week or two; a committee of Carnegie and Caltech astronomers allocates telescope time at Las Campanas (and at Palomar). The dark skies of Las Campanas, the clear nights, the stability to the atmosphere, all have equalled the highest hopes. Superior locations remain available for future telescopes on the Las Campanas ridge and on several nearby peaks. Relations between the astronomers and successive Chilean governments have remained excellent. Carnegie's Las Campanas observatory, along with the Cerro Tololo Inter-American Observatory and the European Southern Observatory at La Silla – all headquartered at La Serena – today make Chile the leading center for observational astronomy arguably of both hemispheres.

In 1980, four years after the dedication of the du Pont Telescope, Cal Tech and the Carnegie Institution ended their 22-year-old arrangement for unified operation of their observatories, called since 1969 the Hale Observatories.[80] Cal Tech owned the 200-inch Hale Telescope on Mount Palomar and the Big Bear Solar Observatory at Big Bear Lake in the San Bernadino Mountains. The Carnegie Institution owned the Mount Wilson Observatory and the Las Campanas Observatory. On June 29, 1984, the Carnegie Institution announced its intention to reduce its operations at the Mount Wilson Observatory and to concentrate its astronomical resources in the operation of the Las Campanas Observatory.[81] This followed a three-year assessment of the Institution's priorities in astronomy.[82] The Mount Wilson Observatory has been operated since 1989 by the Mount Wilson Institute, a non-profit organization created for this purpose.[83]

The Carnegie Institution, the University of Arizona and the Johns Hopkins University formed a mini-consortium on October 1, 1986 to build and operate an 8-meter telescope at the Las Campanas Observatory, calling it the Magellan Project.[84] Hiltner served as Project Manager from the time of his retirement in 1986 until shortly before his death in 1991.[85] Hopkins was forced to withdraw in April 1991 for financial reasons, and Carnegie and Arizona have decided to go ahead with construction of a 6.5-meter telescope.[86] Casting of this mirror took place in early February 1994 at the Steward Observatory Mirror Laboratory in Tucson.[87] The mirror has a very fast focal ratio, f/1.25, and the mounting is of the "alt-azimuth" type. The dome is located on Manqui Peak, about 110 meters higher than the site of the du Pont Telescope.

Part V
Epilogue

20
An active planetary research program replaces the large orbiting telescope, and the Space Division is renamed the Planetary Sciences Division in 1969. The third Director of KPNO takes office in 1971, and the Corporate Office is established in 1972. Inadequate funding for KPNO causes the AURA Board to terminate the rocket program in 1973. The dedication of the 150-inch Mayall Telescope on June 20, 1973. Taking over the management of the Sacramento Peak Observatory (SPO) in 1976, and the merger of the solar programs at KPNO and SPO in 1983 to form the National Solar Observatory (NSO). The subsequent merger of NSO with KPNO and CTIO to form the National Optical Astronomy Observatories (NOAO). The Space Telescope Science Institute (STScI).

Joseph W. Chamberlain was appointed Associate Director – Space Division of KPNO on March 12, 1962,[1] eight and a half months before Cerro Tololo was chosen (Chapter 14) and more than a year before David L. Crawford was appointed Project Manager for the 150-inch telescope (Chapter 15). Chamberlain's letter of acceptance[2] made clear his vision of the role of the Space Division: "The emphasis should always be on the research problems, rather than on instrumentation." He started to build a research staff that would use space vehicles for research in solar system astronomy by making one senior level and four non-tenure appointments.[3] A special AURA–NSF meeting was held on August 13, 1962 to discuss the new goals of the Space Division and "NASA relationships". The minutes of this meeting[4] include comments about the possible use of sounding rockets for research by the KPNO Space Division. Funds for sounding rocket systems and rocket experiments were in the Space Division Budgets for FY62, 63, 64, and 65, but no funds were requested for the orbiting telescope.

Russell A. Nidey, an engineer hired by Meinel in March 1959,[5] became Space Systems Manager and planned the first KPNO Aerobee rocket flight, which took place at the White Sands Missile Range on April 14, 1963.[6] This flight was a failure. "The nose cone did not come off, the rocket spun in the direction opposite to that which had been expected, and the parachute recovery package failed to operate." The next KPNO Aerobee flight, on June 25, was successful, and gave useful air glow data.[7] The third flight, on November 4, was another failure.[8] The fourth flight, on April 7, 1964, was a success.[9] The success of the fifth flight, on June 19, was especially gratifying because it carried "the first instrumentation subsystem to have been designed and built entirely within the Observatory".[10] The sixth flight, on September 15, was a partial success, and concluded Phase I of the rocket program.[11] "Of the six launched to date, two yielded no scientific data, three gave excellent data, and one is expected to yield 25 percent of the hoped-for-data. Thus the flights have so far been somewhat better than 50 percent successful, perhaps slightly better than the national average." The rocket program produced some good science, but it was mostly the work of staff. Very few visiting scientists made use of the KPNO rockets, in contrast with the large demand for time on the KPNO telescopes. The termination of the rocket program in 1973 for financial reasons will be discussed later in this chapter.

In 1968 Chamberlain suggested that the Space Division might change its name to Planetary Sciences Division.[12] This received comprehensive discussion as it went through channels,[13] and was approved by the AURA Board at the 1969 Annual Meeting.[14] Chamberlain resigned as Associate Director effective June 30, 1970,[15] and left the Observatory on May 31, 1971 to become Director of the Lunar Science Institute, in Houston.[16]

Mayall would reach the mandatory retirement age of 65 in 1971, and a Special Nominating Committee to recommend a candidate or candidates for appointment as the third Director of KPNO was appointed in 1970 with members: G.B. Carson, W.A. Hiltner (ex officio), O.C. Mohler, R. Lorenz, G. Wallerstein, R. Wildt, and A.E. Whitford (Chairman). Whitford wrote to directors of observatories and chairmen of Ph.D.-granting astronomy departments asking for nominations and also for comments about: (a) the most desirable direction of development of KPNO and CTIO in the 1970s, and (b) the requisite personal qualities, background and experience of a candidate for Director.[17] Copies of 31 replies received by the Committee, with the recommended names blocked out, were sent to the members of the AURA Board for use in policy discussions at the 1971 Annual Meeting in January.[18] The position of Director of KPNO was offered to Goldberg, and his acceptance, effective September 1, was announced at the March 16, 1971 meeting of the Executive Committee.[19] The Executive Committee handled this by extending Mayall's appointment for 3 months beyond his mandatory retirement date of June 1.

A symposium honoring Mayall had been held on his 65th birthday, May 8, 1971.[20] Invited papers were presented by W.W. Morgan, Margaret Burbidge, R. Minkowski, and A.R. Sandage. A 200-year old Chilean Tinaja from Vicuna was a surprise birthday present from the CTIO staff. Transportation costs by land and sea for the 600-pound earthen jar were paid by the incumbent and past presidents of AURA.

The Executive Committee met a little more than a week after Mayall retired.[21] The second item on the agenda was a resolution of appreciation for Mayall's services as Director, ending: "NOW THEREFORE BE IT RESOLVED, that as a token of appreciation for the dedication of Dr. Mayall to the future of the Observatories, the 158-inch telescope on Kitt Peak shall be known as THE NICHOLAS U. MAYALL

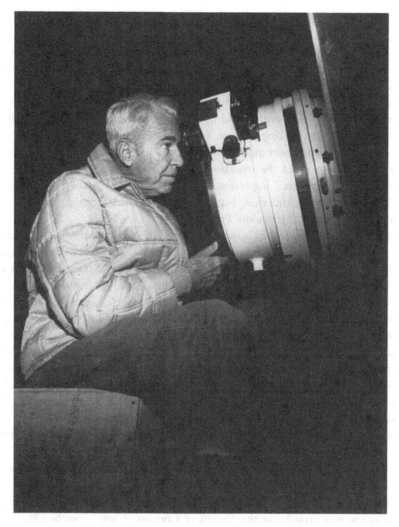

Nicholas U. Mayall observing at the prime focus of the 150-inch (4.0 meter) telescope, 2 March 1973. Following Mayall's retirement, it was dedicated on 20 June 1973 as the Nicholas U. Mayall Telescope. NOAO Photo Archives.

TELESCOPE." It was signed and sealed by Wildt, who began his second term as President of AURA at the 1971 Annual Meeting in January.[22]

Changes that would have a major impact on AURA and the Observatories were already under way at the National Science Foundation. Haworth's six-year term as Director ended a year after Richard Nixon became President, and Nixon did not reappoint him for a second term as Eisenhower had done for Waterman. William D. McElroy, Chairman of the Johns Hopkins Biology Department, was nominated by President Nixon and took office in July 1969.[23] This was coincident with a major reorganization of NSF proposed by the Daddario Committee and passed by Congress.[24] A Deputy Director and four new Assistant Directors, in addition to the Director, would be presidential appointments. KPNO and CTIO would be under the Assistant Director for National and International Programs, and would report to the Office for National Centers and Facility Operations.[25] Dr. Thomas O. Jones, formerly in charge of the NSF Antarctic Program, was appointed Deputy Assistant Director, and served as Acting Assistant Director during the nine-month period before Dr. Thomas B. Owen, a retired Rear Admiral (1970), was appointed Assistant Director in July 1970.[26] During that period Jones "began the long and difficult job of putting [in place] adequate Foundation management of these centers and trying to set their operations up on a slightly different basis than just making a grant to the organizations; the Foundation would actually be involved with them quietly and privately in the management so as to keep a firmer hand and to avoid the objections of both the Congress and OMB on the management of these centers." Representatives of National and International Programs, Daniel Hunt and Gerald Anderson, began to attend AURA meetings with Fleischer or Harold Lane from the Astronomy Program.[27]

Up to this time the relationship between NSF and AURA had been one of colleagues working together to bring about mutually desired results. The bureaucracy created by the Daddario reorganization changed this relationship to one of micromanagement by NSF and "bureaucratically correct" lines of communication. Wildt had met with Hunt the evening before the June 24, 1971 meeting of the AURA Executive Committee.[28] His report to the Executive Committee included: (a) NSF was not willing to recognize the present mode of operation, according to which Mayall and Miller conducted the day to day business with the Foundation; (b) Wildt had explained in vain that Mayall and Miller were Officers of the Corporation; (c) Hunt said that if there were a Chairman of the Board, Dr. Owen, Assistant Director for National and International Programs, would communicate with him. Hunt would communicate with the President, and Anderson would communicate with the Observatory Director and others as he saw fit. In response to a question, Miller said there were 49 people in NSF with whom the Observatory had been communicating.

Miller was asked to prepare a report outlining the problems of communicating with NSF and the administrative problems that had arisen. His report, 19 pages and five diagrams, was discussed by the Organization Committee on July 22, 1971.[29] The report of this meeting was brought to the Executive Committee on November 16.[30] Wildt then reported his discussions with NSF officials on September 24, and stated that it was his firm conviction that AURA could no longer operate without a full-time paid president, and Goldberg concurred. Counsel Richard Bilby distributed a proposed change in the by-laws to create the new position, and it was voted to bring this to the 1972 Annual Meeting of the AURA Board for action. Owen had assured Wildt that NSF would increase the management fee to cover the additional costs of a corporate office with a full-time paid president.

The amendment to the by-laws creating the full-time office of President of AURA was approved by the AURA Board at the 1972 Annual Meeting.[31] Wildt then asked for and received approval to negotiate a five-year contract with Lee, currently the University of Chicago administrative member of the AURA Board. Lee's acceptance was reported to the Executive Committee on March 7.[32] Lee and Goldberg, two of the three incorporators of AURA,[33] now held the two top positions in AURA and KPNO. Wildt and Lorenz were reelected for a second year at the Annual Meeting, but their titles were changed to Chairman of the Board and Vice-Chairman of the

Board. Clarence M. Black, the AURA Assistant Treasurer, was elected Acting Secretary of AURA, and Luella J. Ellis, formerly Mayall's secretary, was elected Acting
Assistant Secretary of AURA. Resolutions thanking Miller and Elliott for their years
of service as Secretary and Assistant Secretary were passed unanimously.

A personality conflict between Goldberg and Miller developed shortly after Goldberg
became a member of the AURA Board in 1966. It strengthened after Goldberg became
KPNO Director on September 1, 1971, and in June 1972 he recommended that Miller's
appointment be terminated.[34] The Executive Committee agreed, and voted to authorize
Goldberg to work out termination arrangements to the extent of full pay for one year,
including the usual staff benefits. This would be charged to contract funds to the extent
agreed by the President and NSF. This was a sad ending for a man who had played
an important and powerful role in building KPNO and CTIO.[35]

Elliott continued on the KPNO staff as Secretary to Harry Albers, Miller's successor,
and left in January 1973 to take a position with Transamerica Corporation in San
Francisco.

1973 was an eventful year. KPNO–CTIO were reorganized, the rocket program was
terminated, and the 150-inch Mayall Telescope was dedicated. The American Astronomical Society chose Goldberg to give the 1973 Henry Norris Russell Lecture,[36] and
he was elected for a 3-year term as President of the International Astronomical Union
(IAU).[37]

Goldberg did not like the divisional organization of KPNO, feeling that the Observatory should be viewed as having a single scientific staff.[38] He proposed that this staff
should be organized into several scientific programs, with the present Associate Directors becoming Program Directors. The AURA Visiting Committee's first report had
expressed a similar point of view.[39] The Organization Committee, with delegated authority, approved the reorganization of KPNO and CTIO as presented by Goldberg and
Blanco on May 2, 1973.[40] This was quickly implemented and was reported at the next
meeting of the Executive Committee.[41] The AURA Board received a report of these
changes at the 1974 Annual Meeting.[42]

The role and cost of the rocket program had been called into question several times
over the years by AURA and by others.[43] An ad hoc Committee on the Rocket Program
recommended to the AURA Executive Committee that the rocket program should be
continued if at all possible, but an attempt should be made to obtain a separate contract
with NSF to support it.[44] This was rejected by NSF.[45] Termination of the program
was discussed by the Scientific Committee[46] and by the Executive Committee,[47] and it
was agreed that the rocket program would be phased out if the FY 1974 KPNO budget
dropped below $8.6 million. The final figure from NSF was $7.8 million, and the
rocket program was eliminated from the FY 1975 budget. The rocket program was
terminated effective October 31, 1973.[48]

Installation of the Nicholas U. Mayall Telescope had been completed in February,
with "first light" seen by Mayall, Crawford, and Hoag on February 27, 1973.[49] A "first
light" press conference was held on the afternoon and evening of March 5, in spite of
cloudy weather. The dedication took place on June 19–20, 1973.[50] The program began
on June 19 with a reception and buffet at the KPNO headquarters, followed by a
symposium ("Copernicus + 500 + ") at the University of Arizona Auditorium. The
speakers were Jesse L. Greenstein, Frank Drake, and Sir Fred Hoyle. The dedication

Pier of 158-inch (4.0 meter) Mayall telescope. The domes of the Steward Observatory 36-inch (0.9 meter) and 90-inch (2.25 meter) telescopes are at the left. NOAO Photo Archives.

ceremony began on Kitt Peak at 11:00 a.m. on June 20, and the dedication address was given by the recently appointed Director of the National Science Foundation, H. Guyford Stever. The 200 persons in attendance included 18 from foreign countries, 11 members of the National Science Board, 22 NSF staff, three NASA staff, and a staff member from the House Committee on Science and Astronautics.

"On July 1, 1973, the formal design and construction project, as a separate Observatory program, came to an official end, and a program of operational testing was begun. Crawford stepped down as Project Manager, and the Mayall Telescope was assigned to Stellar Programs under Hoag. [Roger] Lynds became the astronomer-in-charge of the telescope."[51]

Goldberg's election as President of the IAU was a high honor, but it was partly

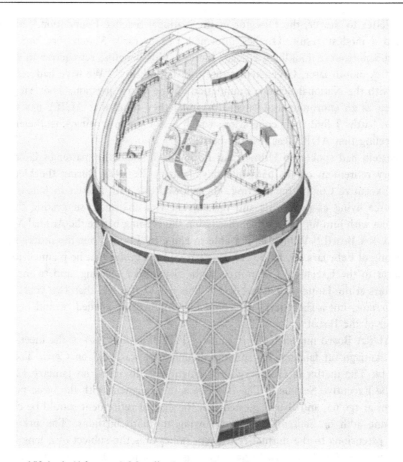

158-inch (4.0 meter) Mayall telescope dome-diagram.

responsible for the conflicts that developed in the last four years of his directorship. The time required by his IAU position prevented him from giving full-time to his duties as Director of KPNO during the period August 1973 to August 1976.[52] The authority he delegated to the Assistant to the Director led to problems with the KPNO user community and with the KPNO staff. Conflicts also developed with Lee and with Greenstein, who was elected Chairman of the Board at the 1974 Annual Meeting.[53]

Salary increases were on the agenda of the June 27, 1974 meeting of the Executive Committee.[54] Lee reported at the next meeting[55] that NSF had not yet approved the proposed salaries for Goldberg or Blanco. "The discussion concluded with the request that the Chairman and the President discuss the matter further and continue working with the National Science Foundation officials." There was a new NSF official for the President to communicate with; Dr. Robert E. Hughes had been nominated to succeed Owen as Assistant Director for National and International Programs.[56] NSF finally approved the salary raise for Blanco, effective July 1, 1974, but took no action on Goldberg's salary. In spite of this, the Executive Committee recommended further increases for Goldberg and Blanco, effective July 1, 1975,[57] but NSF had not acted on either of these as late as January 28, 1976.[58] Greenstein thereupon wrote a strongly worded letter to Stever, the Director of the National Science Foundation,[59] and this produced a modest result. Hughes informed Lee in early March that Stever had approved a portion of Goldberg's salary increase plus benefits, retroactive to January 1, 1976.[60] A month later, Greenstein wrote to Edmondson:[61] "We have had very poor success with the National Science Foundation on the salary question. They view their concession as an enormous and satisfactory one; they think that AURA has won an important battle. I find their concession, which is not yet in writing, insufficient, and have a feeling that AURA has lost the battle."

Greenstein had spoken to Hiltner and Edmondson about the matter of Goldberg's mandatory retirement on his 65th birthday, January 26, 1978, during the December 3, 1974 Executive Committee meeting. He followed this with a letter in January 1975 to the three living ex-presidents and the current vice-chairman, suggesting that this group meet with him for an informal discussion the evening before the Annual Meeting of the AURA Board.[62] Wildt was not able to come to Tucson, but the others met on the evening of February 25, 1975.[63] Greenstein told the group that he planned to bring the matter to the Executive Committee at the September meeting, and to the Board of Directors at the January 1976 meeting.[64] This would give less than two years to find a new director, but a short-term extension for Goldberg, if needed, would be within the power of the Board.

The AURA Board met in Chile in January 1976. The first part of the meeting was held in Santiago on January 12, and the second part was held on Cerro Tololo on January 14. The matter of Goldberg's retirement was discussed on January 12 at the end of the Executive Session.[65] The Board was in agreement with the basic policy of retirement at age 65, and also suggested that the date of retirement should be changed to the June 30th or September 30th following the 65th birthday. The problem of granting extensions to the normal retirement policy was the subject of a lengthy discussion, and the Board expressed unanimous agreement to adhere to the normal retirement policy. The Goldbergs were delayed by bad weather in Cuzco, after a visit to Machu Picchu, but had been present on Cerro Tololo for the final part of the Board meeting.

Lee communicated the sad news that Rupert Wildt had died on Friday, January 9. The Board asked Lee to send an appropriate expression of condolence to Mrs. Wildt.

The festering salary problem combined with the initiation of the search for Goldberg's successor resulted in the harassment of Greenstein and Lee by people who were persuaded by Goldberg or by some of his friends to telephone Greenstein and Lee. Greenstein got fed up and wrote to Goldberg:[66] "This letter is to say that I would appreciate some relief from the rather continuous pressure being generated." Lee said[67] the harassment he had endured had been "unbelievable since returning from Chile." Margaret Burbidge, a member of the AURA Board, had been engaged in discussions with a small group of Board members ("several of us"), and wrote to Greenstein:[68] "I believe that Gil Lee has a five year contract and the Board should decide whether it wishes to reappoint him as President at the end of his first five-year term. I believe that he is in the fourth year of his contract. This being the case, we should be prepared to act at the May meeting." In response, Greenstein called a Special Meeting of the Board of Directors on May 3, 1976 to precede the scheduled meeting of the Executive

Committee on that date.[69] Greenstein wrote to Edmondson:[70] "It is not my intention to reopen to the full Board meeting the question of the retirement date of the present Director. I will, however, moved by various considerations, bring up for discussion what use, if any, the Board wishes to make of the present Director's talents on his retirement from administrative office."

The Special Meeting of the Board of Directors[71] began with approval of an amendment to the NSF contract providing for AURA to operate the Sacramento Peak Observatory (SPO) on an interim basis for a period of 15 months, effective July 1, 1976. Jack B. Zirker was appointed Acting Director. The Board approved the appointment and membership of a Search Committee for the KPNO Observatory Director in the Executive Session. The Board also discussed and unanimously approved extension of Lee's contract beyond May 1, 1977, but only for three years. The Organization Committee presented an interim report, and was asked to continue its study of a long-range corporate structure that could accommodate added responsibilities, such as the permanent management of SPO.

Another Special Meeting of the AURA Board took place on December 3, 1976[72] during a long recess in the scheduled meeting of the Executive Committee.[73] Lee attended the morning part of the Executive Committee meeting, but Greenstein had excused him from attending the Board meeting. The Board meeting began with a discussion of the Report of the Visiting Committee transmitted November 18, 1976, and three members of the Committee were present (James W. Clark, Martin Schwarzschild, E. Joseph Wampler). The Organization Committee had been discussing some of the issues raised by the Visiting Committee, and was in substantial agreement with many, but not all, of the Visiting Committee recommendations. It was agreed that the Executive Committee should be reduced in size, that the President of AURA should be a scientist, and that the required revisions of the by-laws would be presented at the 1977 annual meeting. The Board approved a motion that the Chairman should appoint a Search Committee for a scientist President of AURA. The Executive Committee discussed the manner in which information about the reorganization plan would be given to Lee at the end of the afternoon Executive Session. Greenstein said he would meet with Lee to inform him.

While all of this was happening, there were some important changes at the National Science Foundation. Goetz Oertel replaced Fleischer as Head of the Astronomy Program shortly after the 1975 Annual Meeting, and he attended the June 3 meeting of the Scientific Committee and the June 4 meeting of the Executive Committee.[74] An internal reorganization in the fall of 1975 made a major change in the structure of the Directorates, and Hughes became NSF Assistant Director for Astronomical, Atmospheric, Earth, and Ocean Sciences.[75] This was an improvement because it placed astronomy entirely under one Assistant Director. Oertel thought he had been appointed to be in charge of the NSF Astronomy Program, and he left NSF after Hughes appointed Hunt as Acting Director of the Division of Astronomical Sciences pending the selection and appointment of a Division Director. Stever left the Foundation in August 1976 to become Director of the President's Office of Science and Technology Policy (OSTP),[76] and Richard C. Atkinson served as Acting Director of NSF until he was confirmed as Director in 1977.

Code was elected Chairman of the Board, and Rossi was elected Vice-Chairman at the 1977 Annual Meeting.[77] In attendance was William E. Howard III, the person finally appointed as Director, Division of Astronomical Sciences.

An important item of business was discussion of a letter that Goldberg had written to Chairman Greenstein on December 9, 1976 requesting that he be relieved of his duties as Observatory Director on or before September 30, 1977, a year before his mandatory retirement. The Scientific Committee had passed a resolution asking Goldberg to reconsider, and the Executive Committee tabled this until the Board had received the report of the Search Committee. Hiltner reported that the Search Committee had arrived at a candidate who stood very high on the selection scale set by the Committee, and the candidate had been approved by NSF. The Board approved the recommendation of the Committee that an offer should be made to the candidate.

The tabled motion was then divided into two parts:

(1) The Board of Directors should invite Goldberg to accept an appointment as staff scientist at KPNO from September 30, 1977 to September 30, 1980. The motion passed on a written ballot, 17 in favor, 8 opposed.

(2) The Board of Directors should ask Goldberg to delay his resignation as Observatory Director until a new Director can assume the position, or until September 30, 1978. The motion failed on a written ballot, 10 in favor, 14 opposed, 1 abstention. The Board then voted to accept Goldberg's request.

Lee left the meeting before the Search Committee for President presented its report. Chairman Pings said advertisements had been placed in *The Chronicle of Higher Education*, *The Wall Street Journal*, and *The National Observer*. The Chairman said he had also written letters to the Presidents of AAU Universities, the AURA Board, and the AURA Visiting Committee asking for nominations. 63 applications and 11 nominations had been received.

The Search Committee for President presented its report and a recommendation at a Special Meeting of the AURA Board on September 16, 1977.[78] Pings reported that the position had been declined by three senior distinguished scientists. The committee had then looked at a group of science-administrators, and picked Dr. John M. Teem as their unanimous choice. Following discussion of the proposed employment contract, the Board voted to offer the presidency of AURA, Inc. to Teem effective on or about October 1, 1977.

The Board then discussed Lee's contract with AURA, and it was voted, with five abstentions, to authorize the Chairman to relieve Lee of his responsibilities as President of AURA, Inc. effective October 1, 1977, or to take whatever steps were necessary after consultation with legal counsel to effect termination of his appointment. Teem was then called into the meeting by Chairman Code, who asked for his comments about the challenges of the position.

Nine months before this, the AURA Board had extended Lee's contract to May 1, 1980,[79] and he continued on the AURA payroll in a subordinate position for eight months, until he accepted the position of Assistant Chancellor for Administrative Affairs at the University of Wisconsin-Milwaukee starting in June 1978.[80]

The beginning of Teem's term as President of AURA coincided with the end of Goldberg's directorship of KPNO, and the Search Committee had not yet been successful in finding a new Director. Blanco had reluctantly agreed in June to serve for no longer than one year as Acting Director of KPNO,[81] and he arranged that during his absence John Graham, Patrick Osmer, and Barry Lasker would serve as Acting Director of CTIO on a rotating basis.[82] The Search Committee recommended Geoffrey Burbidge to be KPNO Director, and the AURA Board appointed him for a five-year term, effective November 1, 1978, at the 1978 Annual Meeting of the AURA Board.[83] Fred Gillett served as Interim Acting Director for eight months, after Blanco returned to CTIO following the Board meeting.

Teem did not lack for major organizational problems to deal with as soon as he took office: (1) the change from interim to permanent management of SPO by AURA; (2) reorganization of the management structure of the Observatories to include SPO; (3) reorganization of AURA that was made necessary by enlarging the membership; and (4) the decision to compete for management of a NASA-funded Institute to handle the scientific output of the space telescope.

(1) Management of the Sacramento Peak Observatory by AURA

The Sacramento Peak Observatory, in the mountains to the east of Alamogordo, New Mexico, was founded in 1952 by the United States Air Force. Goldberg reported at the November 19, 1975 meeting of the Executive Committee[84] that the SPO was being "divested of its Air Force support, despite its preeminence in solar physics." He reported to the AURA Board in January 1976 that NSF had been given the responsibility for developing an operating plan for SPO after Air Force support ended on June 30, 1976.[85] He felt that AURA might be asked to manage the SPO facility. This prediction came true, and four months later a Special Meeting of the AURA Board[86] was called to approve an amendment to the NSF contract providing for interim management of SPO by AURA for a 15-month period. Hunt was present and expressed the appreciation of NSF to AURA for this prompt response. The official transfer of SPO from the Air Force to NSF took place in a ceremony at the Observatory on July 1, 1976, and the management by AURA began at that time. Lee and Edmondson represented AURA at the ceremony.

The Board in January had approved a resolution, drafted by Goldberg, urging Stever (at that time both Director of NSF and the President's Science Advisor) to establish a committee to review the overall national effort in solar physics and to make priority recommendations for support of a balanced program by the appropriate government agencies. Goldberg sent the resolution to Stever with a strong cover letter, dated January 23, 1976.[87] The Executive Committee was informed on September 28 that the NSF Astronomy Advisory Committee had created a subcommittee, chaired by Arthur C.B. Walker, to review possible arrangements for the management of SPO.[88] Goldberg attended the first meeting, and Greenstein asked Lee to represent AURA at the second meeting, scheduled for October 6, 1976. The Executive Committee voted "that AURA advise the Walker Committee of its interest in assuming management and operation of the Sacramento Peak Observatory as part of the AURA organization." Discussion of the long-term operation of SPO continued at the December 3 meeting of the Executive Committee,[89] and Greenstein asked the members of the Scientific Committee who were present to vote. Eleven were in favor and five were opposed.

Walker had written to Lee on October 19, 1976, and Greenstein and Lee had prepared a draft reply.[90] Following discussion by the Executive Committee, it was agreed that they would edit the draft and distribute the final version at the meeting of the Walker Committee on December 6, 1976. Walker's letter said the Committee had agreed that a consortium of universities would be the best type of contractor for SPO, and had established a set of six guidelines. The Walker Committee was going to solicit comments from AURA, UCAR, and USRA as to how these guidelines would fit their organizational structures. In particular, the Committee wanted to know how AURA would coordinate the SPO and KPNO solar programs. Lee replied that SPO would be a separate administrative unit.

The Walker Committee was not ready to recommend a contractor for SPO by the time of the 1977 annual meeting of the AURA Board.[91] Howard, attending his first AURA meeting, presented a request from NSF that the interim arrangement be extended for up to one year. The Board reluctantly approved the extension, and also voted that the Board should take steps to prepare a response to the Request for Proposal (RFP) that NSF would issue later in the year. The status of the preparation of the proposal was discussed at the September 16, 1977 special meeting of the AURA Board,[92] and the status of the negotiations with NSF was discussed at the 1978 annual meeting.[93] Zirker was appointed Director of SPO for a 5-year term, beginning on the effective date of the AURA management contract with NSF.

It had been voted to establish a separate Visiting Committee for SPO initially, and to merge it with the KPNO–CTIO Visiting Committee after three years. The first Visiting Committee meeting was held at SPO on October 19–20, 1978,[94] following the selection of AURA to manage SPO. At this point, AURA had separate contracts with NSF to operate KPNO, CTIO, and SPO. This was a logical consequence of the decision by NSF to have separate contracts for KPNO and CTIO.[95]

When the AURA Board agreed to be responsible for interim management of SPO, it asked the Visiting Committee to examine "the structure of AURA and the National Observatories it operates." The strongly worded written report was mild in comparison with the verbal report given by Chairman E.J. Wampler and committee members Martin Schwarzschild and James W. Clark at the December 3, 1976 Special Meeting of the AURA Board.[96] The Committee criticized the present underemphasis of the KPNO service function, the recent handling of personnel problems by the AURA Board, the size of the Executive Committee, the location of the corporate office, and the composition of the AURA Board. With regard to AURA management of SPO, they said: "If the possibility of permanent management of the SPO is to be pursued by AURA, then an adequate representation from the solar physics community must be established on the AURA Board, so the Board can act as an effective advocate for the SPO. . . . A direct line of communication from the Director of the SPO to the AURA Board must be set up, in parallel with the KPNO channel." The AURA Board responded to one of these criticisms at the Special Meeting on September 16, 1977 by approving a revision of the by-laws that would reduce the size of the Executive Committee and allow it to use telephone-conference meetings to handle urgent business.[97] The full Board would meet twice a year to compensate for the smaller size of the Executive Committee.

By the time of the March 6, 1979 meeting of the Executive Committee,[98] several solar astronomers had been added to the AURA Board and Robert F. Howard of the

Jane Pauley interviews A. Keith Pierce on NBC Today Show, 7 April 1978.
NOAO Photo Archives.

Hale Observatories had been appointed a consultant to the Board. Burbidge reported
at this meeting that he had reorganized the KPNO administrative and support services
into three areas: Operations Support, Engineering and Technical Services, and Administrative Services. Furthermore, he had created a unified scientific staff by abolishing
the formal separation of the solar system program and the galactic and extragalactic
program.

(2) Reorganization of the management structure of the observatories to include SPO

The Organization Committee continued its normal review of AURA and the Observatories. A joint meeting with the Screening Committee produced a complicated reorganization plan that was tabled indefinitely at the 1979 Annual Meeting.[99] In addition,
an ad hoc Committee to Consider Observatories' Organization and Budgets (COOB)
had been appointed by Chairman Code in 1980, shortly before Harlan Smith and

Harold Bell were elected Chairman and Vice-Chairman at the 1980 Annual Meeting.[100] COOB gave its first report at the July 14, 1980 meeting of the Executive Committee.[101] The report at the next meeting said:[102] "One of the problems to be resolved is whether AURA or NSF should allocate funds among the three observatories." NSF had been doing this by the process of approving or modifying the separate budgets submitted under the three contracts. The Committee was considering a recommendation that there be a single NSF contract with AURA to cover the three observatories.

COOB submitted a report to the Executive Committee on December 15, 1980.[103] As a first step, COOB recommended that the KPNO and CTIO directors act in concert to present a single integrated Program Plan and a single Long Range Plan. Geoffrey Burbidge and Patrick Osmer, who had been appointed Director of CTIO at the 1980 Annual Meeting, had attended the COOB meeting and were prepared to begin implementation immediately. They also felt that SPO should be included in this arrangement as soon as possible. Jefferies had previously suggested "the possibility of taking the solar component out of the KPNO Program Plan and working it out in conjunction with the SPO plans."[104] Burbidge and Osmer reported their first steps toward implementation at the 1981 Annual Meeting.[105]

COOB, the Organization Committee, and the AURA Board did not move directly from these beginning discussions to the final result. The first step was to consider having the directors of KPNO, CTIO, and SPO report to a Vice-President–Scientific Operations in the corporate office.[106] Before this got very far, NSF asked AURA to study "the pros and cons of combining the SPO and KPNO Solar programs under the Director of SPO."[107] Careful and thorough discussion for more than 18 months by the SPO Standing Committee, the Organization Committee, and the Executive Committee, with input from the SPO and KPNO directors, led the AURA Board at the 1982 Annual Meeting to vote to combine the KPNO solar program with SPO, using the name National Solar Observatory (NSO).[108]

COOB and the Organization Committee had also considered the organizational consequences of replacing the three NSF contracts with a single contract. At first it was envisioned that KPNO, CTIO, and NSO would be combined into a single observatory with a single director. There were strong objections to the loss of identity for the three observatories, and the Board responded by passing the following resolution at the 1982 Annual Meeting:

> RESOLVED THAT the AURA Board adopts as a management objective the appointment of a single Director responsible for operations of all ground-based AURA Observatories. Implementation of a plan to achieve that objective will be subject to detailed evaluation by the AURA Board at its Annual Meeting in February, 1983. The organization will have three units, respectively; Solar, Kitt Peak (non-solar), and CTIO, each headed by an Associate Director. The Associate Directors shall have a large measure of autonomy in the planning and operation of their units, consistent with a single AURA Observatories Program Plan, which has been approved by the Board. Until the new organization is in place, the AURA ground-based Observatory Directors will be encouraged to proceed with steps consistent with this overall policy.

The Organization Committee was asked to reorient its study "to consider only options consistent with the policy guidance in the first sentence of this resolution."

The detailed report of the Organization Committee was presented at the October 1–3, 1982 meeting of the Executive Committee in Chile.[109] It was approved, with the endorsement of the three Observatory Directors, for submission to the AURA Board at the November 3–4, 1982 Special Meeting. The Board voted at the Special Meeting to approve consolidation of all AURA-managed ground-based observatories (KPNO, CTIO, SPO) under a single Director. The new organization structure would comprise the three optical astronomy centers – Kitt Peak National Observatory, Cerro Tololo Inter-American Observatory, and the National Solar Observatory (consisting of SPO and the McMath Observatory), and, as the Director might deem appropriate, centralized administrative and technical services and a separate Advanced Development Programs Division.[110]

Laura P. Bautz, who had replaced William E. Howard III as Division Director for Astronomical Sciences at NSF in March 1982, met with the Executive Committee on February 15, 1983[111] during the time of the AURA 25th anniversary celebration (reported in Chapter 21). Her earlier reluctance to approve the reorganization[112] had diminished, and she seemed to be satisfied by the answers to her questions. She was accompanied by Kurt Riegel, the newly appointed Astronomy Centers Section Head. The Director Search Committee reported that six formal applications had been received plus 15 nominations. The Committee had added names, bringing the total to more than 80. This list had been reduced to 12 highly qualified candidates who were being considered by the Executive Committee.

The Executive Committee met on April 15, 1983 to receive the report of the Search Committee.[113] Five persons had not been interested, and six highly qualified candidates had agreed to be interviewed in Los Angeles on March 23 and 24, or in Chicago on April 5. Teem had informed NSF of the names of the six before the final process of evaluation and selection was begun. On April 6, the Committee selected Dr. John T. Jefferies as the most qualified candidate, and the Executive Committee voted to approve this appointment. Teem received approval from Edward A. Knapp, Director of the National Science Foundation, on May 24,[114] and the AURA press release was issued on June 10,[115] after NSF had also approved the name of the Observatory.

The Executive Committee held a teleconference on May 10[116] to discuss the current status of the name for the combined observatories. At the April 6 meeting the Committee had chosen: National Observatories for Research in Astronomy (NORA). Teem reported that NSF felt this name implied that NORA encompassed ALL of the national observatories, and the potential confusion could be damaging in connection with NSF budgetary requests. Ten names were suggested during the teleconference, and the final choice was: National Observatories for Optical Astronomy (NOOA). Conti, who had been elected Chairman at the 1983 Annual Meeting, reported to the Executive Committee[117] on June 16 that NSF had decided the name should be National Optical Astronomy Observatories (NOAO), and the Executive Committee voted to give formal approval to this name.

R. Kent Honeycutt, the new Indiana University scientific representative, attended the October 3, 1983 meeting of the Executive Committee.[118] The agenda included a report by Jefferies, the Director-designate/NOAO. He said his first draft of the implementation plan for NOAO had been discussed with Teem and Rossi, the Chairman of the Organization Committee. He said he would soon submit a revised plan to the Organization Committee and from there it would go to the Executive Committee in December.

Jefferies also reported that he had been reviewing applications for the new position of NOAO Associate Director/Advanced Development Programs. He said the five-year contract with the current Director of CTIO, Osmer, would be honored for its remaining two and one-half years. By previous action of the AURA Board, the contracts of the current directors of KPNO and NSO would be in effect only until the reorganization had been implemented. The new positions of NOAO Associate Director/KPNO and NOAO Associate Director/NSO were to be advertised immediately. Jefferies said he had encouraged both of the current Directors to apply.

Jefferies discussed the second draft of his implementation plan at the December 12, 1983 meeting of the Executive Committee.[119] He recommended the appointment of Dr. Jacques M. Beckers to the position of NOAO Associate Director/Advanced Development Programs, and this was approved. Jefferies announced at the 1984 Annual Meeting[120] that the Executive Committee had concurred in his recommendation for the appointment of Dr. Sidney Wolff to be Director/KPNO, Associate Director/NOAO and Astronomer with tenure/NOAO. He also gave a detailed report on the progress of implementation of the organization of NOAO. The Board voted to "receive the report of the Director/NOAO with enthusiasm." The appointment of Dr. Robert F. Howard to be Director/NSO, Associate Director/NSO and Astronomer with tenure/NOAO was approved at the next meeting of the Executive Committee.[121]

(3) Reorganization of AURA to accomodate increased membership

The restructuring of AURA had been more talk than action until the 1982 Annual Meeting, when the Board approved an amendment to the by-laws that reduced the size of the Board by cutting the number of institutional representatives from two to one for each member university.[122] This was implemented in 1985, as reported in Chapter 21.

(4) Management of the Space Telescope Science Institute by AURA

NASA's Space Shuttle Program and plans for a Large Space Telescope (LST) were mentioned in Goldberg's report to the AURA Board at the 1974 Annual Meeting.[123] On November 18, 1975 the Scientific Committee[124] discussed possible KPNO participation in the LST instrumentation plans. Goldberg said that many in the astronomical community felt that LST should be operated by a consortium of universities under contract with NASA, and AURA seemed to be a natural vehicle to perform such a function. The Scientific Committee voted to recommend to the Executive Committee that AURA proceed actively toward seeking management of the proposed Space Telescope Scientific Institute. The Executive Committee[125] agreed, and voted to authorize the Chairman and the President to contact NASA and begin discussions concerning AURA's interest in undertaking management of the proposed Institute. The Chairman was instructed to appoint up to three members of the Board to consult with him and the President during their discussions with NASA. Lee reported to the AURA Board on January 14, 1976[126] that he had written to Dr. James C. Fletcher, the Administrator of NASA, after the letter had been reviewed by Pierpont, Wilkinson, and Smith. Fletcher had replied expressing interest in the AURA proposal and stating that he wished to meet with AURA representatives in the next two or three months.

The Scientific Committee discussed the Space Telescope Institute and KPNO involvement in the space program again on May 2, 1976.[127] The following day the Executive Committee[128] voted to create an ad hoc Committee on the Space Telescope Institute to develop a discussion paper on the structure of the Space Telescope Institute, and to submit it to the Space Science Board summer study at Woods Hole. The Committee reported on its activities at the September 28, 1976 meeting of the Executive Committee.[129] The consensus of the meeting was that the ad hoc Committee should be continued, and that the Chairman of the Organization Committee should become an *ex officio* member. By the time of the 1977 Annual Meeting the ad hoc Committee felt its membership should be enlarged for the purpose of making an active study of plans for AURA operation of the Space Telescope Institute.[130]

Smith reported to the Board on September 16, 1977, the day Lee was fired and Teem was hired, that NASA had accepted the Woods Hole report, and had agreed to the establishment of the Space Telescope Institute as a separate entity, not necessarily at the Goddard Space Flight Center although they would prefer that location.[131] He resigned as Chairman of the ad hoc Committee, having accepted the chairmanship of the Committee on Astronomy and Astrophysics of the Space Science Board. During the next two years this Committee would be developing a recommendation to NASA concerning the management of the Space Telescope Institute, and there was a conflict of interest with his chairmanship of the AURA Committee.

The next reported activity of the ad hoc Committee is in the Minutes of the 1979 Annual Meeting of the Board of Directors.[132] The Committee had met on December 5, 1978 and February 6, 1979, and a report was given to the Board. Code, the Chairman of the Committee, said the Committee had concluded that AURA had a responsibility to respond to a Request for Proposal (RFP) from NASA to establish and manage an STSI. The Committee had contacted and visited five universities as potential sites for the STSI, because the proposing organization must specify a site, as well as a management structure. Teem estimated that preparation of a proposal would probably cost as much as $44,000, and said $7,000 had already been spent. The Board voted, with one abstention and one opposed, to approve the process now underway, and instructed the officers and the Committee to develop a proposal for review by the Board at a Fall 1979 meeting.

The ad hoc Committee gave a progress report at the June 5, 1979 meeting of the Executive Committee.[133] Preproposals were being prepared with University of California, San Diego, University of Colorado, University of Chicago, Princeton University, University of Maryland, and Johns Hopkins University. A letter was to be mailed to all AURA Board members giving an outline of the proposed plan, and canceling the proposed July Special Board Meeting. After the review process had been completed, the Committee would recommend a single proposal to the Executive Committee for a yes or no vote. A yes vote would bring the selected preproposal to the Board. Code reported at the 1980 Annual Meeting[134] that the Executive Committee "had received detailed reports on the various sections of the Space Telescope Institute proposal, and that the Committee now recommends it to the Board for final ratification." The Board voted to ratify the action of the Executive Committee concerning submission of the proposal to NASA.

After AURA had chosen The Johns Hopkins University as the site, Code (part time), Teem and Welch (full time) lived in Silver Spring, Maryland, for three months while writing the proposal.[135] They were joined by Barry Lasker, on leave from CTIO, who headed the team which developed the technical and scientific portions of the proposal. Welch was responsible in the business portions. The corporate office word processor was shipped from Tucson, because Computer Sciences Corporation (CSC), the contractor hired by AURA to help prepare the proposal, did not have one. One of the scientific sections of the proposal was limited by NASA to 300 pages, but it turned out to be 320 pages after careful editing. Welch recalled: "Because we had the word processor we were able to reformat that section, put two more lines on a page, add a half inch to the right hand margin, and it came to 300 pages. We didn't have to rewrite a word." The seven volumes of the proposal stood about 18 inches high. Welch delivered four copies to NASA on March 3, and gladly returned to Tucson.

Teem reported to the Executive Committee[136] on July 14, 1980 that NASA was going to make site visits at all of the sites proposed by the bidders for the contract. July 17, 1980 was the date for the Johns Hopkins site visit, and Teem, Code and Welch would participate along with Johns Hopkins personnel. He also reported that JHU had contributed $22,500 to help defray the cost of the proposal.

NASA spent ten months evaluating the four proposals that had been submitted, and selected AURA on January 16, 1981.[137] Code served as Acting Director during the negotiation of the contract and the search for a Director, from January 15 to September 1, 1981. Negotiation of the contract for management of the Space Telescope Science Institute (STScI) took three months, and it was officially signed on April 30.[138]

The reorganization of AURA included creation of a Space Telescope Institute Council (STIC), reporting to the AURA Board, and STIC undertook the search for a Director. The result was reported to the Executive Committee on May 28, 1981.[139] There were initially 60 applications and nominations. This list was reduced to 30, and these were ranked. Members were also asked to list without ranking their 10 to 12 "top choice" candidates. A short list of eight candidates was developed from these two lists. Only four on the short list were interested, and they were interviewed. As a result of the interviews, Dr. Riccardo Giacconi was selected as the prime candidate. The Executive Committee voted to ratify this recommendation, and to recommend that the AURA Board give its final approval to the appointment of Dr. Giacconi. He took office on September 1, 1981, and attended the meeting of the Executive Committee on September 11–12.[140] Six months later, the National Solar Observatory (NSO) was created by combining the Kitt Peak solar program with the Sacramento Peak Observatory (SPO).[141] Giacconi was an active and vocal participant in the discussions that led to the creation of NOAO.[142]

The move of the corporate office to Washington in 1983, reported in detail in Chapter 21, was the inevitable consequence of AURA's contract with NASA in 1981 to establish and manage the Space Telescope Science Institute (STScI) and the formation of the National Optical Astronomy Observatories (NOAO) in 1982.

21 The 25th anniversary celebration of AURA and KPNO on February 14–16, 1983. The move of the corporate office to Washington in the fall of 1983. A re-structured AURA Board and Executive Committee, approved in 1982, takes effect in 1985.

AURA was incorporated in the State of Arizona on October 28, 1957,[1] and Kitt Peak was selected as the site for the National Astronomical Observatory on March 1, 1958.[2] The date chosen for celebrating both of these 25th anniversaries was a convenient time for the 1983 Annual Meeting of the AURA Board of Directors, and coincidentally was the 71st anniversary of Arizona's statehood.

Edmondson was asked to be Chairman of the 25th Anniversary Planning Committee, primarily because he was the only member of the original Board who was still serving.[3] The other members of the Committee were: Albert B. Weaver (University of Arizona) and W. Albert Hiltner (University of Michigan), plus David F. (Kelly) Welch and Muriel Fults from the AURA corporate office. Edmondson and the three Tucson-based members of this group met in Tucson on April 2, 1982, at the time of the AURA Board meeting, and developed a provisional schedule for the two day celebration.[4] The Committee and John Teem, the President of AURA, met at Chicago's O'Hare Airport on June 8 to plan the contents of a special 25th anniversary brochure. Edmondson was to be the editor, and would be responsible for finding qualified individuals to participate in writing the brochure. He made three trips to Tucson for work on the text of the brochure, June 22–29, July 22–30, and September 7–18. Kelly Welch and Muriel Fults arranged for the brochure to be printed by the University of Arizona Press, and deserve full credit for the excellent quality of the finished product.[5] Copies were given to those in attendance at the closing banquet, and were mailed to those who were invited but not able to attend.

The invitation list was put together by mail and telephone between the June 8 meeting of the Committee in Chicago and the end of Edmondson's September trip to Tucson, and a few names were added just before the copy for the program booklet was sent to the printer. Invitations were sent to 339 individuals,[6] and 262 attended the banquet on February 16.[7] It was an enjoyable "family reunion" for the early AURA Board members and retired KPNO and NSF staff members.

A public lecture by an eminent historian of astronomy, Owen Gingerich, on the evening of February 14, began the celebration. His topic was "The Electronic

Revolution: Astronomy from 1957 to 1982." This was preceded by an afternoon press conference. The questions were supposed to be for Gingerich, but one reporter asked Edmondson if there was any truth to the rumor that the AURA Corporate Office was going to be moved to Washington. Such a question had been anticipated, and Edmondson's carefully coached reply was that this was on the agenda of the AURA Board for discussion on February 17 and he could not predict what the action of the Board might be. He added that the purpose of a move to Washington would be to make the AURA Corporate Office more effective in representing the needs of Kitt Peak and Cerro Tololo to NSF and Congress on a day-to-day basis.[8]

The morning of February 15 was devoted to a meeting of the Executive Committee and to two talks for the invited guests at the Westward Look Resort. The Director of the Space Telescope Science Institute, Riccardo Giacconi, spoke about "The Space Telescope Observatory." The Director of the Kitt Peak National Observatory, Geoffrey Burbidge, spoke about "Telescopes and Astronomy in the 1990's." Busses took the participants to Kitt Peak after lunch, and this provided an opportunity to see the observatory's facilities in daylight. Dinner was served in the dining hall, and it took two seatings to accommodate everyone. The Higmans had dinner with the Papago representatives, who were interested to learn that he had worked for the Tribe at the time Kitt Peak was chosen. Views of selected objects through the 50-inch, 84-inch, and 150-inch telescopes were available before the return to Tucson. Short talks about the research being done with these telescopes were given at each site.

There were two more major talks on the morning of February 16. Bradford Smith, University of Arizona planetary scientist, spoke about "An Astronomical Wonderland: The Outer Solar System." Victor Blanco, Director of the Cerro Tololo Inter-American Observatory, spoke about "Crown Jewels of the Southern Skies." Busses took the participants to the picnic grounds of the Catalina Council of the Boy Scouts for a picnic lunch, and a visit to the nearby Arizona–Sonora Desert Museum after lunch. This visit was arranged with the assistance of Ralph Patey, the first Business Manager of KPNO and the Administrative Officer of the Desert Museum in 1983.

The 25th anniversary celebration ended that evening with a banquet at the Westward Look Resort, and the Chairman of the AURA Board of Directors, Harlan Smith, was the master of ceremonies. He began by mentioning that February 14 was the 71st anniversary of Arizona's statehood, and that November 23, 1982 had been the 20th anniversary of the selection of Cerro Tololo. He then introduced several distinguished guests, including Richard Harvill, President Emeritus of the University of Arizona, and asked each to say a few words. He described the important role of Edward Spicer, Professor Emeritus of Anthropology at the University of Arizona, in obtaining permission from the Papago Tribal Council for astronomers to test Kitt Peak as a possible site for the proposed national astronomical observatory, and expressed regret that failing health prevented Dr. Spicer from being present.[9] On behalf of the AURA Board, he presented a framed resolution and a silver clock with an electronic chime to Edmondson in recognition of his twenty-five years of service on the Board. He then introduced Norman Rasmussen, a member of the National Science Board, who presented the NSF Meritorious Public Service Award to Edmondson.[10] At the AURA Board meeting the next day, Edmondson was appointed "Consultant/Historian to the AURA Board" for one year effective on the date of his retirement from Indiana University and resignation from the AURA Board of Directors, July 1, 1983.[11] This appointment has been renewed annually.

The AURA Board and Executive Committee had unusually full agendas to deal with on February 17. The most important action was the decision to move the corporate office to Washington.[12] The move was made during the summer of 1983, and Suite 820 in the Joseph Henry Building became the new location of the AURA corporate office.[13] This building was leased by the National Academy of Sciences from George Washington University, and when the NAS did not renew the lease in 1987 the corporate office moved to Suite 701, 1625 Massachusetts Avenue, N.W.[14] Welch chose to retire and stay in Tucson, and the staff that moved to Washington consisted of Teem, Fults, Phyllis McDowell, and Robert Milkey (the recently appointed Corporate Staff Scientist).

The first mention of the need for a Washington office seems to have been Goldberg's comment at a meeting of an earlier Organization Committee on July 22, 1971:[15] "Dr. Goldberg suggested that perhaps AURA should consider a paid president and an office and small staff, perhaps even in Washington, allowing the Director to give the Observatory scientific direction instead of working on budgets and dealing with the Foundation. Dr. Wildt and Dr. Edmondson agreed that perhaps this should be considered." Nothing happened until the Chairman of the AURA Board, Harlan Smith, reviewed the history of the Corporate Office at the July 14, 1980 meeting of the Executive Committee,[16] and asked the question: "Is Tucson still the best location for the Corporate Office?" The Organization Committee was directed to study the cost and other factors that would be involved in relocating the Corporate Office. The Organization Committee reported at the 1981 Annual Meeting that it was still looking into this question,[17] and presented a preliminary report at the June 19, 1981 meeting of the Executive Committee.[18] Concern about the costs of the move and annual operating cost of the office in Washington led the Executive Committee to conclude that a decision should not be made at this time. There was further discussion at the December 3–4, 1981 meeting of the Organization Committee:[19] "Although the need for an AURA presence in Washington remains high, little work has been done on the Corporate Office location issue due to higher priority issues before the Organization Committee." These included the merger of the solar programs at KPNO and SPO to form the National Solar Observatory (NSO), and the subsequent merger of NSO with KPNO and CTIO to form the National Optical Astronomy Observatories, already discussed in Chapter 20.

The other high priority issue was how should the AURA Board be restructured if additional qualified universities were admitted to membership. The AURA Board had first discussed the matter of increasing AURA's institutional membership at the 1968 annual meeting,[20] and this was quickly followed by an expression of interest from the University of Arizona.[21] The Organization Committee discussed the matter on December 13, 1968, but no action was taken.[22] The AURA Board accepted the report of this meeting at the 1969 annual meeting, but without further discussion.[23] At the 1970 Annual Meeting the AURA Board approved the formation of a Users' Committee reporting to the Director of KPNO, and a Visiting Committee reporting to the AURA Board, but did not discuss increasing the membership.[24]

Fleischer raised the question of increasing the membership of AURA at the

November 19, 1970 meeting of the Executive Committee in Chile.[25] Hiltner told him this was under consideration by the AURA Board and the Organization Committee. Fleischer repeated the question at the 1971 annual meeting,[26] and was told the Organization Committee had been making progress reports to the Executive Committee, and that a final report would be made at the 1972 annual meeting. Fleischer said that for some time there had been strong pressure from several universities to become members of AURA, and the criticisms had been directed to NSF. He felt it would be embarrassing to wait until 1972, and urged quicker action. The Board then voted that a special meeting should be called when the Organization Committee had agreed on a plan. This took only two months, and the special meeting of the AURA Board was held on March 16, 1971.[27] The following recommendations of the Organizing Committee were unanimously approved:

1   That AURA membership be increased.
2   That membership in AURA will be open to Ph.D. degree granting institutions in the United States with a deep commitment to advanced training and research in astronomy.
3   That a committee of three Board members, two scientists and one administrator, be appointed to review applications for membership in AURA, Inc., and make recommendations to the Executive Committee and the Board of Directors.

The meeting lasted 35 minutes, 9:05–9:40 a.m. There was no other business, and a meeting of the Executive Committee followed at 9:40 a.m.[28] Goldberg's acceptance of the directorship of KPNO was announced at the beginning of the Executive Session of this meeting.

Applications for membership in AURA were received from five universities, and these were discussed at the 1972 annual meeting of the AURA Board.[29] It was voted to admit three at this time: University of Arizona, California Institute of Technology, and University of Texas at Austin. The Organization Committee was asked to review the problem of the rate of growth and the ultimate size of the AURA corporation, and report to the Executive Committee not later than the September 1972 meeting. The report was ready for the June 20 meeting,[30] and was presented to the AURA Board at the 1973 annual meeting.[31] The report recommended that any growth in membership should be at a slow rate, and stated that any growth beyond 15 members would require restructuring of the organization.

The University of Colorado was admitted in 1977,[32] the University of Hawaii in 1978,[33] and the University of Illinois in 1980,[34] which brought the membership to 15. The Massachusetts Institute of Technology was admitted in 1981,[35] and the Johns Hopkins University in 1982.[36] Restructuring the AURA Board was clearly overdue. An amendment to the by-laws that would reduce the size of the AURA Board by cutting the number of institutional representatives on the Board from two to one for each member university was proposed by the Organization Committee. This was approved by the AURA Board at the 1982 annual meeting, and was to become effective after the 1985 annual meeting. The President of AURA was authorized to select the university representatives who would continue. Teem presented the list at the 1984 annual meeting.[37] It contained ten astronomers, six administrators, and one not yet determined.

It was just as well that the reorganization of the AURA Board had not become effective before the 1985 annual meeting.[38] Both the wisdom and the combined background of those who had participated in the creation of NOAO were needed to cope with the scientific and political firestorm that was produced by the decision of NOAO Director Jefferies to close SPO in response to a $2.6 million shortfall in NOAO funding caused by a severe reduction in the FY 1986 NOAO budget by NSF.[39]

The Executive Committee met on March 27, 1985, the day before the 1985 annual meeting of the AURA Board of Directors, and received reports from Jefferies and the OAC about the budget-constrained NOAO program options.[40] Jefferies described the strong opposition to the plan to close SPO that he had received from NSF, members of Congress, and the solar community. Based on advice from the SPO Users Committee, he had now decided to keep the SPO Tower Telescope open at a moderate level of operation. Mathis reported that Jefferies had discussed the budgetary basis for his plan to close SPO with the Observatories Advisory Committee (OAC) at a meeting in La Serena in February, and that the OAC supported Jefferies' approach to reallocation of funds in order to strengthen the most promising programs instead of making across the board cuts.

Chairman Conti introduced the discussion the next day at the Board meeting[41] by saying: "Let's try to see how we got into this mess, and what we can do about it." Jefferies said he had intended to advise the SPO staff before, or at least simultaneously with, taking his proposal to the Board. He had not intended to make the proposal public until these two groups and NSF had been advised. Unfortunately, prior to Jefferies' return from Chile, NSO Director Howard had written to the NSO Users Committee advising them of the proposal. Conti said he felt "the well-intentioned but very ill-timed action by Dr. Howard brought about the unfortunate political situation."[42]

When Jefferies discussed with Teem his plan to accelerate informing the SPO staff of his intended recommendation due to the prospective disclosure of this by the Users Committee, Teem told him that NSF must be informed immediately, and that NSF would inform Congress of the correct facts. Teem informed NSF that very day, and NSF advised the New Mexico Congressional delegation the following Monday. Jefferies notified the SPO staff on March 12, 1985 and the Tucson staff on March 14.[43] Immediately after this, members of the solar community began contacting AURA Board members and members of Congress to express their opposition. NSF received a telegram from the Governor of New Mexico as well as communications from the New Mexico Congressional delegation. One of them invited Jefferies to come to Washington to discuss the matter, which he did. He also talked to Bautz at NSF. Subsequently, she wrote to Jefferies: ". . . it is our position that no personnel actions of any kind, including notifications, will be taken which involve staff at NSO Sunspot (SPO) until the NSF formally responds to the AURA proposals. At the earliest this response will be forthcoming in mid-May." This delay meant that termination costs would come from the FY 1986 budget, the very thing Jefferies had tried to avoid by giving timely notice. Hindsight says that NSF should have been notified before the February meeting of the OAC in La Serena, and Teem should have pointed this out to Jefferies.

On June 27–28, the Executive Committee received a report from Teem about the work of the AURA/NSF Committee, chaired by Eugene Parker, that had been set up to study the NOAO Solar Program.[44] Bautz had requested time to talk to the Executive

Committee, and she made some highly critical remarks about the way AURA had handled the SPO problem. She said the FY 1986 Program Plan should not be submitted until the results of the Parker Committee study were known. She also said that NSF would conduct a review of the entire NOAO operation and its management in the fall of 1987.

A special meeting of the Executive Committee on August 14 received a report from the OAC that included:[45] "At Sacramento Peak Observatory, with considerable support of the U.S. Air Force program there, AURA has been able to devise a Program that keeps all of the major facilities operational." This should have been the happy ending, but there was still some bad news to come.

Senator Pete Domenici and Representative Joe Skeen both from New Mexico, had written a three-page letter on September 18, 1985 to Dr. Roland W. Schmitt, Chairman of the National Science Board.[46] They were concerned about the NOAO Program Plan for FY 1986, and wrote: "Just keeping a few telescopes minimally operating is no way to run a National Center." The letter ended with a request for NSF to study "the feasibility of removing the NSO component from NOAO, its management then to be under AURA or even some other university consortium." They wanted the result prior to the next NSF budget submission to the Congress. Teem responded by sending a copy of the Parker Committee Report to Senator Domenici with a 3-page cover letter.[47] Teem's letter plus a letter from Schmitt seemed to pacify the Senator.[48]

NSF accepted the recommendation of the Parker Committee Report, and Bautz wrote to Teem:[49] "We would like to commend the members of the Parker Study for their efforts, the USAF for their increased support of programs at NSO, and AURA for their response to the concerns of the community as embodied in the Parker Report."

The Parker Committee did not share the positive reaction of NSF to the AURA response, and Parker sent Bautz a highly critical "supplementary report" on December 12, 1985, based on mail and telephone communications by members of the Committee.[50] There was strong personal criticism of Jefferies because of some of his recent actions relating to SPO, and the Committee concluded that "there is no alternative to seeking independent management of the NSO." Fortunately, things seemed to have settled down by the time of the 1986 Annual Meeting, except that many SPO staff members continued to be suspicious of the NOAO Director.

Between the 1985 and 1986 Annual Meetings, Teem gave the AURA Board one year's advance notice of his intention to take early retirement, effective September 30, 1986.[51] The election of the University of Maryland, the State University of New York, and the University of Washington at the 1986 Annual Meeting, Teem's last meeting and the first with a smaller Board of Directors under the new system, brought the number of member universities to 20.[52]

A search committee nominated Goetz K. Oertel to succeed Teem as President, and he was elected by a unanimous vote of the AURA Board in a telephone conference meeting on June 27, 1986.[53] The public announcement was released on July 18, and Oertel took office on October 1, 1986.[54]

22 A look back and a look ahead

Goetz K. Oertel, AURA President

This chapter presents a subjective look back, a personal and certainly not disinterested analysis of the present, and a speculative look ahead. It therefore need not meet the high standards of an accurate account of history. Instead, it takes license to look beyond the record and to speculate about underlying causes of the past and present situation, and about what the future may bring. It is further colored by my roots which are in physics and management more than in astronomy, and by indomitable optimism about the future.

Raison d'être

With these cautions in mind, we will start by examining the raison d'être of national observatories and of AURA, and see how it operates and how well. Next, we look at its environment: the community of scientists, and the sponsors and partners in the US and abroad. Accomplishments and challenges are highlighted next. Finally, we look into a crystal ball to get to "the vision thing" as ex-President Bush is said to have put it.

The goal of observatories is to advance astronomy by enabling novel and significant research. In 1953, the observatories in the USA were private. Their owners and operators sought to attract the world's best talent to their institutions where they would have access to telescopes. The strategy worked. Astronomers elsewhere depended for access to telescopes upon the generosity of colleagues at the private observatories. It therefore sat well with them when Leo Goldberg said at an NSF conference (Chapter 4) that ". . . what this country needs is a truly national observatory to which every astronomer with ability and a first class problem can come on leave from (her) university."

This phrase created the environment in which Leo and Robert McMath could create the consortium of universities that became AURA. It also contained seeds for controversy. First, private and national observatories would henceforth compete for ideas and for people, and for sponsors' funds. That should be healthful but can become destructive as it occasionally did.

The last part of Leo's statement can be read to give the national observatory a

251

mission not unlike that of Carnegie's observatories: to be a "center of excellence" where resident staff and researchers on leave from other institutions carry out research to advance astronomy themselves. A contrasting "service" view of the mission of a national observatory is that it enable colleagues elsewhere in the community to visit, take observations, and take the results home to their institutions for analysis. In that view, resident staff ensures that telescopes and instruments work properly, provides service to the user community, and develops plans for future facilities. AURA adopted this latter mission with the important proviso that to provide excellent service, staff must be, and stay, at the forefront of science and technology through a personal program of astronomical research.

What should the national observatories do?

One view is that they should exclusively operate facilities that are unique and beyond the capabilities of private institutions to build and operate. Advocates for that view point out that universities can and do garner funds to construct and operate observatories for their own students and staff, and that more of them would do so if national observatories no longer provide access to facilities that are not unique. Others regard this as an elitist view because not all private universities can provide meaningful observing capabilities to their staffs. This view is also not consistent with Leo's raison d'être for the national observatories.

In a sense, all facilities at the national observatories are unique within the US community, either in capability, or by providing access to the southern skies, or through special instrumentation or exceptional performance as the WIYN telescope on Kitt Peak, or in providing access based on merit without regard to institutional affiliation. Specifically, solar facilities and Cerro Tololo are unique. Even the night-time facilities at Kitt Peak are unique in instrumentation if not in aperture or location.

How does the raison d'être, to enable the best science, relate to whether or not all facilities at the national observatories should be unique? Is uniqueness necessarily more important than the quality and quantity of science enabled? In April 1994, MIT astrophysicist Paul Schechter presented a list of the top ten accomplishments of the national observatories to a panel of the National Research Council as an indicator of the quality of research at the national observatories. He challenged others to list the top ten accomplishments of private observatories. That challenge was not met, though it clearly could be. As to the quantity of research enabled, the national optical astronomy observatories support more than 1,000 visitors annually, many of them students. Time on its night-time telescopes that are not unique is oversubscribed by factors of up to three. By contrast, some unique facilities are not fully subscribed. If quantity and quality of science are the measures, then some non-unique facilities compete well with some unique ones.

If uniqueness is not the sole criterion for merit, and if scientific merit is in the eyes of the beholders, how to set priorities in a budget crunch? Ultimately, the NSF decides, based on recommendations from the observatories director and from AURA, and with independent advice it may seek. As budgets decline, choices involve not only science but also matters of national policy. For example, is it more important to enable research through unique facilities, or to enable research and training by faculty and students, including at institutions that cannot share in the costs to construct and operate private

WIYN 3.5 meter telescope (D864). NOAO Photo Archives.

facilities? The question arises not solely as a social issue but also because large unique facilities tend to be heavily oversubscribed, and successful proposals normally require prior work at smaller, non-unique facilities. Without at least some non-unique facilities at national observatories, proposers from many institutions might lack access to smaller facilities, and therefore also lack the opportunity to lay the foundation for a strong proposal to use large unique facilities.

AURA today

What are our vision, mission, and goals for AURA? As set by the Board of Directors, the vision is to "learn all we can know about the universe and to share knowledge and insights with colleagues and lay audiences." The mission is to "advance astronomy and related sciences, to serve the community and respond to its priorities, to foster educational opportunities, and to enhance international cooperation." The primary goal is to "excel in providing research opportunities at world-class facilities and to grant access based on the scientific merits of proposed work."

AURA manages astronomical and educational programs and projects, large and small, national and international, ground-based and in space. First, we operate the National Optical Astronomy Observatories (NOAO) for the National Science Foundation (NSF). NOAO includes Cerro Tololo Inter-American Observatory, the National Solar Observatory with sites at Kitt Peak and Sacramento Peak and the global "GONG" network, and night-time programs in Tucson with instrumentation programs, Kitt Peak, and support for the US national role in Gemini. The second unit is the International Gemini Project to build modern 8-meter aperture telescopes in Hawaii and in Chile, which AURA manages for a consortium of six national science agencies. The third major unit is the Hubble Space Telescope Science Institute in Baltimore which AURA operates for NASA and the European Space Agency.

How do we in AURA operate? The Directors of the observatories and institutes manage their institutions with broad authority. They report to the board through me as President. The board holds us together, as management, responsible and accountable for the work for our sponsors. The approach of delegation of authority is the opposite of micro-management: the Directors know what their jobs are, they have authority to do them, but they are not told how. They are held accountable for getting things done. While communications are open, and encouraged, all formal direction or guidance from the board and from sponsors flows through one channel: through the President. That keeps signals from getting crossed. In effect, management is left to the people who are closest to the work, not to a remote Czar or to a committee. The approach works because we hire the best people and empower them to do their jobs: people make organizations work, not the other way around.

Good management brings in good people and they, in turn, produce good results. A fine but by no means unique example is Cerro Tololo Inter-American Observatory. There, Chileans and Norte Americanos alike have become legendary for the effective support they provide to visiting astronomers. Their work, often beyond the point of duty, shows their commitment to the mission and makes observing and work at Cerro Tololo productive and pleasant.

The AURA Board is ultimately responsible for everything its employees do. All authority originates with the board. Because of its size–now 41–it meets only once annually, operating more as an assemblage of stake holders than as an operating board. It elects officers and working committees, considers matters of policy, and gives guidance. Between meetings, it functions through an Executive Committee and through councils, one for each observatory or institute. The board described its function, and that of each board member, as that of a "trustee" and advocate for the mission of AURA and its units. That includes oversight and advocacy, advice and support. By relying on committees and councils for most of its work, the board is as effective as it can be at its present size. The board's most important role is to recruit, hire, and retain management, and to set policies that enable us to recruit and retain top quality people at all levels. Those of us in management believe that it does this very well!

Concerns

In a variety of ways, the community has criticized the AURA Board. Fair or unfair, based on fact or on perception, historically correct or overtaken by events, criticism is there. On the one hand, it is fair and factual to say that the board is rather large. On the other hand, some say that "management is doing a great job but the board is bad." That can be regarded as unfair because the quality of management reflects the wisdom of the board that hires and retains it. Another criticism, micro-management, was overtaken by events. It must have been an issue when, reportedly, an observatory director felt that he must take remarks by individual board members as formal guidance or direction. It is not an issue now when all guidance and direction is formally adopted and flows to the Director through a single channel, the President.

The issue of real or perceived conflicts of interest on the parts of board members is a long-standing concern. Could board members put their personal or institutional interests ahead of those of AURA or its centers? Have they done it? First, the potential for such conflicts seems unavoidable unless one excluded from the board and its committees most astronomers with knowledge, competence, interest, and commitment to our science. If one so excluded all colleagues whose interests might somehow compete with those of AURA, one might exclude every competent astronomer from outside AURA. No AURA employee can serve on the board either. Thus, conflicts of interest cannot be avoided, they must be managed. Policy and procedures now require full disclosure of any potential conflicts of interest and bar board and committee members from final discussion and votes on matters in which they may have a conflict.

This policy, adopted in the late 1980s, works well if the potential for conflict of interest is recognized soon enough. That is almost always the case because there are few secrets within the community. When Ben Franklin said that "three may keep a secret if two be dead," he may not have had us in mind but he might have.

A policy, no matter how good and how effective it may be, may be powerless against perceptions. How do they arise? In my tenure of nearly a decade, only individual comments during debates, within the privacy of the boardroom, could have given rise to such perceptions if they had been heard outside that room. The board's actual decisions and resolutions were always within its role as trustee and advocate for the mission of AURA and its Centers, and were constructive and highly motivated. Further, many resolutions sought to further the broader interests of our community which includes private institutions, national centers, and international cooperation. All actions the board took during my tenure reflected statesmanship, not conflicts of interest. I am convinced that we did not suffer but actually benefitted from the candor of discourse within the privacy of the boardroom. Thus, like in the making of laws or sausage, the product is fine even though the process may be bloody.

The full board's review of the mirror decision for the Gemini telescopes is sometimes cited as an example of inappropriate behavior by the board. I disagree. The board served two important purposes here: one in its responsibility to the international project and its sponsors, the other in its responsibility to the US community as the operator of the national observatories, a partner in Gemini. With regard to the first, the board explored if the President, the Director of the international Gemini project, and the Executive Committee had acted appropriately in selecting the mirror. With regard to the second, it heard from the chair of the US Academy committee that had criticized the decision. The board then explored if it should take some action, as the steward for NOAO, an advocate for the US interests in Gemini. The discussion was protracted, heated, candid, and often passionate. Some participants found it objectionable, though on very different grounds. It was almost certainly necessary that it be held, lest a revolution in the US community would end Gemini as we know it, notably Gemini South. In the end, the board formally endorsed the course taken by management and by the international sponsors and gave valuable guidance for the further conduct of the project. It is perhaps ironic that the board is still criticized for what may have been one of its finest hours.

We were fortunate to have Maarten Schmidt chair this historic meeting. His prestige, diplomacy, tact, and firmness enabled the board to pass these shoals, and others, safely. He accepted and internalized his role as chief "trustee and advocate" for the national centers, and he carried through with excellence, style, determination, and a considerable personal investment of time and effort. He fostered not only Gemini and Hubble but all our facilities and plans, whether or not he depended upon them for his own research. A true statesman, he took his stewardship for the national facilities seriously. To return to potential conflicts of interest on the board and how to manage them: Maarten demonstrated that a superb way is to put fine people with high principles in leadership positions and let them work and think things through. During my tenure and well before, we have continued to be blessed with superb leadership from the chairs of the board: Peter Conti, Bob Noyes, and Bob MacQueen preceded Maarten, and Bruce Margon succeeded him.

How do perceptions of conflict of interest arise in spite of the statesmanship of board members and the supportive actions of the board as a whole? They can arise innocently, for example, when a board member makes private comments that appear to promote his own interests over those of AURA's. Even though one can hardly expect a board member not to favor his or her own interests, and even though s/he may preface remarks appropriately, the listener may misunderstand or want to misunderstand. Other ways in which perceptions of conflict of interest could arise would be if board members advocated only those courses of action for AURA which are also in their own institutions' interest, or if they "filibustered" constructive proposals by raising numerous objections, or if they opposed in private that which AURA proposes or advocates in public. It is unfortunate that isolated instances where outsiders believe that they observe such behavior can raise suspicion about a board, regardless of the facts.

Some have suggested that conflicts of interest are a consequence of the by-law that provides that member university presidents appoint the "institutional" members of the AURA board and are therefore entitled to a share in controlling the consortium. Not so: "institutional" board members and their university presidents know that they are to act as "trustees and advocates" for AURA, not as "member representatives," as they serve on the board. As such, they are pledged to work for the best interest of AURA. It is not their role to represent the interests of their institutions over those of AURA.

For the first time in a decade, the board is considering restructuring itself, reducing its size, improving operations, and addressing other issues. It established a working group of former and present leaders of the board and several outsiders which has analyzed the situation and sought input from a large sector of the community. The board will consider the final report and recommendations from this group at a special meeting in January 1996. Any major restructuring that may result would be implemented after the regular annual meeting in April 1996. The board is taking this self-evaluation very seriously. I expect that it will not hesitate to make the changes it may deem appropriate.

Environment in the community and the agencies

Let us now consider the environment in which AURA operates. First and most important is the astronomy community. In advancing astronomy and in operating its Centers, AURA is at once the community's servant, agent, and representative. The community's support is crucial. We seek its input in many ways, through member universities, through the board and its committees, through visiting and users committees, through workshops, and through surveys and other means.

When I was considering joining AURA, John Teem warned: when attacked, astronomers circle the wagons but fire to the inside. Indeed, many astronomers react to reductions in, say, the NSF astronomy budget by attacking their colleagues—and the NSF's wisdom in distributing the reduced budget—rather than by uniting to make a coherent case to increase funding for astronomy.

This type of astro-cannibalism contrasts with astronomers' unquestionable success in making hard choices, through the decadal surveys under Academy auspices, then sticking by their priorities. That approach has been very effective at NASA, though less so at the NSF. There may be many reasons. One may be that astronomers have, on occasion, involved high-level NSF officials in their fights; notably people who are in a position to allocate more resources to astronomy—or fewer!

In AURA, we work hard to settle matters within the community. For example, we encourage, take part, and occasionally sponsor candid debates *within the community*, to explore issues and gain consensus on how to address them together. We listen, we abide by consensus views, and we act—or propose to the Federal agencies—accordingly.

Not everyone necessarily subscribes to the outcomes of consensus-building processes. Whoever feels strongly enough can take their cases to higher authority, for example, to funding agencies such as the Astronomy Division in NSF. That is healthful if it is done collegially and with integrity. Issues in astronomy should, however, be resolved within the community. It can be quite damaging to astronomy to do otherwise, for example, if appeals are taken to officials who allocate resources between astronomy and other disciplines. These officials should hear positive messages about astronomy, why it is good, exciting, interesting, valuable, and worthy of additional support. What they hear from the community should also be consistent with what they are hearing from the advocates for astronomy on their staff, at NSF from the Astronomy Division. If, instead, they are drawn into fights within astronomy, then it would be hardly surprising if they reacted negatively. For example, they might react by reducing the astronomy budget "until the astronomers get their act together." The benefits of the hard work and tough choices in the decadal surveys can be eroded in this way. "Settle astronomy fights among astronomers." "Work together to increase funding for astronomy and for NSF." "Don't fight over a shrinking pie, fight to enlarge it." Such advice is frequently given but not always followed.

The environment for astronomy in NSF and NASA is characterized by the Federal budget process, and by the different missions and characters of the two agencies: NSF *supports* good science; NASA *manages missions* that enable good space science. The difference is reflected in the different relationships between the agencies and science consortia, and even in the contractual vehicle they employ: NASA prefers contracts, NSF prefers cooperative agreements. We respect the needs of the two agencies, as they accept and respect our role as representative of the community, and as its advocate.

Both agencies are Federally funded and subject to annual authorizations and appropriations. The annual budget battles between White House and Congress, and within the Congress, may be more costly in Federal personnel and resources than any other Federal activity. They also take a toll on projects, and on long-term programs and staff, that may not be readily apparent but can be devastating. To cope with starts and ends of fiscal years, to juggle funds from several different years, to be in the dark for several months into the year about what the "real" budget for that year will be, all present challenges. They can be met but take much effort, dedication, time, and money.

A more serious consequence of the budget battles is harder to manage: late decisions on funding levels frequently cause deferrals in procuring needed supplies and parts. Thus, projects slip and staff may need to stay on board–and be paid–longer: costs increase. A comparison between WIYN and GONG illustrates the point. NOAO managed both construction projects during the past decade, both were challenging scientifically and technically. It completed WIYN on schedule and within budget while GONG came in late and above the forecast budget. WIYN was paid for with funds that the participating universities made available when needed. GONG was built with Federal funds that were appropriated annually, but usually at lower levels than were required to remain on schedule and within budget. Thus, GONG cost more and was completed later. It is encouraging that NSF recognized this situation and, in 1995, gave NOAO a budget increment for investment to decrease future costs. NOAO invested one half million dollars to complete GONG as cost-effectively as was then still possible. As a result, GONG was completed one year sooner at a saving of about 1.7 million dollars.

In general, Federal support for astronomy may well have peaked because "discretionary" funding, which provides for NSF and NASA, is declining sharply. The effects could be dramatic. Already, NOAO lost about 30 percent of its funding since 1984 as astronomy's share in NSF funding declined and because priorities within NSF astronomy shifted from centers to grants. NSF asked that NOAO look at budgets at 3, 5, 7, and 9 percent below 1995. That would be in addition to inflation and would therefore amount to 25 to 30 percent below 1995 in buying power. As this is written, the Visiting Committee reports from Chile that Cerro Tololo could not absorb such a reduction without serious damage to its leadership, quality, service, facilities, and staff. NOAO helps Cerro Tololo by continuing to absorb, off the top, the erosion of buying power in Chile from the combined effects of the stronger Peso and Chilean inflation. But it compounds the reduction in funding from the NSF, and escalation within the USA compounds it further. Thus, unless the picture improves, it may become inevitable to close facilities. That could be solar and/or stellar telescopes on Kitt Peak or smaller telescopes everywhere. Should NOAO keep that which is unique or that which is most productive of science in quality and quantity? Either way, NOAO would be quite a different organization than we have known. And it would beg the question how to enable the science that would be lost. NOAO and AURA will wrestle with these questions and answer them, in concert with the NSF.

NSF manages grants and national centers through the Astronomy Division. In recent years, NSF recognized that major projects should not compete at that level and now budgets for them centrally. Grants typically support one or a few people. National centers operate large facilities. Some of them serve more than one thousand astronomers per year and require a staff of hundreds to operate effectively and safely. That includes scientists, engineers, as well as technical, operations, and maintenance staff. The division sets priorities between activities that are as unlike and difficult to compare as individual grants and national centers.

In an effort to reduce the potential for conflicts of interest, NSF is more likely to appoint grantees, or potential grantees, to its advisory committees than people who work at national centers. The advice to NSF has generally been to protect or increase grant funding, if necessary at the expense of national centers. In fact, NSF increased the grants' share of its astronomy funding from 31 percent to 38 percent between 1980 and 1995. In addition, NASA has increased funding for grants substantially. For example, for the Hubble Space Telescope alone, grants for data analysis, Hubble fellows, and archival research increased from nearly zero to $25 million in the past five years. Nevertheless, competition for grants is as tough as ever before. That gives the specter of a grantee community that is increasingly dependent upon "soft" Federal money. At the same time, Federal support is in jeopardy from efforts to balance the budget, and perhaps from waning interest in science in the Congress or at the White House.

The near-term budget outlook for space science is guardedly optimistic. The Hubble Space Telescope program expects to receive a reasonable level of support while it produces superb science and captures the imagination of the public. The Hubble team at Goddard and the Space Telescope Science Institute hope to reduce the cost of the program, including spacecraft operations and engineering support, initially by a factor of two. For the longer term the goal is another factor of 2.5 for the period after the final servicing mission by the Space Shuttle. Such reductions would enable NASA to extend the life of Hubble beyond 2005, the end of the planned fifteen years of orbital life, even as it considers taking a logical next step in the exploration of space: to search for earth-like planets near other stars.

Progress

In 1965, the "Gordon Committee" of the National Academy of Sciences recommended that NASA create a Space Telescope Science Institute to manage Hubble science. The Institute's initial relationship with NASA was turbulent but it succeeded in its mission, thanks to the superb scientific and technical team that its Director Riccardo Giacconi put together and led for a decade. One of its–and his–finest hours came when he accepted software, called the Science Operations Ground System (SOGS), and with it the challenge to make it work with other systems at the Institute and at Goddard. Another such hour came after spherical aberration was discovered, when the Institute assembled and led a joint task group of the entire Hubble team within and outside NASA that recommended how to correct the aberration. NASA accepted and soon carried out the well known successful repair under the leadership of HST project manager Joe Rothenberg, who later became the Goddard Director.

For years, some in NASA had viewed institutes as too independent and had discouraged the idea of similar institutes for other missions. A change in the guard at NASA and at the Institute, and the sustained outstanding performance of the Institute under Giacconi and his successor Bob Williams – former Director at Cerro Tololo – changed that. In December 1995, a high level NASA task force concluded that "independent" institutes are a good idea and that NASA should establish more of them.

Federal funds for science at NSF and NASA come, ultimately, from taxpayers. And taxpayers should benefit through new knowledge, through exploration of the solar system and the universe, through education, and through innovations that might result. We are working with both agencies to help communicate the often esoteric results of research at the frontiers of knowledge to students of all ages through schools, the media, and the World Wide Web. Colleagues at NOAO and at the Institute work with counterparts in museums, schools, and the media to assist them in their missions and to encourage them to use astronomy results. The World Wide Web makes it possible for more people to share in our science and its results, and to do so in greater depth and breadth than ever before. We help by reaching out to them directly, by providing good information, and by offering quality-monitored gateways to other sources.

Science is international; and Cerro Tololo *Interamerican* Observatory, the Hubble Space Telescope, and GONG, have significant and important international roles. The Gemini Project constitutes an additional step toward an international future for our science: costs and benefits are shared proportionately among Argentina, Brazil, Canada, Chile, the United Kingdom, and the United States. Unlike our other observatories, Gemini's primary mission is to equitably serve the six Gemini country communities, not primarily the US community and its associates in the Americas. Thus, the "customer" is the Gemini Board – which includes US astronomers – and not solely the US community. Much confusion ensued within the US community until this distinction was understood or accepted. Gemini is now in good shape scientifically, technically, and structurally. A first rate international staff runs the project, from Tucson, under the leadership of Matt Mountain who succeeded first Director Sidney Wolff.

Funds for science are scarce throughout the world. Most large projects may need international sponsors to proceed. Gemini represents a good way to organize such a project.

Experience with observatories and with Gemini is not the only attribute that prepares AURA well for undertaking international programs and projects. Our board and its councils and committees include international members. Universities in Chile, Mexico, and Canada are international affiliate members. In recognition of Gemini's international nature, the board delegated broader authority to management, and management will propose structural changes to enhance AURA's credibility as manager of international science activities.

In summary, we have much to be proud of. Hubble and its repair succeeded. WIYN and GONG are doing superbly. Gemini is underway, an international alliance forged and supported, and in good technical and management shape. Thanks to Sidney Wolff's leadership, NOAO is doing better than ever, despite the sharp decline in its budget during the past decade. We are dealing well with success and with adversity. NASA rated our performance for Hubble during the past two years at 100 percent and 99 percent, respectively. We made the concept of institutes work for NASA. We became international. Most importantly, we have top quality people at all levels, management and staff. They, of course, deserve the credit.

WIYN 3.5 meter telescope (D865). The WIYN telescope has been jointly funded by the University of Wisconsin, Indiana University, Yale University and NOAO. 40 percent of the time is assigned to NOAO, and the remaining 60 percent is shared by the universities in proportion to their financial contributions. NOAO will provide the funds for operation and maintenance. NOAO Photo Archives.

We are facing ever stiffer challenges. Can we do still more and better for still less? If not, and as budget constraints so require, which facilities should we close and why? How can we attain and maintain consensus among more colleagues on more issues that involve our observatories? Can we improve the climate for science in general, and astronomy in particular, including through a responsible and appropriate outreach program? Can we work even better with our sponsors to serve their needs and those of the communities they represent? Which new opportunities that come our way should we pursue? In addressing these challenges, we will be guided by what we believe is best for our science and for the astronomy community. We will look out for the people

WIYN 3.5 meter telescope dome (D877). NOAO Photo Archives.

who depend upon us and who work for science and the community. To do that, we must be willing to invest in the future: in people, in instruments, and in telescopes, whether or not there may be cause for optimism.

We are reviewing strategies and plans for the future of the centers and of AURA itself. We chartered the "Hubble Space Telescope and Beyond" study under chair Alan Dressler of Carnegie. Its report and recommendations are finding resonance in NASA and could help trigger a Golden Era of exploring space for earth-like planets. We assembled a community workshop on the future of NOAO and are ready to propose that NSF make capital investments to enable better science and reduce future operating costs. We hope to propose a major new solar telescope and enable the National Solar Observatory to consolidate its facilities. We are looking beyond Gemini on the international scene and within the USA. And we are encouraging innovation by our people everywhere. We may accept selected new challenges from NASA or from NSF.

Vision

What of the future? The new generation of private, national, and international telescopes will be built on Mauna Kea, on Cerro Pachon, and elsewhere. They will exceed our expectations. GONG will give us the first glimpse at the interior of the Sun and will help explain its structure, the generation of energy and neutrinos, and the solar cycle. A novel infrared solar telescope will enable new research on the corona and the solar disk. Ground based night-time observatories in both hemispheres will continue to extend their reach as new technology enables them to overcome many limitations of our atmosphere. Hubble will survive the hazards of space, receive new instruments and be boosted into a higher orbit, and perform superbly and cost-effectively far beyond its design life, without requiring Shuttle maintenance or repair after 2005. A 4 to 8-meter aperture space telescope for astronomy will be built and placed in orbit. The search for earth-like planets near other stars requires interferometry from earth orbits and beyond sources of zodiacal light. The limits of interferometry will first be explored on the ground, as a logical stepping stone toward the increasingly powerful space instruments that will follow. They will show if earth-like planets near other stars exist and will look for oxygen and other evidence for life.

Policy-makers will recognize that improvements in modern technologies, from optics to detectors to data processing and enhancement, require that they be challenged to their limits, and that this challenge can uniquely be found by applying them to astronomy and space science. They will also recognize that astronomy is uniquely successful in attracting students to technical study and careers, and that it captures the imagination of the public more than any other science. They will therefore support a stronger astronomy program at the NSF and will set visionary goals for NASA that are exciting to experts and lay audiences alike, and that stimulate technological innovation as they broaden our horizons.

The search for planets near other stars is one such goal. Another would be to establish a beachhead on another body in the solar system, one that would not be just another space ship which can decay in our atmosphere but a place to which humans can return when they are ready, and that they may occupy permanently some day. Still another goal would be to prepare to guard our titanic spaceship Earth against a cometary iceberg or similar threat. These goals seem more important than mere space tourism near earth. They may find resonance with experts and lay audiences alike.

Which goal to adopt is a policy call, more political than scientific. If the Tunguska object had fallen recently instead of in the 1910s, we might soon have found ourselves working on a crash program, perhaps called "space guard" for Earth, to prevent a recurrence. As it is, we can make a more considered choice. By selecting a goal that pushes technology to its limits, and by pursuing it with ingenuity and tenacity, we will advance science and our understanding of the world. Society will also benefit from innovation in technology and education. We will also attract a more diverse community of women and men from different cultures to the pursuit of the great intellectual challenges we humans share.

Notes and references

Chapter 1

1 Letter, Struve to Edmondson, March 19, 1940. Edmondson Personal Files.
2 Letter, Wells to Struve, July 17, 1940. Indiana University Archives.
3 Unpublished, Struve: Plan for Astronomical Collaboration in Connection with the McDonald Observatory, March 16, 1940. Yerkes Observatory Archives.
4 Letter, Hutchins to Rainey, April 9, 1940. Yerkes Observatory Archives.
5 Letter, Struve to Wells, December 27, 1940. Indiana University Archives.

Chapter 2

1 J. Merton England, *A Patron for Pure Science*, National Science Foundation, NSF 82–24, 1982.
2 Milton Lomask, *A Minor Miracle*, National Science Foundation, NSF 76–18, 1976.
3 *Astronomical Journal* **56**, No. 5, October, 1951, pp. 147–148.
4 David H. DeVorkin, "Back to the Future: American Astronomers' Response to the prospect of Federal Patronage, 1947–1955: The origins of the ONR and NSF Programs in astronomy." Unpublished manuscript, 1990.
5 Edmondson heard Klopsteg say this during a conversation that included a few other members of the ONR Committee.
6 Staff Notes on Ad Hoc Meeting of Astronomical Consultants, August 1, 1952.

Chapter 3

1 *Nature* **168**, September 1, 1951, 356.
2 Brief History of Steps Leading to the Establishment of the National Radio Astronomy Observatory. NSF draft, Sept. 10, 1963.
3 Transcript of Seeger interview with Edmondson, April 27, 1979.
4 *Journal of Geophysical Research* **59**, No. 1, March, 1954, pp. 149–201; *Science* **119**, April 30, 1954, 588.
5 Inter-University Radio Astronomy Observatory. MPE Study S-19, Peter van de Kamp.
6 Allan A. Needell: Lloyd Berkner, Merle Tuve, and the Federal Role in Radio Astronomy, Osiris, 2nd series, 1987, 3:261–288.
7 Minutes, NSF MPE Sciences Committees, May 24, 1956, 8:45 p.m.
8 Minutes of Conference on Radio Astronomy Facility, July 11, 1956.

9 Transcript of July 11, 1956 Conference.

10 Minutes of Joint Meeting of NSF Advisory Panel for Astronomical Observatory and NSF Advisory Panel on Radio Astronomy, July 23, 1956.

11 Copies from several sources: NSF History File; Leo Goldberg; Robert R. McMath Collection, Michigan Historical Collections, Bentley Historical Library, University of Michigan.

12 Transcript of Goldberg interview with Edmondson, September 6, 1978.

Chapter 4

1 *SCIENCE*, 115, No. 2983, February 29, 1952, 223–226.

2 Copy provided by Mrs. Ethel Carpenter.

3 Transcript of Irwin interview with Edmondson, August 22, 1978.

4 *Ibid.* Also, transcript of Keller interview with Edmondson, November 4, 1978.

5 It was not easy to find a copy of this proposal. NSF and the three universities did not save copies of proposals that were not funded. The surviving members of the first Astronomy Panel had not saved their copies. However, Leo Goldberg, who attended the August 1, 1952 meeting of the Panel, representing the MPE Divisional Committee, had enough feeling for history to save his copy. His secretary, Carol Gregory, located it in his personal files at KPNO in April 1978.

6 Staff Notes on Ad Hoc Meeting of Astronomical Consultants, August 1, 1952. NSF History Files.

7 Letter Stebbins to Seeger, July 15, 1952. NSF History Files.

8 Letter, Waterman to Carpenter, September 11, 1952. NSF History Files.

9 Letter, Klopsteg to Strand, November 19, 1952. NSF History Files.

10 Letter Klopsteg to McMath, December 16, 1952. NSF History Files.

11 Letter McMath to Whitford, May 14, 1953. Whitford Archives, University of Wisconsin Library.

12 Staff Notes on Meeting of Advisory Panel for Astronomy. February 5–6, 1953. NSF History Files.

13 Circular letter from Struve and Kuiper, March 9, 1953. Robert R. McMath Collection, Michigan Historical Collections, Bentley Historical Library, University of Michigan.

14 Letter, Struve to McMath, May 21, 1953. Robert R. McMath Collection, Michigan Historical Collections, Bentley Historical Library, University of Michigan.

15 Letter, Struve to McMath, June 11, 1953. Robert R. McMath Collection, Michigan Historical Collections, Bentley Historical Library, University of Michigan.

16 Letter, McMath to Klopsteg, June 23, 1953. Robert R. McMath Collection, Michigan Historical Collections, Bentley Historical Library, University of Michigan.

17 Letter, Seeger to McMath, August 5, 1953. Robert R. McMath Collection, Michigan Historical Collections, Bentley Historical Library, University of Michigan.

18 NSF Staff Meeting Notes, March 9, 1953. NSF History Files.

19 Memorandum, The Director to Assistant Director for Mathematical, Physical, and Engineering Sciences, March 16, 1953. NSF History Files.

20 NSF Staff Meeting Notes, May 11, 1953. NSF History Files.

21 Letter, Whitford to NSF (Subject: Grant NSF-0388), October 19, 1953. Whitford Archives, University of Wisconsin Library.

22 Letter, Seeger to Whitford, March 17, 1953. Whitford Archives, University of Wisconsin Library.

23 Letter, Seeger to Whitford, May 1, 1953. Whitford Archives, University of Wisconsin Library.

24 Letter, Putnam to Whitford, May 8, 1953. Whitford Archives, University of Wisconsin Library.

25 Letter, Whitford to Putnam, May 11, 1953. Whitford Archives, University of Wisconsin Library.

26 Letter, Putnam to Whitford, May 13, 1953. Whitford Archives, University of Wisconsin Library.

27 Letter, McMath to Whitford, May 14, 1953. Whitford Archives, University of Wisconsin Library.

28 Letter, Whitford to McMath, May 21, 1953. Whitford Archives, University of Wisconsin Library. The rumor was not unfounded. The NSF Summary Sheet for the Conference (3/4/53) included: "Proposal No. P-295, requesting funds for the establishment of a photo-electric observatory in Arizona, was rejected. It is now proposed to examine the possibility of utilizing the Lowell Observatory with some modernization for this purpose. Mr. R. Putnam, the sole trustee and comptroller of the Lowell Observatory, met with the NSF Ad Hoc Panel for Astronomical Instrumentation and is enthusiastic about the plan. The present conference is designed to determine whether the astronomers believe whether such an arrangement is possible."

29 Letter, Whitford to Struve, May 26, 1953. Whitford Archives, University of Wisconsin Library.

30 Letter, McMath to Whitford, May 27, 1953. Whitford Archives, University of Wisconsin Library.

31 Letter, McMath to Putnam, May 27, 1953. Whitford Archives, University of Wisconsin Library.

32 Letter, Struve to Whitford (longhand), June 1, 1953. Whitford Archives, University of Wisconsin Library.

33 Letter, McMath to Waterman, June 23, 1953. Mentioned in Reference 34.

34 Letter, McMath to Struve, June 23, 1953. Robert R. McMath Collection, Michigan Historical Collections, Bentley Historical Library, University of Michigan.

35 Letter, McMath to Whitford, June 23, 1953. Whitford Archives, University of Wisconsin Library.

36 Letter, Whitford to Seeger, June 28, 1953. Whitford Archives, University of Wisconsin Library.

37 Letter, Whitford to Goldberg. August 12, 1953. Goldberg Personal Files.

38 Letter, Stebbins to Goldberg, August 10, 1953. Goldberg Personal Files.

39 Letter, Goldberg to Stebbins, August 18, 1953. Goldberg Personal Files.

40 *Proceedings of the National Science Foundation Astronomical Photoelectric Conference*, pp. 107–108. Note: The Conference was held August 31 – September 1, 1953. The Proceedings were published October 1955.

41 Transcript of Goldberg interview with Edmondson, September 6, 1978.

42 Op. cit., Reference 40, p. 121.

43 *Ibid.*, pp. 121–122.

44 Transcript of Giclas interview with Edmondson, August 18, 1978.

45 Transcript of Irwin interview with Edmondson, August 22, 1978.

46 MPE Study S-9, November 1953. NSF History Files.

47 *Astronomical Journal*, **60**, No. 1224, January, 1955, 17–32.

48 Op. cit., Reference 40.

49 Letter, Whitford to Goldberg, December 7, 1953. Robert R. McMath Collection, Michigan Historical Collections, Bentley Historical Library, University of Michigan.

50 Letter, Seeger to Goldberg, December 7, 1953. NSF History Files.

51 MPE Study S-18, Peter van de Kamp, May 1955, pp. 5–7. NSF History Files.

52 Diary Note, Seeger, January 15, 1954. NSF History Files.

53 Letter, McMath to Seeger, January 14, 1954. NSF History Files.

54   The charge to the proposed panel would be to advise the NSF of the general astronomical needs that can be met by a specific plan for a national astronomical observatory. In this connection, the Foundation should be informed as to (1) what research and education programs might be undertaken in such an institution,

(2) what facilities would be required, including possible sites, proposed instrument designs, and estimated costs, (3) what organization, including full-time and part-time personnel, would be envisaged, and (4) what fiscal arrangements would be recommended for its permanent operation. May I remind you, also, that the Foundation is interested in a cooperative observatory not merely for photo-electric research as originally proposed, but rather for all associated research. Finally, no time limit would be set on the report of the panel. It is hoped, however, that there would be progress reports from time to time and that every effort would be made to have a complete report rather than a quick one. The Foundation would like the panel to keep the NSF, as well as the American Astronomical Society, informed of its activities.

Letter, Seeger to McMath, January 28, 1954. Robert R. McMath Collection, Michigan Historical Collections, Bentley Historical Library, University of Michigan. Also NSF History Files.

55 Letter, Seeger to McMath, April 19, 1954. NSF History Files.

Chapter 5

1 The information about McMath comes from several sources: Remarks by Leo Goldberg at the 25th Anniversary celebration of the McMath–Hulbert Observatory (Portfolio, Robert R. McMath Collection, Michigan Historical Collections, Bentley Historical Library, University of Michigan); taped interviews with Gilbert L. Lee, Jr. (1978, June 26), Leo Goldberg (1978, September 6 and 1979, February 13), Paul E. Klopsteg (1979, February 28), A. Keith Pierce (1979, March 21), Orren C. Mohler (1979, July 11), Walker L. Cisler (1980, July 31); McMath Obituary, I.S. Bowen, *Year Book*, American Philosophical Society 1962, pp. 149–153.

2 McMath and Waterman became acquainted through their wartime service with the Office of Scientific Research and Development. McMath and Klopsteg became acquainted when they served on the National Research Council's Advisory Committee on Artificial Limbs during the war.

3 Letter, McMath to Struve, May 14, 1953. Robert R. McMath Collection, Michigan Historical Collections, Bentley Historical Library, University of Michigan. The letter reports the successful meeting with Mr. Dodge: "I had the pleasure of intervening with Mr. Dodge, the budget director, who is an old Detroiter. At least, we got the NSF pegged at a decent figure as part of the Eisenhower budget. The Truman budget having become meaningless." The account of the conversation is based on Edmondson's recollection of hearing McMath describe it on several occasions.

4 Letter, Klopsteg to McMath, August 13, 1953. Robert R. McMath Collection, Michigan Historical Collections, Bentley Historical Library, University of Michigan.

5 Letter, Klopsteg to McMath, August 5, 1953. Robert R. McMath Collection, Michigan Historical Collections, Bentley Historical Library, University of Michigan.

6 Letter, McMath to Klopsteg, August 8, 1953. Robert R. McMath Collection, Michigan Historical Collections, Bentley Historical Library, University of Michigan.

7 Minutes, NSF Advisory Panel for NAO, Nov. 4–5, 1954. Peter van de Kamp Personal Files; also Robert R. McMath Collection, Michigan Historical Collections, Bentley Historical Library, University of Michigan.

8 Letter, McMath to Struve, January 30, 1954. Robert R. McMath Collection, Michigan Historical Collections, Bentley Historical Library, University of Michigan.

9 Letter, Struve to McMath, February 16, 1954. Robert R. McMath Collection, Michigan Historical Collections, Bentley Historical Library, University of Michigan.

10 Diary Note. Waterman, undated. NSF History Files.

11 Letter, Waterman to McMath, March 8, 1954. NSF History Files; also Robert R. McMath

Collection, Michigan Historical Collections, Bentley Historical Library, University of Michigan.

12 Letter, McMath to Klopsteg, September 3, 1954. Robert R. McMath Collection, Michigan Historical Collections, Bentley Historical Library, University of Michigan. "Last March 26 I filled out a PSQ for your great organization and some time thereafter an ill-mannered Civil Service investigator appeared here in the office. It would seem to me that with all the clearances from various government agencies now in my possession, it should not take much more than six months to say that I am not a satisfactory consultant for the NSF or possibly – yes."

13 Letter, Bowen to McMath, October 6, 1954. Robert R. McMath Collection, Michigan Historical Collections, Bentley Historical Library, University of Michigan.

> The reason I raise this question is that when I returned from my vacation on Monday I was met by two F.B.I. agents with a curt demand for an explanation of what they interpreted as a discrepancy between my recent personal statement for the present clearance and earlier statements. If I remember correctly my earlier investigation was made by the Navy Intelligence and as nearly as I can figure out the difficulty arose because the Navy Intelligence turned over to the F.B.I. certain data without the accompanying explanation. This combined with a similarity of names made it appear that I had been a member of a subversive organization without having reported it on the present personal statement. Since the actual record is clear I am sure there will be no difficulty in straightening things out but it raises the question as to whether it may not delay clearance beyond the time of the meeting, at least in my case, and consequently make attendance by myself of doubtful advisability.

14 Letter, Struve to McMath, May 10, 1954. Robert R. McMath Collection, Michigan Historical Collections, Bentley Historical Library, University of Michigan.

15 Letter, McMath to NAO Panel, September 30, 1954. Robert R. McMath Collection, Michigan Historical Collections, Bentley Historical Library, University of Michigan.

> "Upon the occasion of my last long visit to Washington, I took up this meeting problem with Drs. Klopsteg and van de Kamp. It turns out that owing to lack of travel funds in the Foundation, it will be necessary for the University of Michigan to apply for a grant so that it can act as host and reimburse you gentlemen for your expenses. It was decided that the University of Michigan, through Dr. Goldberg, would ask for sufficient funds for three meetings at this time, and that the second and third meetings would make provision for our consultants.

16 Letter, Goldberg to van de Kamp, September 24, 1954. NSF Historical Files (NSF/Contracts Michigan 2).

17 Letter, Cummiskey to van de Kamp, September 27, 1954. NSF Historical Files (NSF/Contracts Michigan 2).

18 Letter, Waterman to Hatcher, November 1, 1954. NSF Historical Files (NSF/Contracts Michigan 2).

19 Op cit. Reference 7.

20 Letter, McMath to van de Kamp, December 2, 1954. Robert R. McMath Collection, Michigan Historical Collections, Bentley Historical Library, University of Michigan. The words "and orders" are not in the Minutes of the Panel meeting.

21 Memorandum of conversation between Drs. Leonard Carmichael and Robert R. McMath, Washington, October 25, 1954, in re staff appointments and research program of the Astrophysical Division of the Smithsonian Institution. Robert R. McMath Collection, Michigan Historical Collections, Bentley Historical Library, University of Michigan.

22 Letter, Meinel to McMath, November 29, 1954. Robert R. McMath Collection, Michigan Historical Collections, Bentley Historical Library, University of Michigan. The letter says:

"Bengt telephoned me, etc." Meinel's recollection of what happened is that Stromgren phoned him from Ann Arbor, but told his wife, Marjorie, owing to his absence in Washington for IGY planning meetings. She gave him the message when she met his train at Walworth. [Transcript of Meinel interview with Edmondson, February 5, 1989.].

23 Letter, McMath to Meinel, December 2, 1954. Robert R. McMath Collection, Michigan Historical Collections, Bentley Historical Library, University of Michigan.

24 Letter, McMath to van de Kamp, December 4, 1954. Robert R. McMath Collection, Michigan Historical Collections, Bentley Historical Library, University of Michigan.

25 Letter, van de Kamp to McMath, December 8, 1954. Robert R. McMath Collection, Michigan Historical Collections, Bentley Historical Library, University of Michigan.

26 Letter, Meinel to Bowen, December 15, 1954. Robert R. McMath Collection, Michigan Historical Collections, Bentley Historical Library, University of Michigan.

27 Letter, McMath to Bowen, December 18, 1954. Robert R. McMath Collection, Michigan Historical Collections, Bentley Historical Library, University of Michigan.

28 Letter, McMath to Meinel, December 18, 1954. Robert R. McMath Collection, Michigan Historical Collections, Bentley Historical Library, University of Michigan.

29 Letter, Bowen to Meinel, December 23, 1954. Robert R. McMath Collection, Michigan Historical Collections, Bentley Historical Library, University of Michigan.

30 Letter, van de Kamp to McMath, December 23, 1954. Robert R. McMath Collection, Michigan Historical Collections, Bentley Historical Library, University of Michigan.

31 Letter, McMath to van de Kamp, December 28, 1954. Robert R. McMath Collection, Michigan Historical Collections, Bentley Historical Library, University of Michigan.

32 Letter, Goldberg to van de Kamp, December 30, 1954. Robert R. McMath Collection, Michigan Historical Collections, Bentley Historical Library, University of Michigan.

33 Memorandum, McMath to Members of NAO–NSF Panel, January 25, 1955. Robert R. McMath Collection, Michigan Historical Collections, Bentley Historical Library, University of Michigan.

34 Letter, Watkins to Waterman, February 7, 1955. Robert R. McMath Collection, Michigan Historical Collections, Bentley Historical Library, University of Michigan.

35 Letter, Weddell to Lee, January 28, 1955. Robert R. McMath Collection, Michigan Historical Collections, Bentley Historical Library, University of Michigan.

36 Letter, Whitford to McMath, January 27, 1955. Robert R. McMath Collection, Michigan Historical Collections, Bentley Historical Library, University of Michigan.

37 Letter (longhand), Struve to McMath, January 29, 1955. Robert R. McMath Collection, Michigan Historical Collections, Bentley Historical Library, University of Michigan.

38 Letter, Meinel to McMath, February 3, 1955 with enclosed "Report on Conferences, January 20–26, 1955." Robert R. McMath Collection, Michigan Historical Collections, Bentley Historical Library, University of Michigan.

39 Letter, Babbitt to Harvill, January 27, 1955. Lowell Observatory Archives. This letter is full of misinformation that was given to Babbitt by Meinel, starting with "Executive Scientist of the National Science Foundation" as Meinel's title. The letter continues: "Apparently there is a good chance for Arizona to obtain a large research observatory to be located some place in Arizona, if a sponsoring institution can be secured in this state." The letter concludes: "I, of course, do not know of the financial obligations that would be involved but if these were not too heavy and there are not other phases about this project which I do not understand, the University would be passing up an unusual opportunity if it did not at least investigate this matter very thoroughly. The University of Michigan apparently is already working very aggressively to obtain this project regardless of where location might be. If we are to be successful in getting it, it will take a united front and a great deal of aggressive work by all the parties concerned.

I will greatly appreciate your giving this matter your earnest consideration."

40 Letter, Meinel to McMath, February 19, 1955. Robert R. McMath Collection, Michigan

Historical Collections, Bentley Historical Library, University of Michigan. This letter begins: "I hope that I have not caused you undue worry about tackling things too fast." Meinel was concerned about "the small operational budget" that the NAO Panel had proposed, and felt that outside support would be needed. The letter continues:

A rapid series of events encouraged me to tackle the State of Arizona for the needed support, yet avoid the association of the NAO with the existing state groups. Since such a maneuver would be difficult, if not impossible, within the official Panel channels, I attempted it as an extracurricular activity. My hope was to get a policy decision on the part of the state. I very clearly pointed out that the NAO would be placed where scientific conditions so determined, but pointed out the obvious reasons for the state of Arizona to be the operating agency for the NAO if it were in the state. The strategy was successful, to the extent that an independent campus like Lick was recommended. I expect that President Harvill will drop you a line to set official channels for the next step.

The letter concludes:

Since I have no Panel policy to follow in the above topic I have had a bit of adventure in lining up information for the Panel to consider. The possibility of major new resources for astronomy from Arizona should be a welcomed event aside from the problems it solves for the NAO, so I don't think that you could blame me for attempting to set it up. My chief apprehension is now that the University of Michigan might have had its eye on the NAO, in which case my actions would hardly be pleasant news!

I would very much appreciate it if you would find the opportunity to drop me a reply on your thoughts regarding the questions of title, budget, etc., in view of the opportunity to get solid financial support from Arizona. In conclusion, I want to assure you that the Arizona developments have been quite orderly and are well under control.

41 Letter, McMath to Meinel, February 10, 1955. Robert R. McMath Collection, Michigan Historical Collections, Bentley Historical Library, University of Michigan.

42 Letter, Meinel to McMath, February 14, 1955. Robert R. McMath Collection, Michigan Historical Collections, Bentley Historical Library, University of Michigan.

43 Memorandum, McMath to Members of the NAO–NSF Panel, February 16, 1955. Robert R. McMath Collection, Michigan Historical Collections, Bentley Historical Library, University of Michigan.

44 Letter, McMath to Bowen, February 16, 1955. Robert R. McMath Collection, Michigan Historical Collections, Bentley Historical Library, University of Michigan.

45 Letter, Bowen to McMath, February 21, 1955. Robert R. McMath Collection, Michigan Historical Collections, Bentley Historical Library, University of Michigan.

46 Letter, McMath to Bowen, February 24, 1955. Robert R. McMath Collection, Michigan Historical Collections, Bentley Historical Library, University of Michigan.

47 Letter, Harvill to Seeger, February 14, 1955. Robert R. McMath Collection, Michigan Historical Collections, Bentley Historical Library, University of Michigan. Also in NSF History Files.

48 Letter, Harvill to McMath, February 24, 1955. Robert R. McMath Collection, Michigan Historical Collections, Bentley Historical Library, University of Michigan. Also in NSF History Files.

49 Memorandum, McMath to Members of NAO-NSF Panel, February 26, 1955. Robert R. McMath Collection, Michigan Historical Collections, Bentley Historical Library, University of Michigan. Also in NSF History Files.

50 Letter, McMath to Harvill, February 26, 1955. Robert R. McMath Collection, Michigan Historical Collections, Bentley Historical Library, University of Michigan. Also in NSF History Files.

51 Letter, Harvill to McMath, March 7, 1955. Robert R. McMath Collection, Michigan Historical Collections, Bentley Historical Library, University of Michigan.

52 Diary Note, ATW, March 3, 1955. NSF History Files.

53 Transcript of Harvill interview with Edmondson, August 30, 1978.

54 Letter, McMath to van de Kamp, March 3, 1955. Robert R. McMath Collection, Michigan Historical Collections, Bentley Historical Library, University of Michigan.

55 Letter, McMath to van de Kamp, March 3, 1955. Robert R. McMath Collection, Michigan Historical Collections, Bentley Historical Library, University of Michigan.

56 Diary Note, van de Kamp, March 4, 1955. NSF History Files.

57 Letter, Bowen to McMath, March 8, 1955. Robert R. McMath Collection, Michigan Historical Collections, Bentley Historical Library, University of Michigan.

58 Letter, McMath to Bowen, March 12, 1955. Robert R. McMath Collection, Michigan Historical Collections, Bentley Historical Library, University of Michigan.

59 Letter, McMath to Meinel, March 12, 1955. Robert R. McMath Collection, Michigan Historical Collections, Bentley Historical Library, University of Michigan.

60 Letter, Meinel to McMath, March 14, 1955. Robert R. McMath Collection, Michigan Historical Collections, Bentley Historical Library, University of Michigan.

61 Memorandum, McMath to Panel, March 15, 1955. Robert R. McMath Collection, Michigan Historical Collections, Bentley Historical Library, University of Michigan.

62 Letter, McMath to Bowen, March 17, 1955. Robert R. McMath Collection, Michigan Historical Collections, Bentley Historical Library, University of Michigan.

63 Letter, McMath to Bowen and Struve, March 19, 1955. Robert R. McMath Collection, Michigan Historical Collections, Bentley Historical Library, University of Michigan.

64 Letter, Bowen to McMath, March 25, 1955. Robert R. McMath Collection, Michigan Historical Collections, Bentley Historical Library, University of Michigan.

65 Letter, McMath to van de Kamp, April 12, 1955. Robert R. McMath Collection, Michigan Historical Collections, Bentley Historical Library, University of Michigan.

66 Letter, Struve to McMath, April 2, 1955. Robert R. McMath Collection, Michigan Historical Collections, Bentley Historical Library, University of Michigan. Struve promised to write "a few paragraphs concerning the educational value of the proposed observatory."

67 Letter, Meinel to McMath, April 8, 1955. Robert R. McMath Collection, Michigan Historical Collections, Bentley Historical Library, University of Michigan.

68 Letter, McMath to Meinel, April 9, 1955. Robert R. McMath Collection, Michigan Historical Collections, Bentley Historical Library, University of Michigan. The U.S. Postal Service gave quick service in 1955.

69 Letter, Goldberg to Meinel, April 11, 1955. Robert R. McMath Collection, Michigan Historical Collections, Bentley Historical Library, University of Michigan.

70 Transcript of Meinel interview with Edmondson, February 5, 1979.

71 Transcript of Abt interview with Edmondson, March 27, 1979.

72 Letter, Meinel to McMath, May 5, 1955. Robert R. McMath Collection, Michigan Historical Collections, Bentley Historical Library, University of Michigan.

73 Log Book, National Astronomical Observatory Site Survey, Helmut A. Abt. This is a 123-page notebook containing Abt's longhand notes, and some by Meinel, for the period May 7, 1955 to June 1, 1956. Archives, Kitt Peak National Observatory Library, Tucson.

74 Memorandum, McMath to NAO–NSF Advisory Panel, April 30, 1955. Robert R. McMath Collection, Michigan Historical Collections, Bentley Historical Library, University of Michigan.

75 MPE Study S-18, Third Draft, National Science Foundation, Peter van de Kamp, May 3, 1955. NSF History Files. The final draft, dated May 1955, was entitled "Inter-University Astronomical Observatory." There were now five recommendations for NSF action. Operation by a group of universities was still recommended, but the question of ownership was not mentioned.

76 Memorandum, McMath to NAO–NSF Advisory Panel, May 10, 1955. Letter, Whitford to McMath, May 12, 1955. Letter, Stromgren to McMath, May 16, 1955. Robert R. McMath Collection, Michigan Historical Collections, Bentley Historical Library, University of Michigan.

77 Memorandum, McMath to NAO–NSF Advisory Panel, May 21, 1955. Robert R. McMath Collection, Michigan Historical Collections, Bentley Historical Library, University of Michigan.

78 Memorandum, Program Director for Engineering Sciences to Program Director for Astronomy, May 6, 1955. NSF History Files.

79 Diary Note, Seeger, May 25, 1955. NSF History Files.

80 Two memoranda, McMath to NAO–NSF Advisory Panel, June 2, 1955. Robert R. McMath Collection, Michigan Historical Collections, Bentley Historical Library, University of Michigan.

81 Op. cit. Reference 11.

82 Letter, Whitford to McMath, June 7, 1955. Robert R. McMath Collection, Michigan Historical Collections, Bentley Historical Library, University of Michigan.

83 Letter, Bowen to McMath, June 7, 1955. Robert R. McMath Collection, Michigan Historical Collections, Bentley Historical Library, University of Michigan.

84 Letter, McMath to Bowen, June 9, 1955. Robert R. McMath Collection, Michigan Historical Collections, Bentley Historical Library, University of Michigan.

85 "Inter-university Observatory" was the name used in the final draft of MPE Study S-18. NSF History Files. The Annual Reports of the National Science Foundation for FY 54 and 55 used the title: "Advisory Panel for National Astronomical Observatory." The Reports for FY 56 and 57 used the title: "Advisory Panel for Astronomical Observatory."

86 Memorandum, McMath to NAO–NSF Panel, June 9, 1955. Robert R. McMath Collection, Michigan Historical Collections, Bentley Historical Library, University of Michigan.

87 Letter, Struve to McMath, June 13, 1955. Robert R. McMath Collection, Michigan Historical Collections, Bentley Historical Library, University of Michigan.

88 Letter, Stromgren to McMath, June 23, 1955. Robert R. McMath Collection, Michigan Historical Collections, Bentley Historical Library, University of Michigan.

89 Memorandum, McMath to NAO–NSF Panel, June 22, 1955. Robert R. McMath Collection, Michigan Historical Collections, Bentley Historical Library, University of Michigan. Also, Diary Note, van de Kamp, June 17, 1955, and Diary Note, Waterman, June 21, 1955. NSF History Files.

90 Memorandum, Goldberg to NAO–NSF Panel, June 24, 1955. Robert R. McMath Collection, Michigan Historical Collections, Bentley Historical Library, University of Michigan.

91 Letter, Watkins to Waterman, July 6, 1955. Robert R. McMath Collection, Michigan Historical Collections, Bentley Historical Library, University of Michigan.

92 Letter, Seeger to McMath, July 19, 1955. Robert R. McMath Collection, Michigan Historical Collections, Bentley Historical Library, University of Michigan.

93 Letter, McMath to Seeger, July 21, 1955. Robert R. McMath Collection, Michigan Historical Collections, Bentley Historical Library, University of Michigan.

94 Letter, McMath to Seeger, July 21, 1955. Robert R. McMath Collection, Michigan Historical Collections, Bentley Historical Library, University of Michigan.

95 Letter, McMath to Waterman, July 28, 1955. Robert R. McMath Collection, Michigan Historical Collections, Bentley Historical Library, University of Michigan.

96 Letter and Proposal, Cummiskey to van de Kamp, July 27, 1955. NSF History Files.

97 Cable, Waterman to McMath, August 26, 1955. Robert R. McMath Collection, Michigan Historical Collections, Bentley Historical Library, University of Michigan.

98 Letter, Watkins to Waterman, September 1, 1955. Robert R. McMath Collection, Michigan Historical Collections, Bentley Historical Library, University of Michigan.

99 Letter, McMath to Waterman, September 15, 1955. Robert R. McMath Collection, Michigan Historical Collections, Bentley Historical Library, University of Michigan.

100 Minutes, Third Meeting of National Astronomical Observatory Panel, Dublin, Ireland, August 31, 1955. NSF History Files.

101 Letter, McMath to Waterman, September 15, 1955. [Not the same as 99)]. Robert R. McMath Collection, Michigan Historical Collections, Bentley Historical Library, University of Michigan.

102 Letter, Waterman to McMath, September 26, 1955. Robert R. McMath Collection, Michigan Historical Collections, Bentley Historical Library, University of Michigan.

103 Letter, Waterman to McMath, November 4, 1955. Robert R.McMath Collection, Michigan Historical Collections, Bentley Historical Library, University of Michigan.

104 Letter, McMath to Nier, September 20, 1955. Robert R. McMath Collection, Michigan Historical Collections, Bentley Historical Library, University of Michigan.

105 Letter, Goldberg to Nier, October 7, 1955. Robert R. McMath Collection, Michigan Historical Collections, Bentley Historical Library, University of Michigan.

106 Transcript of Goldberg interview with Edmondson, September 6, 1978.

107 Op. cit. Reference 73, p. 61.

108 Letter, Meinel to McMath, May 12, 1955; Letter and Report, Meinel to McMath, June 15, 1955; and Letter, McMath to Meinel, July 21, 1955. Robert R. McMath Collection, Michigan Historical Collections, Bentley Historical Library, University of Michigan.

109 Letter, Meinel to McMath, July 27, 1955. Robert R. McMath Collection, Michigan Historical Collections, Bentley Historical Library, University of Michigan.

110 Letter, Meinel to McMath, September 6, 1955. Robert R. McMath Collection, Michigan Historical Collections, Bentley Historical Library, University of Michigan.

111 Memorandum, McMath to NAO–NSF Advisory Panel, September 20, 1955. Robert R. McMath Collection, Michigan Historical Collections, Bentley Historical Library, University of Michigan.

112 Memorandum, McMath to NAO–NSF Advisory Panel, October 4, 1955. Robert R. McMath Collection, Michigan Historical Collections, Bentley Historical Library, University of Michigan.

113 Letter, Seeger to McMath, September 13, 1955. Robert R. McMath Collection, Michigan Historical Collections, Bentley Historical Library, University of Michigan.

114 Memorandum, Hogg to "Dear Fellow Member of the American Astronomical Society," October 18, 1955. NSF History Files.

115 Memorandum, Waterman to Hall, November 3, 1955; and Memorandum, Hall to Waterman, November 16, 1955. NSF, Office of the Director Subject Files, 1949–56, Box 9, Division of MPE Sciences, Astronomy (Federal Records Center).

116 Letter, Meinel to McMath, October 26, 1955. Robert R. McMath Collection, Michigan Historical Collections, Bentley Historical Library, University of Michigan.

117 Op. cit. Reference 96.

118 Letter, Meinel to McMath, August 19, 1955. Robert R. McMath Collection, Michigan Historical Collections, Bentley Historical Library, University of Michigan.

119 Letter, Meinel to Haverland, August 19, 1955. Robert R. McMath Collection, Michigan Historical Collections, Bentley Historical Library, University of Michigan.

120 Letter, Meinel to Waterman, August 19, 1955. Robert R. McMath Collection, Michigan Historical Collections, Bentley Historical Library, University of Michigan.

121 Letter, Waterman to Emmons, September 13, 1955. Robert R. McMath Collection, Michigan Historical Collections, Bentley Historical Library, University of Michigan.

122 Letter, Hawley to Haverland, September 12, 1955. Robert R. McMath Collection, Michigan Historical Collections, Bentley Historical Library, University of Michigan.

123 Letter, Carpenter to Meinel, October 17, 1955. Robert R. McMath Collection, Michigan Historical Collections, Bentley Historical Library, University of Michigan.

124 Transcripts: Harvill interview with Edmondson, August 30, 1978; Spicer interviews with Edmondson, January 25 and February 8, 1979; Haury interview with Edmondson, March 16, 1979.

125 Letter, Rosamond B. Spicer to Edmondson, September 25, 1989.

126 Transcript: Meinel interview with Edmondson, February 3, 1979.

127 Letter, Gilmore to Haverland, December 13, 1955. NSF History Files; Robert R. McMath Collection, Michigan Historical Collections, Bentley Historical Library, University of Michigan.

128 Letter and enclosure, Haverland to Abt, December 15, 1955. Robert R. McMath Collection, Michigan Historical Collections, Bentley Historical Library, University of Michigan.

129 Letter, Meinel to Haverland, December 20, 1955. Robert R. McMath Collection, Michigan Historical Collections, Bentley Historical Library, University of Michigan.

130 Letter, McMath to Carpenter, December 27, 1955. Robert R. McMath Collection, Michigan Historical Collections, Bentley Historical Library, University of Michigan.

131 Resolution No. 860 of the Papago Council, January 6, 1956. Robert R. McMath Collection, Michigan Historical Collections, Bentley Historical Library, University of Michigan.

132 Letter, Meinel to McMath, January 3, 1956. Robert R. McMath Collection, Michigan Historical Collections, Bentley Historical Library, University of Michigan.

133 Letter, Meinel to Hogg, March 16, 1956. NSF History Files.

134 Letter Meinel to McMath, October 11, 1956. NAO Panel Archives, Kitt Peak National Observatory Library, Tucson.

Chapter 6

1 Letter, McMath to Bowen, August 17, 1954. Robert R. McMath Collection, Michigan Historical Collections, Bentley Historical Library, University of Michigan.

2 Report, National Observatory, August 1954. Longhand draft in Bowen's NAO Panel Files, in storage at KPNO. Found by Muriel Fults and sent to Edmondson November 20, 1989.

3 Memorandum. McMath to Drs. Bowen, Stromgren, Struve, and Whitford, Members of the National Observatory Committee of the National Science Foundation, May 12, 1954. Robert R. McMath Collection, Michigan Historical Collections, Bentley Historical Library, University of Michigan.

4 Memorandum, McMath to Drs. Stromgren, Struve, and Whitford, Members of the National Observatory Committee, August 18, 1954. Robert R. McMath Collection, Michigan Historical Collections, Bentley Historical Library, University of Michigan.

5 Letter, McMath to Struve, March 24, 1953. Robert R. McMath Collection, Michigan Historical Collections, Bentley Historical Library, University of Michigan. This was in response to the March 9, 1953 Struve–Kuiper circular letter to observatory directors. Op. cit. 4, Reference 13.

6 Letter, McMath to Bowen, August 21, 1954. Robert R. McMath Collection, Michigan Historical Collections, Bentley Historical Library, University of Michigan.

7 Op. cit. Reference 1.

8 Letter, McMath to Struve, August 30, 1954. Robert R. McMath Collection, Michigan Historical Collections, Bentley Historical Library, University of Michigan.

9 Letter Struve to McMath, August 24, 1954. Robert R. McMath Collection, Michigan Historical Collections, Bentley Historical Library, University of Michigan. Struve proposed that NSF be asked immediately for a substantial grant to start a site survey and engineering work on a 60-inch telescope. He recommended that John Irwin be employed to conduct the site survey. McMath agreed that "John Irwin would make a splendid 'appointee' of our committee." However, Irwin's name was not one of the five on the list considered at the first meeting of the NAO Panel, according to van de Kamp's notes. Edmondson personal files.

10 Op. cit. 5, Reference 7.

11 Minutes, NSF Advisory Panel for NAO, March 29–31, 1955. Robert R. McMath Collection. Michigan Historical Collections, Bentley Historical Library, University of Michigan.

12 Op. cit. 5, References 90 and 91.

13 Proposal to the National Science Foundation from Association of Universities for Research in Astronomy, Inc. for the construction and operation of a Cooperative Astronomical Observatory, page 4, October 29, 1957. Edmondson personal files.

14 Letter, Meinel to McMath, October 26, 1955. Robert R. McMath Collection, Michigan Historical Collections, Bentley Historical Library, University of Michigan.

15 Letter, McMath to Meinel, November 5, 1955. Robert R. McMath Collection, Michigan Historical Collections, Bentley Historical Library, University of Michigan.

16 Transcript of Abt interview with Edmondson, March 27, 1979. Minutes, AURA Scientific Committee, March 1, 1959, p. 2. Edmondson personal files.

17 Letter, Meinel to McMath, May 21, 1956. Lists the four Arizona sites as having been chosen for the erection of the 60-foot towers. NAO Panel Archives, Kitt Peak National Observatory Library, Tucson. Undated list entitled "Astronomical Observatory, Testing Sites." NSF History Files, NSF/Contracts, Michigan. NSF Press Release: "National Science Foundation Grants $545,000 to the University of Michigan for Studies relating to an Optical Astronomy Observatory." For release Sunday papers August 26, 1956. NSF History Files. *New York Herald Tribune*, September 5, 1956; *Seattle Post-Intelligencer*, date missing. NAO Panel Archives, Kitt Peak National Observatory Library, Tucson.

18 The need for a large solar telescope at a site with superb daytime seeing was discussed at each of the first three meetings of the NAO Panel. The panel voted at the third meeting to instruct Meinel to consider the solar daytime seeing problem and a possible site for a large solar telescope in addition to the search for a site with excellent night-time stellar seeing. Minutes, Third Meeting of National Astronomical Observatory Panel, August 31, 1955, Dublin, Ireland. NSF Historical Files.

19 Helmut A. Abt: Log Book, National Astronomical Observatory Site Survey, pp. 91–94. Archives, Kitt Peak National Observatory Library, Tucson.

20 Leon E. Salanave, Mars Observing Expedition in California, Sky and Telescope, August 1956, p. 469. Leon E. Salanave, Observing at Junipero Serra Peak, Sky and Telescope, May 1957, pp. 320–321. Salanave also presented reports at the August 24–28, 1956 and December 26–29, 1956 meetings of the American Astronomical Society, Astronomical Journal 62, pp. 31–32, February 1957 and p. 98, May 1957.

21 Op. cit. Reference 19, pp. 100–105, 113–116, 120–122. He had previously visited this area on December 7–8, 1955 and the Sierra Ancha Mountains are mentioned in the Log, p. 69.

22 Letter, Barker to Weddell, July 10, 1956; Letter, Meinel to McMath, July 11, 1956; Letter, McMath to Meinel, July 19, 1956; Letter, Harold A. Beelar (attorney-at-law) to Meinel, August 7, 1956; Letter, Meinel to McMath, August 14, 1956. Robert R. McMath Collection, Michigan Historical Collections, Bentley Historical Library, University of Michigan. Meinel's July 11 letter to McMath described it as "accidentally setting foot on a mining claim." His August 14 letter more accurately called it "the accidental casting of a tower foundation on a mining claim."

23 Letter, McMath to Meinel, August 18, 1956. Robert R. McMath Collection, Michigan Historical Collections, Bentley Historical Library, University of Michigan.

24 Letter, McMath to Goldberg, August 18, 1956; Memo, KW to LG, August 20, 1956. Robert R. McMath Collection, Michigan Historical Collections, Bentley Historical Library, University of Michigan. McMath thought Mr. Beelar was "the lessors' lawyer." Mrs. Weddell described him as the "attorney employed by Meinel 'to secure surface clearance on McFadden Peak, Sierra Ancha Mts.'" Both may have been right because Abt's recollection is: "It turned out they picked the same lawyer as the rancher was using." Transcript, Abt interview with Edmondson, March 27, 1979. It should be noted that the NAO Panel had discussed the legal problem of mining claims and grazing rights a year before this event

and had concluded it was "advisable to avoid any detailed site examinations for even Phase I until this question is answered." Minutes, NAO Panel, Second Meeting, March 29–31, 1955, Ann Arbor. NSF History Files. Furthermore, Meinel wrote to McMath on August 19, 1955 that "the Sierra Ancha region is definitely excluded since the extent and quality of the Uranium deposits is leading to the erection of a processing plant in the area." Robert R. McMath Collection, Michigan Historical Collections, Bentley Historical Library, University of Michigan. His May 21, 1956 letter to McMath says: "Excavations for the pier are being made this week at the Sierra Ancha site since it is easiest to reach." Op. cit. Reference 17. No record has been found to explain the change in the status of this site.

25 A.B. Meinel, Final Report on the Site Selection Survey for the National Astronomical Observatory, March 1, 1958. Edmondson personal files.

26 Minutes, Fourth Meeting of National Astronomical Observatory Panel, February 25–26, 1956, Pasadena, California. NSF History Files.

27 Memorandum, Ruth S. Wallace to Drs. Bowen, Stromgren, Struve, Whitford, members of NAO–NSF Advisory Panel, April 4, 1956. Bowen NAO Panel Files.

28 Letter, Meinel to McMath, September 12, 1956. NAO Panel Archives, Kitt Peak National Observatory Library, Tucson. "It looks to be harder to test than even Kitt Peak because it is completely inaccessible during the months November to April because of the seasonal rains."

29 Memorandum, McMath to Drs. Bowen, Stromgren, Struve, Whitford, Members of NAO–NF Advisory Panel, October 11, 1956 and October 23, 1956. NAO Panel Archives, Kitt Peak National Observatory Library, Tucson.

30 Letter, Salanave to McMath, January 15, 1957. NAO Panel Archives, Kitt Peak National Observatory Library, Tucson.

31 Minutes, Fifth Meeting of Advisory Panel for Astronomical Observatory, February 25–26, 1957. NSF History Files.

32 Leon Salanave, Part I of Report on Pacific Islands Survey, May 1, 1957. Solar Technical Reports, Kitt Peak National Observatory Library, Tucson.

33 Letter, Meinel to McMath, September 12, 1956. [Note: Not the same letter as Reference 28.] NAO Panel Archives, Kitt Peak National Observatory Library, Tucson. Note: The official University of Michigan appointment used the title "Research Engineer" because it was listed as an academic appointment in the budget. Letter, Goldberg to Robert L. Williams, September 20, 1956. College, Box 108, Michigan Historical Collections, Bentley Historical Library, University of Michigan.

34 Letter, Salanave to McMath, January 28, 1957. NAO Panel Archives, Kitt Peak National Observatory Library, Tucson.

35 Letter, Pierce to Meinel, March 27, 1957. NAO Panel Archives, Kitt Peak National Observatory Library, Tucson. Note: A. Keith Pierce, a staff member of the McMath–Hulbert Observatory, had been assigned by McMath to work with Meinel on the solar seeing site survey. He visited several European observatories in October and November 1956. Report: "Trip of A.K. Pierce to Investigate Solar Instrumentation and Seeing Problems of Several European Observatories: Cambridge, Oxford, Gottingen, Paris, Pic-du-Midi, December 5, 1956." NAO Panel, Solar Technical Reports, Kitt Peak National Observatory Library, Tucson.

36 Letter, Meinel to McMath, July 22, 1957. NAO Panel Archives, Kitt Peak National Observatory Library, Tucson. Thompson reported that Junipero Serra Peak is much harder logistically than Kitt Peak, with 26 miles of primitive road compared with 3 miles for Kitt Peak.

37 Letter, Meinel to McMath, August 22, 1957. Robert R. McMath Collection, Michigan Historical Collections, Bentley Historical Library, University of Michigan. Also Bowen NAO Panel files.

38 Transcript of Meinel interview with Edmondson, February 5, 1979.

39 Op. cit. Reference 31.

40 Letter, Meinel to McMath, May 6, 1957. This letter also acknowledged that Kitt Peak was "the only site with routine testing" at that time. Bowen NAO Panel Files.

41 Memorandum, McMath to Drs. Bowen, Goldberg, Stromgren, Struve, Whitford, Members of the NAO–NSF Advisory Panel, May 9, 1957. Bowen NAO Panel Files.

42 Letter, McMath to Bowen, May 9, 1957. Bowen NAO Panel Files.

43 Letter, Bowen to McMath, May 15, 1957. Bowen NAO Panel Files.

44 Letter, Whitford to McMath, May 15, 1957. Bowen NAO Panel Files.

45 Letter, Struve to McMath, May 21, 1957. Bowen NAO Panel Files.

46 Letter, Meinel to McMath, May 17, 1957. Bowen NAO Panel Files.

47 Letter Meinel to Pierce, May 20, 1957. KPNO Solar Project Files (Pierce).

48 Letter, McMath to Meinel, May 24, 1957. Letter, Meinel to McMath, June 3, 1957. Letter, Whitford to McMath, June 7, 1957. Letter, Meinel to McMath, June 10, 1957. Bowen NAO Panel Files.

49 Letter, Meinel to McMath, June 27, 1957. KPNO Solar Project Files (Pierce), Kitt Peak National Observatory Library, Tucson.

50 Letter, Meinel to McMath, May 6, 1957. Bowen NAO Panel Files.

51 Letter, Meinel to McMath, May 27, 1957. Bowen NAO Panel Files.

52 Letter, Meinel to McMath, June 10, 1957. Bowen NAO Panel Files.

53 Transcript of Golson interview with Edmondson, March 15, 1979.

54 Transcript of Olson interview with Edmondson, November 24, 1978.

55 Letter, Meinel to McMath, July 31, 1957. Bowen NAO Panel Files.

56 Letter, Whitford to McMath, August 2, 1957. Robert R. McMath Collection, Michigan Historical Collections, Bentley Historical Library, University of Michigan.

57 Letter, Bowen to McMath, August 3, 1957. Robert R. McMath Collection, Michigan Historical Collections, Bentley Historical Library, University of Michigan.

58 Letter, McMath to Bowen, August 10, 1957. Bowen NAO Panel Files.

59 Memorandum, McMath to Drs. Goldberg, Bowen, Stromgren, Struve and Whitford, Members of the NAO–NSF Advisory Panel only. August 10, 1957. Robert R. McMath Collection, Michigan Historical Collections, Bentley Historical Library, University of Michigan.

60 Letter, Johnson to Whitford, August 23, 1957. Bowen NAO Panel Files. Note: Harold L. Johnson was employed by Meinel to work in the Phoenix office on a part-time basis from September 1956 to May 1957, and he served as a consultant for a few months after that. Meinel wanted to give him a full-time research position in the NAO, but this was not possible before AURA came into being, and the AURA contract with NSF was signed.

61 A.B. Meinel, Technical Report No. 12, Observatory Sites, Environment and Weather Patterns, August 15, 1957. Bowen NAO Panel Files.

62 Letter, Meinel to McMath, August 14, 1957. Bowen NAO Panel Files.

63 Letter, McMath to Meinel, August 17, 1957. Memorandum, McMath to Drs. Bowen, Goldberg, Stromgren, Struve, Whitford, Members NAO-NSF Advisory Panel only, August 17, 1957. Letter, McMath to Whitford, August 17, 1957. Letter, Whitford to McMath, August 19, 1957. Robert R. McMath Collection, Michigan Historical Collections, Bentley Historical Library, University of Michigan.

64 Letter, Meinel to McMath, August 20, 1957. Robert R. McMath Collection, Michigan Historical Collections, Bentley Historical Library, University of Michigan.

65 Letter, McMath to Meinel, August 20, 1957. Bowen NAO Panel Files.

66 Letter, Meinel to McMath, August 22, 1957. Bowen NAO Panel Files.

67 Letter, Whitford to McMath, September 3, 1957. Bowen NAO Panel Files.

68 Letter, McMath to McGregor Fund, February 7, 1949. Robert R. McMath Collection, Michigan Historical Collections, Bentley Historical Library, University of Michigan.

69 Transcript of Orren C. Mohler interview with Edmondson, July 11, 1979.

70 Op. cit. Reference 3.

71 Letter, McMath to J.E. Luton (NSF), June 23, 1960. Robert R. McMath Collection,

Michigan Historical Collections, Bentley Historical Library, University of Michigan. This letter implies that a University of Michigan truck was used. Pierce's recollection is different: "He immediately sent his truck at his expense from Motors Metal Manufacturing Company of Detroit to the Lynn plant to pick up the blanks. They were brought to Michigan, where they were stored behind the Observatory in Ann Arbor for many years." Transcript of A. Keith Pierce interview with Edmondson, March 21, 1979.

72 Op. cit. 5, Reference 100.

73 Letter, Struve to McMath, September 29, 1955. Robert R. McMath Collection, Michigan Historical Collections, Bentley Historical Library, University of Michigan. Letter, Meinel to Bowen, September 30, 1955. Bowen NAO Files. Letter, McMath to Struve, October 13, 1955. Robert R. McMath Collection, Michigan Historical Collections, Bentley Historical Library, University of Michigan.

74 Letter, McMath to Hogg, March 20, 1956. NSF History Files.

75 Op. cit. Reference 35, Note.

76 Diary Note, Hogg, May 14, 1956. NSF History Files. Letter, Hogg to McMath, May 17, 1956. NSF History Files. Memorandum and Recommendations, Hogg to Waterman, May 16, 1956. NSF History Files.

77 Proposal to NSF from University of Michigan, "Continuation of Studies Leading to the Establishment of an American Astronomical Observatory," May 17, 1956. NSF History Files.

78 Diary Note, Hogg, June 2, 1956. NSF History Files. Memorandum, McMath to NAO–NSF Panel Members only (Drs. Bowen, Stromgren, Whitford), June 5, 1956. Robert R. McMath Collection, Michigan Historical Collections, Bentley Historical Library, University of Michigan. The House version of the budget bill ordered NSF to spend no money for large-scale facilities until the entire funding was in hand. The NSB deleted the 80-inch disk and other items that were considered to be for permanent installation as an interim solution.

79 Proposal Review History, G2622, $545,000, 15 months, June 1956. NSF History Files. Note: The reviews were returned to Hogg's successor, the author of this book. Several Panel members asked why they were being asked to review this. "I do not understand this application. If it is tied in with the National Observatory project, then why is our Panel being canvassed?" Other comments included: "It appears on the surface, however, that the decision to include elaborate solar apparatus reflects the personal preferences of the panel members, rather than the needs for which the National Observatory was originally intended to cater." "In fact most think of the NAO as primarily a facility for photoelectric observations." "I have been told the money has already been granted."

80 Letter, Waterman to Hatcher, August 2, 1956. Letter, Watkins to Waterman, August 6, 1956. Robert R. McMath Collection, Michigan Historical Collections, Bentley Historical Library, University of Michigan.

81 Op. cit. Reference 17.

82 Op. cit. Reference 29.

83 Op. cit. Reference 35, Note.

84 Op. cit. 3, Reference 10.

85 Memorandum, McMath to NAO–NSF Advisory Panel Members Only (Drs. Bowen, Stromgren, Struve, Whitford, Goldberg), November 27, 1956. Bowen NAO Files. Note: Goldberg became a member of the Panel in mid-November. Letter, McMath to Pierpont, November 13, 1956. Robert R. McMath Collection, Michigan Historical Collections, Bentley Historical Library, University of Michigan.

86 McMath, Proposal: "Preliminary Engineering Study of a Very Large Thermally-Controlled Solar Telescope," $60,000, November 26, 1956. NAO Panel Archives, Kitt Peak National Observatory Library, Tucson.

87 Op. cit. Reference 85.

88 Telegrams, Struve, Stromgren, and Bowen to McMath, November 29, 1956. Letter, Whitford to McMath, November 29, 1956. NAO Panel Archives, Kitt Peak National Observatory Library, Tucson.

89 Memorandum, McMath to Drs. Bowen, Goldberg, Stromgren, Struve, Whitford. Members of NAO–NSF Advisory Panel, December 4, 1956. NAO Panel Archives, Kitt Peak National Observatory Library, Tucson.

90 Memorandum, Sherwood to McMath, Nier, Osborne, and Seeger. October 29, 1956. Letter, McMath to Sherwood, November 13, 1956. Robert R. McMath Collection, Michigan Historical Collections, Bentley Historical Library, University of Michigan.

91 Letter, McMath to Edmondson, December 4, 1956. NAO Panel Archives, Kitt Peak National Observatory Library, Tucson.

92 Letter, McMath to Seeger, December 4, 1956. NAO Panel Archives, Kitt Peak National Observatory Library, Tucson.

93 Op. cit. Reference 89.

94 Letter, Waterman to Hatcher, February 26, 1957. NAO Panel Archives, Kitt Peak National Observatory Library, Tucson.

95 Letter, McMath to Edmondson, March 14, 1957. NSF History Files.

96 Op. cit. Reference 85.

97 Letter, Edmondson to McMath, October 18, 1956. Robert R. McMath Collection, Michigan Historical Collections, Bentley Historical Library, University of Michigan.

98 Op. cit. Reference 94.

99 Letter, McMath to Seeger, July 5, 1957. Robert R. McMath Collection, Michigan Historical Collections, Bentley Historical Library, University of Michigan. Note: A detailed breakdown was sent to Edmondson. Letter, McMath to Edmondson, August 9, 1957. NAO Panel Archives (Solar), Kitt Peak National Observatory Library, Tucson.

100 Op. cit. 2, Reference 1, page 214.
     Letter, Edmondson to Bok, June 15, 1957. Edmondson personal files.

101 Minutes, Advisory Panel for Astronomy, August 17–18, 1957. NSF History Files.

102 Letter, Eckhardt to Goldberg, August 16, 1957. NSF History Files.

103 Letter, Eckhardt to McMath, September 6, 1957. NSF History Files. Note: Eckhardt had alerted Goldberg and McMath that this might happen in telephone conversations on September 5. McMath told him another $50,000 would finish the solar site survey, and Eckhardt promised to discuss the question with Klopsteg and Waterman. Memorandum, McMath to Drs. Bowen, Goldberg, Stromgren, Struve, and Whitford, Members of NAO–NSF Panel only, September 5, 1957, Robert R. McMath Collection, Michigan Historical Collections, Bentley Historical Library, University of Michigan.

104 Letter, McMath to Eckhardt, September 12, 1957. NSF History Files.

105 Letter, Eckhardt to McMath, September 19, 1957. NSF History Files.

106 Letter, McMath to Eckhardt, September 24, 1957. NSF History Files.

107 Letter, Pierce to Salanave, September 13, 1957. NAO Panel Archives (Solar), Kitt Peak National Observatory Library, Tucson.

108 Letter, Pierce to Meinel, October 11, 1957. NAO Panel Archives (Solar), Kitt Peak National Observatory Library, Tucson.Telegram, Salanave to Pierce, October 19, 1957. NAO Panel Archives (Solar), Kitt Peak National Observatory Library, Tucson.

109 Letter, Pierce to Salanave, October 22, 1957. Note: This letter was written six days before AURA was incorporated. NAO Panel Archives (Solar), Kitt Peak National Observatory Library, Tucson.

110 Letter, Salanave to Aden and Keith, November 11, 1957. NAO Panel Archives (Solar), Kitt Peak National Observatory Library, Tucson.

111 Letter, Pierce to Salanave, December 3, 1957. Letter, Salanave to Pierce, December 6, 1957. Letter, Salanave to Pierce, January 10, 1958. NAO Panel Archives (Solar), Kitt Peak National Observatory Library, Tucson.

112 Letter, Pierce to Salanave, February 19, 1958. Letter, Salanave to Pierce, February 24, 1958. NAO Panel Archives (Solar), Kitt Peak National Observatory Library, Tucson.

113 Letter, Meinel to Salanave, March 3, 1958. Letter, Pierce to Beecher, March 12, 1958. Letter, Pierce to Salanave, March 12, 1958. Note: There are negative comments about Salanave written by Meinel on a copy of this letter which he sent to Pierce. Letter, Pierce to Salanave, March 15, 1958. Letter, Pierce to Meinel and Ralph Patey, March 18, 1958. Letter, Pierce to Salanave, March 18, 1958. Letter, Patey to Pierce, March 20, 1958. Letter, Pierce to Salanave, March 28, 1958. Letter, Patey to Pierce, April 9, 1958. Letter, Pierce to Patey, April 11, 1958. NAO Panel Archives (Solar), Kitt Peak National Observatory Library, Tucson.

114 Letter, Pierce to Shane, July 4, 1958. Letter, Pierce to Patey, August 8, 1958. Letter, Salanave to Pierce, August 19, 1958. Letter, Salanave to Pierce, August 31, 1958. Letter, Salanave to Beecher, September 5, 1958. NAO Panel Archives (Solar), Kitt Peak National Observatory Library, Tucson.

Chapter 7

1 Op. cit. 4, Reference 54.

2 Op. cit. 5, Reference 7.

3 Op. cit. 5, Reference 65.

4 Minutes, NSF Advisory Panel for Astronomy. January 14–15, 1955. NSF History Files.

5 Struve, *The General Needs of Astronomy*, Publications of the Astronomical Society of the Pacific, Vol. 67, pp. 214–223, 1955.

6 Russell, "Some Problems of Sidereal Astronomy", Proceedings of the National Academy of Sciences, Volume V, Issue 10, 1919. Also printed as a separate Bulletin of the National Research Council.

7 *New Horizons in Astronomy* (ed. Fred L. Whipple), Smithsonian Contributions to Astrophysics, Volume 1, Number 1, 1956.

8 Ground-based Astronomy, A Ten-year Program. A Report prepared by the Panel on Astronomical Facilities [A.E. Whitford, Chairman] for the Committee on Science and Public Policy of the National Academy of Sciences. National Academy of Sciences - National Research Council, Washington, D.C. 1964. This has been followed by the Greenstein Report (1970s), the Field Report (1980s), and the Bahcall Report (1990s).

9 Letter, Irwin to McMath, May 17, 1954. Robert R. McMath Collection, Michigan Historical Collections, Bentley Historical Library, University of Michigan.

10 Letter, McMath to Irwin, May 22, 1954. Robert R. McMath Collection, Michigan Historical Collections, Bentley Historical Library, University of Michigan.

11 Letter, Irwin to McMath, June 10, 1954. Robert R. McMath Collection, Michigan Historical Collections, Bentley Historical Library, University of Michigan.
McMath's letter made it clear that he did not understand what Irwin was offering to send to him. Irwin's diplomatic reply overlooked this and explained the background for the letters. It is interesting to note that Struve proposed to McMath that "we try to secure the services of John Irwin of Indiana University, who was particularly interested in the project of a photoelectric telescope and whose cooperation would be valuable in view of his enthusiasm and experience. He would also be one of the most interested astronomers from the midwest who might later make effective use of such a telescope."
Letter, Struve to McMath, August 24, 1954. Robert R. McMath Collection, Michigan Historical Collections, Bentley Historical Library, University of Michigan.
McMath's reply said: "I quite agree that John Irwin would make a splendid 'appointee' of our committee." Letter, McMath to Struve, August 30, 1954. Robert R. McMath Collection, Michigan Historical Collections, Bentley Historical Library, University of Michigan. Irwin's name was not on the list considered by the NAO Panel at its first meeting, November 4–5, 1954.

12 Op. cit. 4, Reference 47. Also see 4, Reference 40.

13 The Plan for a New American Observatory, A.E. Whitford. (First draft of remarks at dinner of the AAS, Troy, New York, November 10, 1955.) Whitford Archives, University of Wisconsin Library. A condensed version of the talk was subsequently printed in the *Publications of the Astronomical Society of the Pacific*, **68**, 115–117, 1956, and *Sky and Telescope*, **XV**, 107 and 111, January 1956 contained a summary. A news story from the Associated Press about the opening of the Phoenix site survey office also mentioned Whitford's talk. *Arizona Daily Star*, November 15, 1955.

The news stories that appeared following Whitford's talk also attracted the attention of members of Congress and others. Examples are: Letter, January 30, 1956, Representative Clifton Young (Nevada) to Waterman [Suggested Mount Charleston in southern Nevada.] Letter, February 3, 1956. Waterman to Young. Letter, March 23, 1956. Sunderlin to Young. Memorandum, February 3, 1956. Hogg to Sunderlin. [The secretary of Senator Clinton Anderson (New Mexico) phoned to find out if the site was already selected.]

Diary Note, May 10, 1956, Hogg. [Rear Admiral Byrd talked to Hogg, after trying to contact Waterman. He mentioned that his friend, George Getz, owns some land which might be donated, but did not specify the location. He sent his Chief of Staff to get the list of names of the NAO Panel members.] NSF History Files.

14 Letter, Whitford to Seeger, October 10, 1955. NSF History Files.

Letter, Whitford to Hogg, October 20, 1955. NSF History Files.

Diary Note, Seeger, October 31, 1955. NSF History Files. This summarized a telephone conversation between Dr. Seeger and Dr. Whitford regarding announcement of the four NAO Panel grants to the University of Michigan and the work of the Panel at the AAS Troy meeting, November 1955. Helen S. Hogg was on the third phone. "Dr. Seeger said it was highly desirable that something be said to the AAS, in fact Dr. McMath was asked to keep in touch with the AAS." NSF History Files.

15 Letter, McMath to Hogg, November 30, 1955. NSF History Files.

16 Diary Note, Hogg, Report on AAS Meeting, November 9–12, 1955. NSF History Files.

17 Note: The author was present at this meeting in his capacity as AAS Treasurer. He was not aware at that time of the political undercurrents and tensions.

18 Letter, Whitford to McMath, January 6, 1956. Robert R. McMath Collection, Michigan Historical Collections, Bentley Historical Library, University of Michigan.

19 Diary Note, Hall, January 31, 1956. NSF History Files. This described a meeting to inform the NSF Public Information Office about the NAO site survey. Participants were: Meinel, Hogg, Hall, and Embrey.

20 Letter, Embrey to Meinel, February 10, 1956, and enclosure: Press Release Announcing Erection of Seeing Tower on Kitt Peak in the Quinlan Mountains. NSF History Files.

21 Letter, Meinel to Embrey, February 16, 1956. NSF History Files.

22 Memorandum, Embrey to Hall, March 5, 1956, and press release: National Science Foundation Sponsors Site Survey for New National Observatory. NSF History Files.

23 Letter. Embrey to Meinel, March 15, 1956. NSF History Files.

24 Op. cit. 5, Reference 53. Dr. Harvill said one of the things that always worried him a bit about the way Aden Meinel operated was the way Aden seemed to favor the evening newspaper, the *Tucson Citizen*, and this did cause Harvill some minor problems because a member of the Board of Regents was connected with the morning newspaper.

25 Letter, Meinel to Embrey, March 20, 1956. NSF History Files.

26 Letter, Hogg to Meinel, March 12, 1956. NSF History Files. Meinel told Hogg "the end of April" and his letter to Embrey said "the last week in May."

27 Op. cit. 5, Reference 133.

28 Letter, Hall to McMath, March 23, 1956. NSF History Files.

29 Memorandum, McMath to NAO–NSF Panel, April 3, 1956. NSF History Files.

30 Letter, McMath to Hall, April 3, 1956. NSF History Files.

31 Memorandum, Hall to Waterman, April 10, 1956. NSF History Files.
32 Letter, Meinel to McMath, August 14, 1956. NAO Panel Archives, Kitt Peak National Observatory Library, Tucson.
33 Letter, Embrey to McMath, August 21, 1956. NAO Panel Archives, Kitt Peak National Observatory Library, Tucson.This was the cover letter with the news release about the $545,000 grant. Referring to Meinel's August 14 letter she said: "I was entirely unaware myself of the date on which it was scheduled to take place. I thought Dr. Meinel was aware of our special interest . . ., so I cannot understand why we were not given some notice as to when this particular event would take place." She also said she made no arrangements with *LIFE* to cover the event.
34 Letter, Bowen to McMath, April 11, 1956. NAO Panel Archives, Pitt Peak National Observatory Library, Tucson. "New Giant Telescope to Be Set Up" was the headline of the enclosed article from the April 11, 1956 *Los Angeles Times*.
35 Letter, Embrey to McMath, August 1, 1956. NSF History Files.
36 Letter, McMath to Embrey, August 8, 1956. NSF History Files. Embrey passed it on to Hall with the comment: "I don't think McMath sees what the difficulty is, but I think we should make a fresh start – as he suggests – with my going to Phoenix to see for myself." Hall replied: "I think you ought to go." Edmondson's response, written in Flagstaff, was similar: "I get the impression that McMath has not seen the point, but at least he seems to be cooperative. My reaction to his suggestion that you should make a trip to Phoenix to see the setup there is that you should do it if at all possible." Memorandum, Edmondson to Embrey, August 19, 1956. NSF History Files.
37 Letter, Meinel to McMath, August 13, 1956. NSF History Files.
38 Letter, Embrey to McMath, August 9, 1956. NAO Panel Archives, Kitt Peak National Observatory Library, Tucson.
39 Letter, McMath to Embrey, August 15, 1956. NSF History Files.
40 Op. cit. Reference 33, and 6, Reference 17.
41 Meinel requested McMath's permission to abandon the Sierra Ancha site (McFadden Peak) on August 14, and McMath approved on August 18. See 6, References 22 and 23.
42 Letter, McMath to Embrey, August 23, 1956. NSF History Files.
43 Diary Note, Embrey, October 3, 1956. NSF History Files.
44 Letter, Embrey to McMath, October 9, 1956. NSF History Files.
45 Op. cit. Reference 43.
46 Letter, Embrey to Meinel, October 5, 1956. NSF History Files.
    Letter, Embrey to Meinel, October 9, 1956. NSF History Files.
    Op. cit. Reference 44.
    Letter, McMath to Embrey, October 16, 1956. NSF History Files.
47 Letter, Edmondson to McMath, October 18, 1956. Robert R. McMath Collection, Michigan Historical Collections, Bentley Historical Library, University of Michigan.
48 Diary Note, Seeger, March 1, 1957. NSF History Files. "Dr. Carpenter and I discussed the plan of the University to move its own Observatory, which had been one of the first requests made to the Foundation. It is their present plan to locate it not too distant from Kitt Peak. I suggested to Dr. Carpenter that serious consideration should be given to the possibility of locating it on Kitt Peak provided suitable arrangements could be made."
    Letter, Carpenter to Seeger, March 14, 1957. NSF History Files.
    Carpenter responded: "I have discussed with my colleagues your suggestion of the possibility of the establishment of the 36-inch Steward reflector on a site common to or contiguous with the National Observatory site if this should be chosen at Kitt Peak. We are agreed on the excellent scientific advantages of such a location and will explore the matter administratively with Dean Patrick and President Harvill."
49 Letter, Meinel to Embrey, February 20, 1957. NSF History Files.
50 Press Release, Eugene H. Kone, for release Tuesday, March 5, 1957. NSF History Files.

The source of Kone's misinformation is not identified in the NSF files. Meinel gave Kone correct information about the site survey and the NSF support through the University of Michigan but without mentioning his title; this was too late for the press release. Letter, Meinel to Kone, March 5, 1957. NSF History Files.

51 Letter, Embrey to Kone, March 7, 1957. NSF History Files. Letter, Kone to Embrey, March 8, 1957. NSF History Files. Kone said the information from Embrey had "arrived in time for use in conjunction with a press conference involving Dr. Meinel." He enclosed a copy of the corrected press release.

52 Op. cit. 5, Reference 66. Letter, Struve to McMath, with enclosure, April 15, 1956. Robert R. McMath Collection, Michigan Historical Collections, Bentley Historical Library, University of Michigan.

53 Op. cit. Reference 13.

54 Letter, Irwin to Whitford, December 4, 1955. Robert R. McMath Collection, Michigan Historical Collections, Bentley Historical Library, University of Michigan.

55 Op. cit. 5, Reference 7.

56 Op. cit. 5, Reference 65.

57 Op. cit. 5, Reference 100.

58 Op. cit. 6, Reference 31.

59 Op. cit. 5, Reference 65.

60 Letter, Meinel to McMath, May 21, 1956. NAO Panel Archives, Kitt Peak National Observatory Library, Tucson.

61 Letter, McMath to Meinel, June 2, 1956. NAO Panel Archives, Kitt Peak National Observatory Library, Tucson.

62 Letter, Struve to McMath, June 4, 1956. NAO Panel Archives, Kitt Peak National Observatory Library, Tucson.

63 Letter, Meinel to McMath, June 11, 1956. Bowen NAO Panel Files. Letter, McMath to Meinel, July 19, 1956. Bowen NAO Panel Files. Letter, Meinel to Edmondson, July 25, 1956. NAO Panel Archives, Kitt Peak National Observatory Library, Tucson. Letter, Meinel to McMath, August 7, 1956. Bowen NAO Panel Files. Letter, McMath to Meinel, August 16, 1956. NAO Panel Archives, Kitt Peak National Observatory Library, Tucson. Letter, McMath to Whitford, August 23, 1956. NAO Panel Archives, Kitt Peak National Observatory Library, Tucson.

64 Memorandum, McMath to Panel, August 23, 1956. NAO Panel Archives, Kitt Peak National Observatory Library, Tucson. Charles Jones was recommended by Bowen, Bruce Rule, and Meinel to replace Joseph Nunn for the 36-inch telescope engineering design.

65 Op. cit. Reference 33.

66 Letter, Meinel to McMath, October 11, 1956. NAO Panel Archives, Kitt Peak National Observatory Library, Tucson.

67 Letter, McMath to Meinel, October 11, 1956. NAO Panel Archives, Kitt Peak National Observatory Library, Tucson. Letter, McMath to Meinel, October 16, 1956. NAO Panel Archives, Kitt Peak National Observatory Library, Tucson.

68 Letter, Bowen to Meinel, October 16, 1956. NAO Panel Archives, Kitt Peak National Observatory Library, Tucson. Bowen also criticized the reasoning that led to the choice of the size of the 36-inch telescope.

69 Two letters, McMath to Meinel, October 18, 1956. NAO Panel Archives, Kitt Peak National Observatory Library, Tucson. Letter, McMath to Bowen, October 18, 1956. NAO Panel Archives, Kitt Peak National Observatory Library, Tucson.

70 Letter, Bowen to McMath, October 23, 1956. NAO Panel Archives, Kitt Peak National Observatory Library, Tucson. Bowen objected to Meinel's suggestion that the 16-inch telescopes could be permanent instruments for the observatory. He said: "It may have been worthwhile in any case if it was helpful to make Meinel realize that he should inform the committee of changes before going too far ahead."

71 Letter, Meinel to McMath, October 25, 1956. NAO Panel Archives, Kitt Peak National Observatory Library, Tucson.

72 Memorandum, McMath to Panel, October 26, 1956. NAO Panel Archives, Kitt Peak National Observatory Library, Tucson.

73 Letter, Bowen to McMath, November 2, 1956. NAO Panel Archives, Kitt Peak National Observatory Library, Tucson.

74 Letter, Bowen to McMath, November 6, 1956. NAO Panel Archives, Kitt Peak National Observatory Library, Tucson.

75 Letter, McMath to Bowen, November 13, 1956. NAO Panel Archives, Kitt Peak National Observatory Library, Tucson.

76 Memorandum, McMath to Panel, November 13, 1956. NAO Panel Archives, Kitt Peak National Observatory Library, Tucson.

77 Letter, Whitford to McMath, November 16, 1956. NAO Panel Archives, Kitt Peak National Observatory Library, Tucson.
Letter, Whitford to McMath, November 20, 1956. NAO Panel Archives, Kitt Peak National Observatory Library, Tucson.

78 Letter, Bowen to McMath, November 19, 1956. NAO Panel Archives, Kitt Peak National Observatory Library, Tucson.

79 Op. cit. 6, Reference 3.

80 Op. cit. 6, Reference 2.

81 Letter, Stromgren to McMath, September 24, 1954. Robert R. McMath Collection, Michigan Historical Collections, Bentley Historical Library, University of Michigan.

82 Op. cit. 5, Reference 7.

83 Diary Note, Seeger, June 2, 1955. NSF History Files.

84 Op. cit. 5, Reference 80.

85 Memorandum, McMath to Panel, October 4, 1955. Robert R. McMath Collection, Michigan Historical Collections, Bentley Historical Library, University of Michigan.

86 Memorandum, McMath to Panel, January 3, 1956. Robert R. McMath Collection, Michigan Historical Collections, Bentley Historical Library, University of Michigan.

87 Letter, Whitford to McMath, January 6, 1956. Robert R. McMath Collection, Michigan Historical Collections, Bentley Historical Library, University of Michigan.

88 Letter, Bowen to McMath, January 6, 1956. Robert R. McMath Collection, Michigan Historical Collections, Bentley Historical Library, University of Michigan.

89 Memorandum, McMath to Panel, January 18, 1956. NSF History Files.
Memorandum, McMath to Panel, February 9, 1956. NSF History Files.
Seeger told McMath that NSF "would prefer that the Trustee Group not be dominated by the scientists." McMath responded by proposing that each university would "designate two from their top administrative group and one scientist, presumably the director or the chairman of their department of astronomy." McMath also proposed that "membership in the Operating Group would be on a rotation basis in order to give all the 'have not' institutions a chance at the management of the new observatory. Whether or not the Director of the new observatory would have full charge of the programming is a matter for the decision of the Operating Group."

90 Letter, Hogg to McMath, March 19, 1956, and Letter, McMath to Hogg, April 12, 1956. Robert R. McMath Collection, Michigan Historical Collections, Bentley Historical Library, University of Michigan. Hogg's 14 page draft minutes were shortened to 5.2 pages by McMath.

91 Letter, McMath to Hogg, March 3, 1956. Robert R. McMath Collection, Michigan Historical Collections, Bentley Historical Library, University of Michigan.

92 Op. cit. 5, Reference 86.

93 Memorandum, McMath to Panel, March 10, 1956. Robert R. McMath Collection, Michigan Historical Collections, Bentley Historical Library, University of Michigan. This is a report about the March 8–9 meeting of the MPE Divisional Committee. Dr. Paul Gross, Vice Chairman of the NSB and Chairman of the Board's MPE Committee, spoke about NSF support of large-scale facilities. "He twice stated that the [NSB] understood that the NAO would need support 'in perpetuity'."

94 Letter, Goldberg to McMath, March 30, 1956. Robert R. McMath Collection, Michigan Historical Collections. Bentley Historical Library, University of Michigan. "I have just about come to the conclusion that the plan of organization proposed by the Panel is unworkable, and if I am not sold on it myself, it is going to be exceedingly difficult for me to convince the various university administrators of its workability." Goldberg's proposals about ownership and administration of NAO were passed on in McMath's letter to Hogg cited below.

95 Memorandum, McMath to NAO–NSF Advisory Panel, April 3, 1956, Robert R. McMath Collection, Michigan Historical Collections. Bentley Historical Library, University of Michigan.

96 Memorandum, McMath to NAO–NSF Advisory Panel, April 7, 1956. Robert R. McMath Collection, Michigan Historical Collections. Bentley Historical Library, University of Michigan.

97 Letter, McMath to Hogg, April 28, 1956. NSF History Files.

98 Letter, McMath to Seeger, March 10, 1956. Robert R. McMath Collection, Michigan Historical Collections, Bentley Historical Library, University of Michigan. McMath's conversation with Dr. Paul Gross, Vice Chairman of the NSB, seems to be the first use of the term "Letter of Intent." Gross suggested this was something "which Leo Goldberg could use as an entering wedge in his effort to get various universities to join the trustee group."

99 Letter, McMath to Waterman, July 3, 1956, NSF History Files.

100 Letter, Waterman to McMath, October 16, 1956. NSF History Files.
 FY 1957 was a good year for astronomy. The NSF appropriation was $40 million, an increase by a factor of 2.5 from the previous year. $5.0 million of this was for astronomy: $4.0 million to start the National Radio Astronomy Observatory, $345,000 for the final Michigan site survey grant, and $455,000 for university research grants. This was one-eighth of the NSF budget, or one-fifth of the increase.

101 Memorandum, Goldberg to Members and Academic Consultants of NAO–NSF Panel: Doctors McMath, Bowen, Stromgren, Struve, Whitford, Menzel, and Shane. January 28, 1957. Bowen NAO Panel Files.

102 Op. cit., 6, Reference 31. Cal Tech, Case, Pennsylvania, Virginia, and Yale each had three or more staff, but were not invited. Dr. Bowen had indicated that Cal Tech probably would not wish to be included in the operating organization. It was agreed, at McMath's request, to make Goldberg (previously a Consultant) Vice Chairman of the Panel. Presumably this was to give him official status to convene the meeting of university representatives.

103 Memorandum, T.E. Randall to J.A. Franklin, April 3, 1957. Indiana University Archives, Office of Vice-President and Treasurer. This is a report about the meeting. Indiana University was represented by John B. Irwin, Acting Chairman of the Department of Astronomy, and T.E. Randall, Assistant Treasurer. Edmondson represented NSF. Goldberg wrote to Edmondson on April 2, saying: "I hope you are as pleased as I am with the outcome of last Friday's meeting." He sent a summary of the results of the meeting to Waterman on April 3, and told him: "The meeting was called by the University of Michigan in the aftermath of your letter of October 16th to Dr. McMath."

104 Memorandum, Goldberg to 19 persons, April 12, 1957. Indiana University Archives, Office

of Vice-President and Treasurer. The name of the proposed Association of Universities for Research in Astronomy (AURA) was first suggested in July 1956 by William G. Pollard, Director of the Oak Ridge Institute for Nuclear Studies (ORINS). Goldberg was one of the twelve persons invited by Pollard to incorporate under this name, and he recalled many years later: "I saw at once that Association of Universities for Research in Astronomy spelled AURA, and I think I proposed that name at one of the meetings of the organizing committee, but the idea came unwittingly from Pollard at Oak Ridge, Tennessee." Op. cit. 3, Reference 12.

105 Letter, Goldberg to Waterman, May 3, 1957. NSF History Files. He also enclosed a copy of the proposed plan of organization that would be discussed at a meeting on July 1. Waterman included this letter and the enclosure in the agenda material for the May 20 meeting of the National Science Board. Edmondson was invited to "attend the discussion" of this agenda item, and was shocked when Waterman, without prior warning, asked him to make the presentation to the Board.

106 Minutes, National Science Board, May 20, 1957. NSF History Files.

107 Minutes, Goldberg and NSF officials, June 27, 1957. NSF History Files.

108 Minutes, Organizing Committee for AURA, Inc., July 1, 1957. Edmondson personal files. J.A. Franklin represented Indiana University at this meeting.

109 Letter, Goldberg to Eckhardt, September 12, 1957. NSF History Files. Eckhardt phoned Goldberg on September 11 to inform him that the NSB had approved the AURA plan. The major change in the material presented to the NSB was a revised budget, sent by Goldberg on August 28. Geoffrey Keller became the NSF Program Director for Astronomy on September 1, 1957. Edmondson stayed through September 6, the date of the National Science Board Meeting.

110 Memorandum, Goldberg to Organizing Committee for AURA, Inc., September 12, 1957. Edmondson personal files.

111 Memorandum, Goldberg to Organizing Committee for AURA, Inc. September 17, 1957. Edmondson personal files.

112 Memorandum, Goldberg to Organizing Committee for AURA, Inc. October 11, 1957. Edmondson personal files.

Chapter 8

1 Op. cit. 7, Reference 109.

2 Letter, Waterman to Goldberg, September 19, 1957. NSF History Files.

3 Letter, Strouss to Cummiskey, September 12, 1957. Edmondson personal files.

4 Op. cit. 7, Reference 111.

5 Memorandum, Goldberg to Organizing Committee, September 30, 1957. Edmondson personal files.

6 Op. cit. 7, Reference 112.

7 The NSF General Counsel, William J. Hoff, had ruled that neither Edmondson nor McMath should sign the Contract owing to their previous connections with NSF. See Reference 33 below. For many years Edmondson believed the articles he signed were "lost" for the same reason. Edmondson raised this question with Goldberg during a conversation in the early 1980s. Goldberg replied: "I goofed. The Articles that we signed read 'Frank K. Edmondson, Jr.' and the Phoenix attorneys said they could not change this to match the signature because it would invalidate the notarized document." Hindsight says it was probably just as well that this happened.

8 Minutes, Incorporators of AURA, October 28, 1957, Ann Arbor, Michigan. Edmondson personal files.
The directors designated by the seven Member Universities were:

Harvard University
D.H. Menzel, Director, Harvard College Observatory
Edward Reynolds, Administrative Vice-President

Indiana University
F.K. Edmondson, Chairman, Department of Astronomy
J.A. Franklin, Vice-President and Treasurer

Ohio State University
P.C. Keenan, Professor of Astronomy
C.F. Miller, Controller

University of California
J.M. Miller, Assistant Vice-President – Business Affairs
C.D. Shane, Director, Lick Observatory

University of Chicago
W.B. Harrell, Vice-President – Business Affairs
G.P. Kuiper, Director, Yerkes Observatory

University of Michigan
G.L. Lee, Jr., Controller
R.R. McMath, Director, McMath–Hulbert Observatory

University of Wisconsin
A.W. Peterson, Vice-President
A.E. Whitford, Director, Washburn Observatory

9 Minutes, Board of Directors of AURA, Inc., October 28, 1957, Ann Arbor, Michigan. Edmondson personal files.

As a result of its meeting on October 27 at 10 P.M., the Nominating Committee presented the following report:

For Officers of AURA:
President   Robert R. McMath
Vice-President   Frank K. Edmondson
Secretary   James M. Miller

For Membership on the Executive Committee of AURA:
Robert R. McMath   Term of three years
A.W. Peterson   Term of three years
Frank K. Edmondson   Term of three years
D.H. Menzel   Term of two years
C.F. Miller   Term of two years
W.B. Harrell   Term of one year
C.D. Shane   Term of one year

For Director-at-large
Peter van de Kamp, Swarthmore College (3 years)
Carl K. Seyfert, Vanderbilt University (2 years)

For Consultants:
I.S. Bowen
Otto Struve
Leo Goldberg

For Engineering Consultants:
Bruce Rule
Donald Hendrix

10 Op. cit. Reference 5.
   Letter, Hoff to Goldberg, October 8, 1957. Edmondson personal files.
11 This is the author's vivid recollection of what happened at the meeting. The Minutes simply say "After discussion it was voted, etc."
   [Note: McMath wrote to Meinel a year and a half before this: "Speaking for myself as an individual, I would be delighted to consider you as the first director of the NAO." Letter, McMath to Meinel, June 2, 1956. Robert R. McMath Collection, Michigan Historical Collections, Bentley Historical Library, University of Michigan.]
12 Memorandum, Reynolds to Administrative Subcommittee of the Board of Directors of AURA, Inc. (J.A. Franklin, C.F. Miller, J.M. Miller, W.B. Harrell, G.L. Lee, Jr., A.W. Peterson), November 1, 1957. J.A. Franklin files, Indiana University Archives.
13 Letter, Peterson to Reynolds, November 8, 1957. J.A. Franklin files, Indiana University Archives. The positions were: Business Manager, Chief Engineer, Purchasing Agent, Accountant, and Personnel Officer. A part-time Treasurer might be needed if it was not possible to have a bank perform the services of a Treasurer.
14 Memorandum, Reynolds to Administrative Subcommittee of the Board of Directors of AURA, Inc., November 25, 1957. J.A. Franklin files, Indiana University Archives.
15 Memorandum, Reynolds to Administrative Subcommittee of the Board of Directors of AURA, Inc., December 4, 1957. J.A. Franklin files, Indiana University Archives.
16 Transcript of Franklin interview with Edmondson. November 6, 1979. Patrick had been Director of the Indiana University Memorial Union Building, and Mrs. Patrick had been Franklin's personal secretary. They moved to Arizona to cope with her serious health problem. This did some good because she outlived him by several years. Patrick eventually became President of the VNB and Chairman of the Board.
17 Letter, Franklin to Patrick, July 2, 1957. J.A. Franklin files, Indiana University Archives.
18 Letter, Patrick to Franklin, July 8, 1957. J.A. Franklin files, Indiana University Archives.
   Letter, Franklin to Patrick, July 11, 1957. J.A. Franklin files, Indiana University Archives.
   Letter, Franklin to Cummiskey, July 11, 1957. J.A. Franklin files, Indiana University Archives.
   Letter, Patrick to Franklin, July 17, 1957. J.A. Franklin files, Indiana University Archives.
19 Op. cit., 7, Reference 111.
20 Letter, Franklin to Patrick, October 29, 1957. J.A. Franklin files, Indiana University Archives.
21 Letter, Patrick to Franklin, November 7, 1957. J.A. Franklin files, Indiana University Archives.
22 Letter Franklin to Patrick, November 25, 1957. J.A. Franklin files, Indiana University Archives.
23 Letter, Reynolds to Franklin, November 25, 1957. J.A. Franklin files, Indiana University Archives.
   Letter, Harrell to Franklin, November 26, 1957. J.A. Franklin files, Indiana University Archives.
   Letter, Lee to Franklin, November 27, 1957. J.A. Franklin files, Indiana University Archives.
24 Letter, Shane to McMath, December 3, 1957. Edmondson personal files. This was a report summarizing the work of the Organizing Committee, including seven conclusions and recommendations.
25 Diary Note, Keller, December 3, 1957. NSF History Files. An additional sentence said: "I am to call McMath this afternoon and advise him that present Foundation plans call for sending of a contract, already signed by the Foundation, to him before the next AURA Board Meeting on December 11–12."
26 Letter, Luton to Lee, December 9, 1957. Copy attached to Minutes of AURA Board of Directors, December 13, 1957, 2:25 p.m., Agenda Item 15. Edmondson personal files.

27 Letter, Edmondson to Embrey, November 4, 1957. Edmondson personal files.
   Letter, Embrey to Edmondson, November 6, 1957. Edmondson personal files.
   Letter, Embrey to Edmondson, November 18, 1957. Edmondson personal files. A draft of her proposed press release was enclosed.
   Memorandum, Embrey to Waterman, December 2, 1957. Edmondson personal files.
28 Letter, McMath to Embrey, November 27, 1957. Edmondson personal files.
29 Op. cit., 2, Reference 1. A photograph of the signing ceremony is on the fourth of six pages of photographs preceding Chapter 14, p. 279. An additional reference to the signing is on p. 289.
30 PRESS RELEASE NSF-57-149. National Science Foundation Contracts with Newly Organized Group of Universities for Construction and Operation of Optical Astronomy Observatory. For Press, Radio, and T.V. For Release – Noon, MST, Friday, December 13, 1957. Edmondson personal files.
31 Letter, Embrey to Edmondson, December 23, 1957. Edmondson personal files.
32 Letter, Eckhardt to McMath, December 9, 1957. NSF History Files.
33 Minutes, Board of Directors of AURA, Inc., December 12, 1957, Phoenix, Arizona. Edmondson personal files.
34 Letter, McMath to Eckhardt, December 18, 157. NSF History Files.
35 Letter, Miller to Eckhardt, December 16, 1957. NSF History Files.
36 Letter, Luton to Miller, January 2, 1958. NSF History Files. Note: J.E. Luton, Assistant Director for Adminstration was designated by Waterman as the person in the Foundation to be responsible for the adminstration of the AURA contract. See Reference 51 below.
37 Op. cit. Reference 33. "ACTION: The following recommendation was unanamously approved by the Board of Directors: That Dr. A.B. Meinel be appointed Observatory Director at an initial basic salary of $15,000 exclusive of fringe benefits, subject to the approval of the National Science Foundation for a period of three years [sic] terminating June 3, 1960, the appointment to be effective at the date of turn-over of funds from the University of Michigan to AURA."
38 Kuiper attended only two Board meetings: October 28, 1957, and December 8, 1958. He did not attend the meetings of the Scientific Committee in 1958 (March 1, July 9, Dec. 7) and the Annual Meeting of the Board on March 3, 1959. At this point the AURA officers asked Harrell to have the University of Chicago replace him with someone who would come to meetings and play an active role in the work of the Board and the Scientific Committee. Hiltner was appointed six months before Kuiper's two year term would have ended.
39 Minutes, Board of Directors of AURA, Inc., December 13, 1957, Phoenix, Arizona. Edmondson personal files. Peterson attended the meeting of the Administrative Subcommittee on the evening of December 11 and the first half-hour of the Board meeting on the morning of December 12. He had to return to Milwaukee for a meeting of the Wisconsin Board of Regents on the evening of December 12.
   Memorandum. Peterson to members of the Administrative Subcommittee. December 2, 1957. J.A. Franklin files, Indiana University Archives. Op.cit. Reference 33.
40 Eckhardt told McMath on the telephone that this was not satisfactory. "The NSF prefers that neither Dr. Edmondson nor I act in that capacity. Apparently prompt action is required and it looks to the writer as though Mr. Gilbert Lee was just going to have to function." Memorandum. McMath to Members of the AURA Board of Directors and Scientific Consultants, January 2, 1958. Edmondson personal files.
41 Letter, McMath to Edmondson. December 5, 1957. Edmondson personal files.
42 Op.cit. Reference 26.
43 Letter, Edmondson to McMath, December 4, 1957. Edmondson personal files.
   Letter, McMath to Edmondson, December 5, 1957. Edmondson personal files.
   Letter, Edmondson to McMath, December 7, 1957. Edmondson personal files.
   Comment: Yale had requested NSF funding for a duplicate of the Lick astrograph to be

placed in the Southern hemisphere, and Eckhardt told McMath that he would like to have this under AURA. Edmondson told McMath: "If Eckhardt is serious about placing the Southern hemisphere astrograph under AURA, then I feel it would be a serious mistake to delay an invitation to Yale too long. In fact, it makes me a bit sorry they were not in the organizing group." Yale was the strongest institution among those not invited to join in the incorporation of AURA.

44 Letter, McMath to Eckhardt, December 18, 1957. NSF History Files.
45 Memorandum, Keller to Eckhardt, December 18, 1957. NSF History Files. Keller reported: "The purpose of Dr. Meinel's call was to pass on a request from Dr. McMath (President of AURA) that as soon as possible $300,000 be deposited to AURA's account in Phoenix." He also reported that Minton Moore had been appointed Treasurer of AURA, and that Edmondson had been designated as "contact man between AURA and the Foundation." NSF History Files.
46 Memorandum, Eckhardt to Luton, December 19, 1957. NSF History Files.
47 Memorandum, Luton to Hoff, December 24, 1957. NSF History Files.
48 Diary Note, Waterman, December 26, 2957. NSF History Files. McMath asked about the availability of funds for AURA and the status of the contract. Waterman told him: "I learned that it would come up for my signature shortly and told him so."
    McMath had not heard from Waterman in response to his recommendation that Meinel be appointed Director. Waterman told him that he approved and would make sure the letter was sent. [Note: It was mailed on December 31.] Finally, McMath had asked one more time about continuation of his panel. Waterman replied that "we plan to continue a panel on astronomy which would cover the entire field and discontinue the one on AURA."
49 Letter, McMath to Waterman, December 26, 1957. NSF History Files. This letter did not mention the decision to discontinue the NAO panel.
50 Op.cit. Reference 36.
51 Letter, Waterman to McMath, January 8, 1958. NSF History Files.
52 Letter, Luton to Lee, June 9, 1958. NSF History Files.
53 Minutes, Executive Committee of AURA, Inc., September 13, 2958, Ann Arbor, Michigan. Edmondson personal files.
54 Contract NSF-C63, Amendment No. 1, November 22, 1958. This included $4,000,000 for the solar telescope, and increased the funds provided to $7,545,000. Robert R. McMath Collection, Michigan Historical Collections, Bentley Historical Library, University of Michigan.
55 The NSF appropriation had gone from $225,000 in FY 1951 to $8,000,000 in FY 1954, and to $40,000,000 in FY 1957. The FY 1958 appropriation was also $40,000,000, cut from President Eisenhower's request of $65,000,000. The FY 1959 request was $140,000,000, but this had been reduced to $115,000,000 in the appropriations bill that was ready for action by the Congress.
56 National Science Foundation, Eighth Annual Report for the Fiscal Year ended June 30, 1958, p. 116. Edmondson personal files.
57 National Science Foundation, Ninth Annual Report for the Fiscal Year ended June 30, 1959, p. 134. Edmondson personal files.
58 The first appropriations for FY 1958 and FY 1959 added up to 75.6 percent of President Eisenhower's requests. The final appropriations added up to 90.6 percent of the requests.
59 Letter, Goldberg to Waterman, October 21, 1957. Robert R. McMath Collection, Michigan Historical Collections, Bentley Historical Library, University of Michigan. The other needs were: (a) restoration of capital equipment funds for Greenbank in FY 59; (b) acceleration of design studies for a very large radio telescope for Greenbank; (c) support of immediate design studies for the X-inch telescope for the NAO, X being 300–400 inches; and (d) additional travel funds for US astronomers attending the Moscow General Assembly of the IAU in August 1958.

60 Letter, McMath to Struve, October 19, 1957. Robert R. McMath Collection, Michigan Historical Collections, Bentley Historical Library, University of Michigan.

61 Letter, Struve to McMath, October 21, 1957. Robert R. McMath Collection, Michigan Historical Collections, Bentley Historical Library, University of Michigan.

62 Letter, McMath to Struve. October 26, 1957. Robert R. McMath Collection, Michigan Historical Collections, Bentley Historical Library, University of Michigan.

63 Letter, McMath to Bowen, October 26, 1957. Robert R. McMath Collection, Michigan Historical Collections, Bentley Historical Library, University of Michigan.

64 Letter, McMath to Waterman, October 26, 1957. Robert R. McMath Collection, Michigan Historical Collections, Bentley Historical Library, University of Michigan.

65 Letter, McMath to E.F. Osborne, November 30, 1957. Robert R. McMath Collection, Michigan Historical Collections, Bentley Historical Library, University of Michigan. McMath reported the reinstatement of the solar telescope and his resignation from the MPE Divisional Committee to the AURA Board at the Phoenix meeting. Op.cit. Reference 33. He submitted his resignation to Waterman in a telegram, December 2, 1957. Robert R. McMath Collection, Bentley Historical Library, University of Michigan.
Goldberg had informed the NAO Panel that the solar telescope had been reinstated for FY59 the day after McMath told him about Keller's phone call. Memorandum, Goldberg to Members of the NAO–NSF Advisory Panel and Academic Consultants, November 8, 1957. Robert R. McMath Collection, Bentley Historical Library, University of Michigan.

66 Memorandum, J.M. Miller to Members of the Executive Committee, AURA, Inc., January 15, 1958. Edmondson personal files.

67 Proposal to the National Science Foundation from the Association of Universities for Research in Astronomy, Inc. for the Construction of the World's Largest Solar Astronomical Telescope (60-inches aperture, 300-feet focal length), January 28, 1958. Edmondson personal files.

68 Minutes, Executive Committee of AURA, Inc., February 10–11, 1958, Pasadena, California. Edmondson personal files.

69 Independent Offices Appropriation Bill, 1959, June 5, 1958. Report [to accompany H.R. 11574]. Edmondson personal files.

70 Memorandum, Waterman to Members of the National Science Board, June 11, 1958. NSF History Files.

71 Longhand letter, Pierce to McMath, undated (June 1958). Robert R. McMath Collection, Michigan Historical Collections, Bentley Historical Library, University of Michigan.

72 Associated Press and other news reports. Edmondson personal files.

73 Telegram, Luton to Patey, July 29, 1958. Edmondson personal files.

74 J.M. Miller to Members of the Executive Committee of AURA, Inc., January 15, 1958. Edmondson personal files. This report has nine items based on Miller's January 9–10, 1958 trip to Phoenix, including: "4. Business Manager – I learned that the President instructed the Director to employ Mr. Patey as Business Manager for AURA. This appointment is effective January 13, 1958, and although the action has no endorsement from the Executive Committee, it is presumed that the urgency justified the action. Apparently, Mr. Patey has not had too much experience in the type of business management involving AURA and a Government contract, and it is contemplated, as part of the appointment, to employ as needed Western Business Consultants (Hiram Davis and J.L. Turner). Mr. Patey was in Michigan during my visit so that I had no opportunity to meet him and review the details of the job." The Executive Committee ratified this action on September 13, 1958.
Note: Patey was manager of the Mohave County Chamber of Commerce in Kingman from 1955 to early 1957, when he took a position as assistant manager with the Tucson Chamber of Commerce. He happened to walk past the Phoenix Headquarters soon after it opened and saw the sign "National Astronomical Observatory." He entered the building, introduced himself to Meinel and suggested that a site in the Hualapai mountains near Kingman should

be considered. His office in Tucson provided a useful base for the site survey before Kitt Peak was chosen.

75 Telegram, Goldwater to George Rosenberg, Tucson Daily Citizen, July 22, 1958. Edmondson personal files.

Diary note, Waterman, July 29, 1958. NSF History Files. McMath phoned to report that Meinel had received a telegram and also a personal letter from Senator Hayden stating clearly that the project carries $5,000,000 for the new telescope, plus $1,000,000 for the road. This was contrary to the NSF understanding.

76 Memorandum, Luton to Waterman, Sept. 5, 1958. NSF History Files. $5,000,000 covered both the telescope and the road according to information from the office of Congressman Albert Thomas.

Pierce recalls being in McMath's office when someone from NSF called and asked if the solar telescope could be built for the reduced figure. McMath answered that it could. Transcript of Pierce interview with Edmondson. March 21, 1979.

77 Minutes, Executive Committee of AURA, Inc., September 13, 1958, Ann Arbor. Edmondson personal files.

78 See Reference 67. A copy of the revised proposal is also in Edmondson's personal files.

79 Op.cit. 6, Reference 35.

Op.cit. Reference 76. "Keith Pierce appears to be Dr. McMath's right hand man in all matters of interest to him and certainly will be the person with primary responsibility for the construction of the solar telescope."

80 Letter, McMath to Waterman, July 23, 1958. NAO Panel Archives (solar), Kitt Peak National Observatory Library, Tucson.

81 Memorandum, Waterman to Luton, July 29, 1958. NSF History Files.

Luton returned the memo to Waterman with his agreement in longhand written at the bottom.

82 Memorandum, Waterman to Members of the National Science Board, October 2, 1958. NSF History Files.

83 Memorandum, McMath to Members of the Board of Directors of AURA, Inc., October 14, 1958. Edmondson personal files.

The breakdown was:

Solar telescope	$4,000,000
Operations	300,000
University of Michigan return of previous grants	95,000
Eckhardt's promise to McMath	50,000
(36" telescope supplement)	
	$4,445,000

84 NSF-58-169. National Science Foundation Announces Fund Allocations for National Astronomical Observatory. FOR RELEASE November 9, 1958. NSF History Files.

85 Op.cit. Reference 54.

The solar telescope was added to the equipment approved for the observatory, and the phrase "3.1 million dollars ($3,100,000) was replaced by $7.545 million dollars ($7,545,000)."

Chapter 9

1 Op.cit. 7, Reference 13.

2 Op.cit. 6, Reference 55.

3 Op.cit. 6, Reference 67.

4 Proposal to the National Science Foundation from Association of Universities for Research in Astronomy, Inc. for the Construction and Operation of a Cooperative Astronomical Observatory, pp. 6–8. October 29, 1957. Edmondson personal files.

5 Op.cit. 8, Reference 39, p. 7. The pressure for an early decision about the site came from NSF verbally, never in writing, as the author remembers it. A delay in the site selection after the incorporation of AURA would cause a loss of momentum in Congress for the support of NAO and other NSF research facilities, according to the NSF perception.

6 Minutes, Executive Committee of AURA, Inc., February 10, 1958, Pasadena, California. Edmondson personal files.

7 Memorandum, Meinel to Members of the Board of Directors AURA, Inc. (cc: L. Goldberg), January 27, 1958. Edmondson personal files.

8 This is the way the author remembers it.

9 Op.cit. 6, Reference 25.

10 Minutes, Scientific Committee of the Board of Directors of AURA, Inc., March 1–2, 1958, Pasadena, California. Edmondson personal files. Minutes were written by C.D. Shane (Chairman) and by Ralph Patey (Acting Recording Secretary) for the morning of March 1 (site selection), and minutes were written by Peter van de Kamp for the afternoon of March 1 and the morning of March 2. Patey left after the morning session to go to Washington to report the selection of Kitt Peak to the National Science Foundation.
   Note: The Scientific Committee was established at the first meeting of the AURA Board. It was referred to as a subcommittee in parallel with the Administrative Subcommittee during the first two Board meetings, and also in the minutes of the Executive Committee, Reference 6 above. McMath objected to this, and Shane's memorandum with the call to the March 1–2 meeting, dated February 14, 1958, was addressed to "All Members of the Scientific Committee of AURA." Edmondson personal files.

11 Op.cit. 8, Reference 38.

12 Letter, Luton to McMath, February 4, 1958; and Letter, Luton to Strouss, February 4, 1958. Edmondson personal files.

13 Letter, McMath to Luton, March 8, 1958. Edmondson personal files.

14 DRAFT–NOT FOR RELEASE. Kitt Peak chosen as Site for New National Astronomical Observatory, March 5, 1958. Edmondson personal files.

15 Letter, Embrey to McMath, March 5, 1958. Robert R. McMath Collection, Bentley Historical Library, University of Michigan.

16 Letter, McMath to Luton, March 11, 1958. Edmondson personal files.

17 Letter, McMath to Luton, March 12, 1958. Edmondson personal files.

18 Letter, McMath to Shane, March 13, 1958. Edmondson personal files.

19 Letter, Luton to McMath, March 18, 1958. Edmondson personal files.

20 FOR PRESS, RADIO & TV, NSF-58-116. FOR RELEASE–Friday, March 14, 1958. "Kitt Peak chosen as site for new Astronomical Observatory." Edmondson personal files.

21 For example, the story was on the first page of Section C of the *New York Times*, March 15, 1958. Edmondson personal files.
   The Mohave County Miner, a weekly newspaper published on Thursday was given special permission to publish the story on March 13, because the paper would not be on the newstands until March 14. Letter, McMath to Embrey, March 27, 1958. Edmondson personal files. Letter, Embrey to McMath, March 31, 1958. Edmondson personal files. Patey spoke at a Chamber of Commerce dinner in Kingman on the evening of the 13th to let them know in advance that the Hualapai site had not been chosen, and that the selection of Kitt Peak would be announced the next day. A reporter from a Phoenix newspaper who was present ignored the NSF press release and thoroughly garbled the story. Letter, Embrey to Harvill, March 14, 1958. Edmondson personal files.

22 Minutes, Executive Committee of AURA, Inc., April 4–5, 1958, Phoenix, Arizona. Edmondson personal files.

23 Letter, Meinel to McMath, March 11, 1958. Edmondson personal files.

24 Resolution of the Papago Council, RES. No. 976, March 7, 1958. Edmondson personal files.

25 Op.cit. 5, Reference 131.

26 Senator Goldwater told Counsel Strouss that "he was quite sure the local Indian Agent was wrong in his statement that new legislation would be required. However, he is immediately requesting an opinion from the Department of the Interior, Bureau of Indian Affairs, advising as to the steps necessary for the government to acquire the site."
Letter, Strouss to Edmondson, December 18, 1957. Edmondson personal files.

27 84th Congress, 1st session, Bill Number S.34, Public Law No. 255, 8/9/55.
Goldwater Papers, Arizona Historical Foundation, Hayden Library, Arizona State University, Tempe, Arizona. Goldwater's bill was limited to Indian lands in Arizona, but was amended to include all Indian lands before passage. A similar bill (H.R. 2681) was introduced in the House of Representatives by Congressman Stewart Udall.

28 Op.cit. Reference 22, page 3. Letter, Miller to Jennings, Strouss, Salmon, and Trask, April 28, 1958. Edmondson personal files.

29 Letter, Strouss to Luton, January 15, 1958. NSF History Files.
Letter, Strouss to Meinel, January 20, 1958. AURA Archives, Kitt Peak National Observatory Library, Tucson. This letter mentions a phone call from Professor John Denton of the University of Arizona, who served as the Attorney for the Papago Tribe.
Letter, Luton to Strouss, February 4, 1958. NSF History Files.
Draft, AURA proposal to Papago Tribe, undated [see Reference 23]. Edmondson personal files.

30 Letter, Luton to Meinel, March 19, 1958. AURA Archives, Kitt Peak National Observatory Library, Tucson.
Letter, Patey to McMath, March 19, 1958. AURA Archives, Kitt Peak National Observatory Library, Tucson. Patey wrote: "Mr. Jennings is currently drawing a rough draft of the Lease Agreement."

31 Transcript of Shoenhair interview with Edmondson, March 23, 1979.

32 Letter and enclosure, Luton to Patey, June 11, 1958. Edmondson personal files.

33 Letter, Patey to McMath, June 17, 1958. Edmondson personal files.

34 Telegram, Luton to Jim Hart (*Tucson Daily Citizen*), July 22, 1958. Edmondson personal files.

35 Congressional Record – Senate, July 31, 1958, p. 14346.

36 The lease bill was passed, but the appropriations bill was vetoed. The revised appropriations bill was passed on August 23, 1958 [see 8, Reference 72]. Both bills were probably signed the same day.

37 Letter, Patey to Luton, May 16, 1958. AURA Archives, Kitt Peak National Observatory Library, Tucson.

38 Letter, Luton to Patey, July 18, 1958. Edmondson personal files.

39 Letter, Patey to Luton, September 8, 1958. Edmondson personal files.

40 Minutes, Executive Committee of AURA, Inc., September 13, 1958. Edmondson personal files. Keller promised to "do all he could to expendite this matter in Washington immediately upon his return."

41 Minutes, Papago Tribal Council, October 3, 1958. Edmondson personal files (copy supplied by C.J. Higman).
Minutes, Staff Meeting, AURA, Inc., Tucson, Arizona, October 6, 1958. Edmondson personal files.

42 Lease, dated 24th day of October, 1958, by and between the PAPAGO TRIBE, ARIZONA and the UNITED STATES OF AMERICA, acting through the National Science Foundation. Kitt Peak National Observatory Operations files, Tucson.

43 Op.cit. 5, References 38–40, and 47–53.

44 Op.cit. 5, References 124–130.

45 Memorandum, Luton to Sheppard, April 22, 1958. NSF History Files.
Letter, Meinel to McMath, April 29, 1958. Edmondson personal files.

Letter, Meinel to Shane, April 29, 1958. Edmondson personal files.

46 Letter, Meinel to McMath, March 20, 1958. AURA Archives, Kitt Peak National Observatory Library, Tucson.

47 Letter, Meinel to McMath, December 23, 1957. AURA Archives, Kitt Peak National Observatory Library, Tucson. A 99-year lease from the University was the favored idea at this time. Harvill told Meinel the University could acquire the land immediately upon a decision by AURA.

48 Op.cit. 8, Reference 74.

49 Op.cit. 8, Reference 68.

50 Letter, Meinel to McMath, February 26, 1958. Edmondson personal files. This letter also reported that Arizona State University was interested in offering a site on the Tempe campus for the city offices if either the Hualapai or Mormon Mountain site were chosen.

51 Letter, Luton to Meinel, March 19, 1958. AURA Archives, Kitt Peak National Observatory Library, Tucson.

52 Letter, Luton to McMath, March 25, 1958. AURA Archives, Kitt Peak National Observatory Library, Tucson.

53 Op.cit. 7, Reference 48.

54 Op.cit. Reference 22.

55 Letter, Harvill to Edmondson, March 28, 1958. Edmondson personal files.

56 Letter, Meinel to Luton, March 31, 1958. Edmondson personal files.

57 Letter, Luton to McMath, April 8, 1958. Edmondson personal files.

58 Memorandum, Shane to Members of Executive Committee of AURA, April 23, 1958. Edmondson personal files.

59 Letter, Miller to Shane, April 28, 1958. AURA Archives, Kitt Peak National Observatory Library, Tucson.

60 Letter, Luton to Meinel, April 22, 1958, NSF History Files.

61 Letter, Meinel (signed by Patey) to McMath, May 14, 1958. Edmondson personal files.
Letter, McCormick (Univ. of Arizona) to Meinel, May 15, 1958. AURA Archives, Kitt Peak National Observatory Library, Tucson.
Op. cit. Reference 37.

62 Minutes, Executive Committee of AURA, Inc., September 13, 1958, Ann Arbor, Michigan. Edmondson personal files.

63 Letter, Meinel to Luton, October 24, 1958. AURA Archives, Kitt Peak National Observatory Library, Tucson.

64 Letter, McMath to Luton, October 27, 1958. AURA Archives, Kitt Peak National Observatory Library, Tucson.

65 Letter, Klaffter to Luton, October 29, 1958. AURA Archives, Kitt Peak National Observatory Library, Tucson.

66 Telegram, Miller to Luton, November 21, 1958. AURA Archives, Kitt Peak National Observatory Library, Tucson.

67 Telegram, Luton to Miller, November 25, 1958. AURA Archives, Kitt Peak National Observatory Library, Tucson.

68 Letter, Luton to Miller, November 26, 1958. AURA Archives, Kitt Peak National Observatory Library, Tucson.

69 Letter, Meinel to Luton, November 26, 1958. AURA Archives, Kitt Peak National Observatory Library, Tucson.

70 Telegram, Luton to Meinel, December 4, 1958. Copy with minutes, Reference 71.

71 Minutes, Scientific Committee of AURA, Inc., December 7, 1958. Edmondson personal files.

72 Letter, Harvill to McMath, December 2, 1958. Copy with Minutes, Reference 71.

73 Op. cit. Reference 70.

74 Minutes, Board of Directors of AURA, Inc., December 8, 1958. Edmondson personal files.

75 Telegram, Luton to Lee, December 8, 1958. Edmondson personal files.
76 Letter, Lee to Luton, December 9, 1958. Edmondson personal files.
   Memorandum, Lee to McMath, Shane, J.M. Miller, Peterson, Patey, and Shoenhair, December 9, 1958. Edmondson personal files.
   Letter, Luton to Lee, December 15, 1958. Edmondson personal files.
77 Diary Note, Shane, December 15, 1958. Edmondson personal files.
78 Letter, Shoenhair to Luton, December 20, 1958. Edmondson personal files.
   Letter, Luton to McCormick, December 26, 1958. Edmondson personal files.
   Letter, Shoenhair to McCormick, January 23, 1959. Edmondson personal files.
79 Minutes, Executive Committee of AURA, Inc., June 2, 1959, Ann Arbor, Michigan. Edmondson personal files.

Chapter 10

1 Op. cit. 5, Reference 87.
2 Op. cit. 4, Reference 40.
3 Op. cit. 4, Reference 50.
4 Letter, Waterman to McMath, January 29, 1954. Robert R. McMath Collection, Michigan Historical Collections, Bentley Historical Library. The University of Michigan.
5 Op. cit. 4, Reference 51.
   Op. cit. 5, Reference 75.
6 Memorandum (p. 2, last paragraph), McMath to Drs. Bowen, Stromgren, Struve, Whitford. Members of the NAO–NSF Panel. June 2, 1955. Robert R. McMath Collection, Michigan Historical Collections, Bentley Historical Library. the University of Michigan.
   Note: McMath did not mention this in his May 21 memo to the Panel. Op. cit. 5, Reference 77.
7 Op. cit. 5, Reference 96.
8 Op. cit. 6, Reference 31.
   Note: McMath kept the old name the previous year when he wrote the minutes of the February 25–26, 1956 meeting.
9 Op. cit. 8, Reference 9, pp. 4–5, and attachment.
10 Memorandum, Waterman to Members of the National Science Board, Nov. 19, 1957. NSF History Files.
11 Op. cit. 8, Reference 30.
12 Memorandum, J. Miller to Members of the Board of Directors, AURA, Inc. November 3, 1958. Edmondson personal files. The names were: National Optical Astronomical Observatory, Kitt Peak Observatory, Kitt Peak National Observatory, National Optical Astronomical and Solar Observatory, and National Observatory for Optical and Solar Astronomy.
13 Letter, McMath to J.M. Miller, November 6, 1958. Edmondson personal files.
14 Minutes, AURA Board of Directors, December 8, 1958, page 8, Ann Arbor Michigan. Edmondson personal files. Geoffrey Keller, NSF Program Director for Astronomy, reported that "National Optical Astronomy Observatory," which is symmetrical to "National Radio Astronomy Observatory," would be acceptable to Dr. Waterman. During the discussion it was pointed out that optical astronomy is not the name of a recognized field in astronomy, as is radio astronomy. Edmondson proposed "Kitt Peak National Observatory," pointing out the precedents, such as Mount Wilson Observatory (for using the name of the site) and Argonne National Laboratory (for the use of the word "National" to indicate US Government funding). Randal Robertson, the new NSF Assistant Director for MPE, agreed with this name. Harrell moved, Seyfert seconded, and the Board unanimously approved "Kitt Peak National Observatory," subject to its acceptance by the National Science Foundation.
15 Letter, McMath to Waterman, December 22, 1958. Edmondson personal files.
16 Letter, Waterman to McMath, February 3, 1959. Edmondson personal files.

17 Minutes, Special Meeting of the Board of Directors of AURA, Inc., March 3, 1959, Tucson, Arizona. Edmondson personal files. Waterman's letter had been sent to McMath in Pasadena. McMath wrote to Secretary Miller on February 24: "I found the original letter confirming the official name of our observatory in my brief case, and I am enclosing it and a Thermo-fax copy for your office files." Edmondson personal files.
Miller's report to the Board is on p. 5 of the Minutes.

18 Memorandum, Shane to Members of the Executive Committee of AURA, January 13, 1958. Edmondson personal files. "We also examined some plans for facilities located in a city near the site. It was the general agreement that the plans should be somewhat enlarged as regards space."

19 Op. cit. 9, Reference 6, p. 4.

20 Letter, Meinel to McMath, March 26, 1958. Edmondson personal files.

21 Letter, Shane, Meinel and Patey to Luton, May 13, 1958. Edmondson personal files. The breakdown was:

	FY58	FY59
City land	$ 90,000	-0-
Equipment & library	$ 20,000	$ 71,000
Engineering & Architect	$ 10,000	$ 10,000
Building	-0-	$185,000
Optical tunnel	-0-	$ 30,000
TOTAL	$120,000	$296,000

The amount for city land would be determined by negotiations with the University of Arizona. The expenditure for the building was $40,000 less than estimated needs. The library room and some laboratory space were omitted from these plans.

22 Minutes, Scientific Committee of AURA, Inc., July 9, 1958, Phoenix, Arizona. Edmondson personal files.

23 Op. cit. 9, Reference 62, pp. 4–5.

24 Op. cit. 9, Reference 71, p. 4.
Luton had proposed that the University of Arizona acquire the land and sell it to AURA in a letter to Meinel dated April 22, 1958 [9, Reference 60]. Almost 8 months passed before he approved the transaction on December 8 [9, References 75 and 76].

25 Letter and enclosures, Meinel to Shane, February 12, 1959. Edmondson personal files.

26 Operations Report, Month of April 1959, p. 3. Edmondson personal files.

27 NSF Summary Sheet S-6, FJC/PHK:wgt, 8/5/55. NSF History Files.
Op. cit. 5, References 96–99.

28 Op. cit. 6, References 77–80.

29 Op. cit. 9, Reference 6, pp. 2–3. Approval was also given to using Kitt Peak as the temporary site of this telescope. This became moot when Kitt Peak was selected as the site of the NAO three weeks later.

30 Op. cit. 9, Reference 22, pp. 7–8. The Boller and Chivens bid was $10,000 lower than the next bid, and $132,178 was the high bid. Engineering estimates were $70,000 to $80,000. The high bid on the building was $138,718. The next to the lowest bid, $118,000, was from Sundt Construction Co. of Tucson. This firm was the general contractor for the Kitt Peak 150-inch building.

31 Op. cit. 8, Reference 32. Amendment No. 1 later added $4,445,000 to the contract, including $95,000 returned by the University of Michigan and the $50,000 supplement that Eckhardt had promised to McMath. See 8, Reference 83 for the detailed breakdown.

32 Op. cit. Reference 22, p. 5.

33 Op. cit. 9, Reference 42.

34 Minutes, Staff Meeting, AURA, Inc., September 29, and October 13. Edmondson personal files.

Note: (1) Staff Meetings Minutes were issued on a weekly basis September 29, 1958 – January 12, 1959. These are a valuable resource for anyone who is interested in the details of what happened in this period.

(2) AURA Operations Reports were issued on a monthly basis March 1, 1959 – May 3, 1960.

(3) Kitt Peak National Observatory Monthly Reports were issued June 1960 – February 1966, Bi-Monthly Reports March 1966 – December 1969, Quarterly Reports in 1970–71, and Quarterly Bulletins in 1972–78. Newsletters with varying formats and schedules have been issued since 1978.

35 Operations Report, Month of December 1959. Edmondson personal files.

36 Operations Report, Month of February 1960. Edmondson personal files.

37 Operations Report, Month of March 1960. Edmondson personal files.

38 Op. cit. References 27 and 28.

39 Op. cit. 6, Reference 78.

40 Op. cit. Reference 29, p. 5.

41 Op. cit. 9, Reference 10.

42 Op. cit. Reference 30, p. 5.

43 Op. cit. Reference 22.

44 Minutes, Scientific Committee of AURA, Inc., December 7, 1958. Edmondson personal files.

45 Op. cit. Reference 14.

46 Memorandum, Code to Members of the Scientific Committee of AURA, Inc., January 5, 1959. Edmondson personal files. The members of the Committee were: Code – Chairman, Baustian, Bowen, Carpenter, Jones, Mayall, Patey, Pierce, Rule, and Shane.

47 Letter and enclosures, Patey to Shane, April 2, 1959. Edmondson personal files.

48 Op. cit. 9, Reference 79, pp. 11–12.

49 Notes on a Meeting of the Solar Telescope Review Committee, October 7, 1959, Tucson, Arizona. Edmondson personal files.

50 Op. cit. Reference 37, p. 7.

51 Op. cit. 5, Reference 65. 30-inch and 80-inch telescopes were proposed at the first meeting of the Panel, and the problem of solar seeing was also discussed [5, Reference 7].

52 Op. cit. 5, Reference 100.

53 Op. cit. 8, References 75 and 76.

54 Op. cit. 8, Reference 54.

55 Op. cit. 8, Reference 77, Executive Session, p. 1.

56 Op. cit. Reference 24, p. 2.

57 Op. cit. Reference 14, p. 4.

58 Minutes, Special Meeting of the Board of Directors of AURA, Inc., March 3, 1959, Tucson, Arizona. Edmondson personal files.

59 The information in this paragraph comes from the Weekly Staff Meetings Minutes for January 12, 1959, and the Operations Report for March 8–31 and April, 1959. [See Reference 34, Note.] Edmondson personal files.

60 Operations Report, Week of March 1–7, 1959, p. 3. Edmondson personal files.

61 Operations Report covering the period of March 8–31, 1959, p. 3. Edmondson personal files.

62 Operations Report, Month of June 1959, p. 4a. Edmondson personal files.

63 Operations Report, Month of August, 1959, p. 12. Edmondson personal files.

64 Operations Report, Month of July 1959, p. 4, and Operations Report, Month of August 1959, p. 3. Edmondson personal files.

65 Operations Report, Month of September 1959, p. 5. Edmondson personal files.

66 Operations Report, Month of December 1959, pp. 3–4. Edmondson personal files.

67 Operations Report, Month of January 1960, pp. 4–5. Edmondson personal files.

68 Operations Report, Month of February 1960, pp. 7, 8. Edmondson personal files.

69 *Ibid*, p. 4.

70 Operations Report, Month of March 1960, p. 5. Edmondson personal files.

71 Minutes, Weekly Staff Meeting, AURA, Inc. January 12, 1959, pp. 1–2. Edmondson personal files.

Note: Weekly was included in the heading of these minutes starting October 27, 1958.

72 Letter, Patey to Colonel Moynahan, December 29, 1958. Robert R. McMath Collection, Michigan Historical Collections, Bentley Historical Library, University of Michigan.

73 Letter, Colonel Moynahan to Patey, December 31, 1955. Robert R. McMath Collection, Michigan Historical Collections, Bentley Historical Library, University of Michigan.

74 Letter, Meinel to Shane, January 8, 1959. Robert R. McMath Collection, Michigan Historical Collections, Bentley Historical Library, University of Michigan.

75 A second letter, Meinel to Shane, January 8, 1959. Robert R. McMath Collection, Michigan Historical Collections, Bentley Historical Library, University of Michigan.

76 *Tucson Daily Citizen*, January 9, 1959. Edmondson personal files.

77 Letter, Meinel to Shane, January 14, 1959. Robert R. McMath Collection, Michigan Historical Collections, Bentley Historical Library, University of Michigan.

78 Letter, Shane to McMath, January 14, 1959. Robert R. McMath Collection, Michigan Historical Collections, Bentley Historical Library, University of Michigan.

79 Op. cit. Reference 71.

80 Op. cit. Reference 75.

81 Letter, McMath to Shane, January 17, 1959. Robert R. McMath Collection, Michigan Historical Collections, Bentley Historical Library, University.

82 Op. cit. Reference 17, p. 1.

83 Letter, Meinel to Colonel Moynahan, January 23, 1959. Robert R. McMath Collection, Michigan Historical Collections, Bentley Historical Library, University of Michigan.

84 Op. cit. Reference 26, pp. 1 and 4.

Letter General Moorman to Shane, April 15, 1959. Robert R. McMath Collection, Michigan Historical Collections, Bentley Historical Library, University of Michigan.

85 Operations Report, Month of May 1959, p. 1. Edmondson personal files. The author recalls that Colonel Moynahan was transferred to another post shortly after this.

86 Letter and enclosure, Shane to Waterman, April 16, 1959.

87 Minutes, Executive Committee of AURA, Inc., June 2, 1959, p. 1. Edmondson personal files.

88 Letter, Waterman to Brucker, June 22, 1959. NSF History Files.

Note: History did repeat itself in June 1981 when the Arizona Air National Guard and the US Air Force announced plans to build on auxiliary airfield in the Altar Valley, just nine miles to the southeast of Kitt Peak. It was to be used primarily for training flights to practice takeoffs and landings. [*Arizona Republic*, June 18, 1981; *Arizona Daily Star*, June 19, 1981. Edmondson personal files.]

The Air Force took a hard line at the beginning, but decided in November to put the facility at Libby Army Airfield, Sierra Vista, near Fort Huachuca. [Letter, Col. H.N. Campbell to John B. Slaughter, Director of NSF, December 1, 1981. Edmondson personal files.] Senator Goldwater had withdrawn his support for the Altar Valley site in September, and this must have had a strong influence on the Air Force decision. [*Arizona Daily Wildcat*, September 2, 1981. Edmondson personal files.]

89 Minutes, Staff Meeting, AURA, Inc., p. 4, October 6, 1958. Edmondson personal files.

90 Memorandum, Embrey to Hall, October 15, 1958. NSF History Files.

91 Memorandum, Luton to Embrey, October 20, 1958. NSF History Files.

92 Letter, Keller to Meinel, October 23, 1958. NSF History Files.

93 Memorandum, Embrey to Luton, October 21, 1958. NSF History Files.

94 Minutes, Staff Meeting, AURA, Inc., p. 2, October 20, 1958. Edmondson personal files.

95 Letter and enclosure, Patey to Edmondson, October 22, 1958. Edmondson personal files. Considering the date, this must have been sent only for information. Edmondson's reply (October 31) corrected errors in the titles Patey used for Keller, Robertson, Embrey, and Brode. "I think it is of the utmost importance that you make every effort to be correct in matters of this kind." Edmondson personal files.

96 Letter, Edmondson to Meinel, January 22, 1959. Edmondson personal files.

97 Op. cit. Reference 58, p. 11, Agenda item 17. In early February Shane had talked to Harvill "about inviting the N.S.F. Board to a meeting in Tucson October 12 with AURA and the University of Arizona as joint hosts." [Letter, Shane to McMath, February 5, 1959. Edmondson personal files.] No reason was given for choosing this date.

98 Letter, Edmondson to Waterman, March 20, 1959. Edmondson personal files.

99 Letter, Waterman to Edmondson, March 30, 1959. Edmondson personal files.

100 Letter, Harvill to Edmondson, April 27, 1959. Edmondson personal files.
    Memorandum, Carpenter to Harvill, April 29, 1959. Edmondson personal files.
    Letter, Edmondson to Harvill, May 7, 1959. Edmondson personal files.

101 Letter, Edmondson to Waterman, May 4, 1959. Edmondson personal files.

102 Letter, McMath to Edmondson, May 7, 1959. Edmondson personal files.

103 Letter, Edmondson to McMath, May 13, 1959. Edmondson personal files.

104 Letter, Waterman to McMath, October 22, 1959. Edmondson personal files.
    Letter, McMath to Waterman, October 24, 1959. Edmondson personal files.

105 Letter, Patey to McMath, October 26, 1959. Edmondson personal files.

106 Letter, McMath to Harvill, October 27, 1959. Edmondson personal files.

107 Letter, Harvill to Waterman, November 4, 1959. Edmondson personal files.

108 Letter, Harvill to McMath, November 4, 1959. Edmondson personal files.

109 Letter, McMath to Waterman, November 10, 1959. Edmondson personal files.

110 Letter, Patey to Edmondson, November 4, 1959. Edmondson personal files.

111 Letter, McMath to Edmondson, November 17, 1959. Edmondson personal files.

112 Letter, J.M. Mitchell (NSF) to McMath, September 16, 1958. Edmondson personal files.
    Letter, McMath to Mitchell, September 19, 1958. Edmondson personal files.
    Mitchell's letter seems to be the first suggestion that an informational brochure should be prepared, similar to the brochure for the Green Bank ground breaking. McMath suggested that Embrey and Edmondson could collaborate to produce a brochure in "draft" form.
    Letter, Embrey to McMath, October 10, 1958. NSF History Files. His letter to Mitchell had been turned over to her. She had recently been transferred to the Office of the Director (Research Assistant to the Director), but had been asked to continue to be responsible for certain public relations aspects of the major astronomy projects. She would be interested in helping to prepare a brochure.
    Memorandum and attachment, Edmondson to Lee Anna Embrey, R.R. McMath (for information), C.D. Shane, and Tucson (A.B. Meinel, A.K. Pierce, R. Patey), January 11, 1959. Edmondson personal files. This listed ten subject headings and assigned the responsibilites for writing. Dr. Shane and Mrs. Shane would prepare the material for the printer.

113 Memorandum and enclosures, Edmondson to Embrey, Patey and Shane, November 7, 1959. Edmondson personal files.

114 Letter, Embrey to Edmondson, November 16, 1959. Edmondson personal files. It is worth noting that Embrey wrote the talk that Dr. Waterman gave as part of the Kitt Peak dedication ceremony. Her new position in the office of the Director included writing some of his speeches.

115 Letter, Edmondson to Paine, December 11, 1959. Edmondson personal files.

116 Letter, Shane to McMath, November 11, 1959. Edmondson personal files.

117 Letter, AURA Secretary Miller to Mrs. C.D. Shane, April 14, 1960. Edmondson personal files.

118 Letter, Edmondson to Harvill, November 19, 1959. Edmondson personal files.

119 Letter, Harvill to Edmondson, November 24, 1959. Edmondson personal files.

120 Report of AURA Committee for Dedication Ceremony, 10 December, 1959. Edmondson personal files.

121 Letter, Edmondson to McMath, December 11, 1959. Edmondson personal files.

122 Harvill and Waterman were also in direct communication about logistic and other details. Two examples are: Letter, Waterman to Harvill, December 3, 1959. Edmondson personal files. Letter, Harvill to Waterman, December 10, 1959. Edmondson personal files.

123 Letter, Mohler to Edmondson, December 18, 1959. Edmondson personal files.

124 Letter, Edmondson to McMath, December 23, 1959. Edmondson personal files.

125 Two letters, McMath to Edmondson, December 22, 1959. Edmondson personal files.

126 Letter, McMath to Morgan, January 14, 1960. Edmondson personal files.

127 Letter, Morgan to McMath, January 20, 1960. Edmondson personal files.

128 Op. cit. Reference 120.

129 Letter, Mulders to Edmondson, February 2, 1960. Edmondson personal files. This confirmed an oral suggestion made on January 25.

130 Letter, Patey to Edmondson, January 29, 1960. Edmondson personal files. This reported a telephone conversation with Vernice Anderson, Executive Secretary of the NSB.
Letter, Patey to Mulders, February 5, 1960. Edmondson personal files.
Letter, Mulders to Patey, February 10, 1960. Edmondson personal files.

131 Letter and enclosures, Edmondson to Waterman, January 11, 1960. Edmondson personal files. The enclosures included the list of suggested guests and a copy of the revised program for the Dedication. Copies were sent to McMath, Shane, Carpenter, J.M. Miller (AURA Secretary), and Patey.

132 Letter, Waterman to Edmondson, January 20, 1960. Edmondson personal files.

133 Report, Papago Pottery Cache on Kitt Peak, Emil W. Haury, Director, Arizona State Museum, May 6, 1960. Edmondson personal files. The author also has a photocopy of Dr. Haury's longhand field notes. They include a sketch and verbal description: "Crevice opens up amidst jumble of granite boulders into a triangular pocket about a meter on each side. Pots were cached under overhanging boulder on N side. To reach cache one had to lie on stomach and reach in from the top." The Report reads: "To recover them, it was necessary to lie on one's stomach and reach them from above."

134 Dr. Haury had conducted a survey the previous month along the right of way of the public highway to Kitt Peak as mapped by the Bureau of Public Roads.
Report of Archaeological Survey Along Right of Way for Kitt Peak Observatory Road, February, 1960. Edmondson personal files.

135 Schedule, March 12–15, 1960, and Dedication Program, March 15, 1960. Edmondson personal files.

136 Transcript of Harvill interview with Edmondson, August 30, 1978. By lucky chance, Harvill was seated next to Enos Francisco during the dedication speeches. He whispered to Harvill: "They don't seem to have any place for me on this program. I wish they did, because I have something I would like to say." Harvill told him they were probably going to ask him to speak at the luncheon. When the ceremony ended Harvill got in touch with a member of the AURA group, probably Meinel or Shane, and told him that Enos Francisco should be asked to speak at the luncheon. This was done, and potential embarrassment was avoided.

137 Sky and Telescope, XIX, No. 7, May, 1960. This magazine had published an earlier article about the Observatory, written by Meinel, five months after Kitt Peak was chosen. XVII, No. 10, August, 1958.

138 Minutes, Annual Meeting of the Board of Directors of AURA, Inc., Executive Session pp. 1–2, March 14, 1960, Tucson, Arizona. Edmondson personal files.
Tucson Citizen, March 15, 1960.

139 Memorandum, Shane to All Employees, AURA, Inc., March 16, 1960. Edmondson personal files.

Chapter 11

1 Homer E. Newell, *Beyond the Atmosphere*, Chapters 7 and 8. NASA History Series, NASA SP-4211. U.S. Government Printing Office, 1980.

2 Op. cit. 10, Reference 22.

3 Letter, Shane to McMath, July 10, 1958. Solar Telescope Archives, Kitt Peak National Observatory Library, Tucson.

4 Letter, Meinel to Shane, January 21, 1959. Edmondson personal files. This letter was written following a verbal discussion on January 19 about future plans for the observatory. Copies were sent to McMath and Arthur Code, Chairman of the Scientific Committee.

5 The pending projects were at Princeton and Harvard. The author recalls that Meinel had investigated the possibility of joining the Princeton ultraviolet space astronomy program.

6 Longhand letter, Shane to McMath, January 21, 1959. Robert R. McMath Collection, Michigan Historical Collections, Bentley Historical Library, University of Michigan.

7 Letter, Shane to Keller, February 2, 1959. Edmondson personal files.

8 Letter, Keller to Shane, February 9, 1959. Edmondson personal files.

9 Op. cit. 7, Reference 103.

10 Transcript of Spitzer interview with Edmondson, February 16, 1983.

11 Letter, Spitzer to Edmondson, January 24, 1983. Edmondson personal files. An enclosure (Comments on F.K. Edmondson Questions on AURA History) gave short answers to the questions that were discussed during the interview on February 16.

12 L. Spitzer, Jr., "Astronomical Advantages of an Extra-Terrestrial Observatory," Douglas Aircraft Company, September 1, 1946. Reprinted in *The Astronomy Quarterly*, 7, 1990, 131–142.

13 Spitzer attended the March 14, 1960 Board Meeting, but did not stay for the Dedication on March 15.

14 Minutes, Scientific Committee of AURA, Inc., March 1, 1959. Edmondson personal files.

15 Minutes, Special Meeting of the Board of Directors of AURA, Inc., March 3, 1959, Tucson, Arizona. Edmondson personal files.

16 *Tucson Daily Citizen*, March 4, 1959.

17 Letter, J.M. Miller to Robert F. Goheen, March 19, 1959. Robert R. McMath Collection, Michigan Historical Collections, Bentley Historical Library, University of Michigan.

18 Letter, Goheen to Miller, April 8, 1959. Robert R. McMath Collection, Michigan Historical Collections, Bentley Historical Library, University of Michigan.

19 Letter, McMath to Goheen, April 16, 1959. Robert R. McMath Collection, Michigan Historical Collections, Bentley Historical Library, University of Michigan.

20 Proposal to NSF for a Research Grant. PHASE I. "Preliminary Conceptual Design, and Experimental Studies for a Large Aperture Orbital Telescope." April 1959.
Note: The author's file of AURA Board documents about the AURA–KPNO space telescope project were given to the Space Telescope History Project, Space History Department, National Air and Space Museum, Washington, D.C. in June 1984. Working duplicates of selected items have been retained in Bloomington.

21 Letter, Waterman to Association of Universities for Research in Astronomy, Inc., June 9, 1959. Edmondson personal files.

22 Minutes, Executive Committee of AURA, Inc., June 2, 1959, Ann Arbor, Michigan. Edmondson personal files.

23 Transcript of Roman interview with Edmondson, November 17, 1982.

24 Op. cit. Reference 10.

25 Op. cit. 10, References 71–88.

26 Letter, Meinel to Shane, April 15, 1959. Edmondson personal files.

27 PRESENTATION, Department of the Army, Pentagon, Washington 25, D.C., April 24, 1959. Edmondson personal files.

28 Letter, Meinel to Shane, May 6, 1959. Edmondson personal files.

29 Two letters, Meinel to Shane, May 28, 1959. Edmondson personal files.

30 DISPOSITION FORM, Unclassified, File No. ORDAB, DSPE, SUBJECT: AURA–ABMA meeting, SATELLITE TELESCOPE. Edmondson personal files.

31 The author's copy is the source for the illustration on p. 45 of *"The Space Telescope"* by Robert W. Smith, Cambridge University Press, 1989. Another copy is in the National Air and Space Museum as part of an exhibit called "Telescopes that Never Were."

32 Minutes, Scientific Committee of AURA, Inc., October 9, 1959, Kitt Peak National Observatory. Edmondson personal files.

33 Letter, Luton to Meinel, February 13, 1959. NSF History Files.

34 The author clearly remembers learning about this in McMath's room at the HiWay House the day before the meeting of the Scientific Committee. Chairman McMath and President Shane did most of the talking, with Vice-President Edmondson listening. The three of us agreed that Meinel should not continue as Observatory Director. We may have been under some pressure from NSF, but this would not have been put in writing.

35 Minutes, Executive Committee of AURA, Inc., March 13, 1960, Tucson, Arizona. Edmondson personal files.

36 Op. cit. 10, Reference 138.

37 Op. cit. 8, Reference 37. Meinel's appointment was for a three-year term ending June 3, 1960. It would have been simpler to wait two months and not reappoint him, but no one thought of this at the time.

38 Letter, Luton to Mayall, October 14, 1960. Edmondson personal files.

39 Letter and enclosure, Mayall to Luton, October 27, 1960. Edmondson personal files.

40 Letter, Nidey to Spitzer, November 16, 1960. Edmondson personal files. A draft of the minutes was sent with this letter.

41 Letter, Shane to Keller, November 12, 1960. Edmondson personal files (from Spitzer).

42 Op. cit. Reference 23.

43 Minutes, Joint meeting of the Executive Committee and the Scientific Committee of AURA, Inc., March 11, 1961, Tucson, Arizona. Edmondson personal files.

44 Minutes, Board of Directors of AURA, Inc., March 13, 1961, Tucson, Arizona. Edmondson personal files.

45 Letter, Mayall to McMath, April 5, 1961. Robert R. McMath Collection, Michigan Historical Collections, Bentley Historical Library, University of Michigan.

46 Letter, Meinel to Mayall, June 21, 1961. Robert R. McMath Collection, Michigan Historical Collections, Bentley Historical Library, University of Michigan.
Note: Meinel's resignation was a feature news story in the June 13, 1961 *Tucson Daily Citizen*.

47 Diary Note, Waterman, March 14, 1961. NSF History Files. "Shane said he realized this proposal would have to be approved by the NSF. I told him that I would want to talk with our people because such a radical move would create problems for us."

48 Diary Note, Waterman, March 20, 1961. NSF History Files.

49 Draft of the transcript of the meeting in Washington, 10:05 a.m. March 26, 1961, Regarding the AURA Board action on the space program. Edmondson personal files.

50 Draft of the transcript of the meeting in Washington 11:00 a.m. March 26, 1961, regarding the AURA Board action on the space program. Edmondson personal files.

51 Letter, McMath to Mayall, March 30, 1961. Robert R. McMath Collection, Michigan Historical Collections, Bentley Historical Library, University of Michigan.

52 Letter, McMath to Miller, April 6, 1961. Robert R. McMath Collection, Michigan Historical Collections, Bentley Historical Library, University of Michigan.

53 Op. cit. Reference 45.

54 Minutes, Special Meeting of the Board of Directors of AURA, Inc., June 14, 1961, Board Room of National Science Foundation headquarters, Washington, D.C. Edmondson personal files.

55 Letter, Mayall to Van Allen, June 19, 1961. Robert R. McMath Collection, Michigan Historical Collections, Bentley Historical Library, University of Michigan.

56 Letter, Van Allen to Mayall, July 11, 1961. Robert R. McMath Collection, Michigan Historical Collections, Bentley Historical Library, University of Michigan.

57 Letter, Mayall to McMath, July 19, 1961. Robert R. McMath Collection, Michigan Historical Collections, Bentley Historical Library, University of Michigan.

58 Letter, Mayall to Friedman, July 27, 1961. Robert R. McMath Collection, Michigan Historical Collections, Bentley Historical Library, University of Michigan.

59 Longhand letter, Mayall to McMath, July 31, 1961. Robert R. McMath Collection, Michigan Historical Collections, Bentley Historical Library, University of Michigan.

60 Letter, Mayall to Scherer, September 18, 1961. Edmondson personal files.

61 Letter, Mayall to Friedman, September 18, 1961. Edmondson personal files.

62 Letter, Shane to McMath, September 27, 1961. Robert R. McMath Collection, Michigan Historical Collections, Bentley Historical Library, University of Michigan.

63 Minutes, special meeting of the Board of Directors of AURA, Inc., October 27, 1961, Tucson, Arizona. Edmondson personal files.

64 Letter, Mayall to Friedman, October 31, 1961. Edmondson personal files.

65 Letter, Shane to Carson *et al.*, November 2, 1961. Edmondson personal files.
    Note: The Board had voted that a new ad hoc committee should be appointed at the June 19, 1961 meeting (Reference 54), but this was never done. The old committee continued to function.

66 Memorandum, Edmondson to ad hoc committee to search for Associate Director for KPNO Space Division (C.S. Gage, L. Goldberg, R. Leighton, N.U. Mayall, D.H. Menzel–Chairman, C.D. Shane, L. Spitzer), and others in attendance at November 15, 1961 meeting in Washington (J.M. Miller, A.K. Pierce, W.G. Whaley), November 22, 1961. Edmondson personal files.

67 Letter, Mohler to Menzel, September 19, 1961. Robert R. McMath Collection, Michigan Historical Collections, Bentley Historical Library, University of Michigan.

68 Letter, Mohler to Edmondson, November 25, 1961. Edmondson personal files.

69 The author and Lee were the only members of the AURA Board to attend the funeral service on January 5. Others who might have attended were prevented by a severe ice storm in the midwest.

70 Letter, Friedman to Mayall, November 30, 1961. Edmondson personal files.

71 Letter, Mayall to Board of Directors, AURA, Inc., December 5, 1961. Edmondson personal files.

72 Edmondson's brief notes about the phone call. Edmondson personal files.

73 Letter, Menzel to Code, November 27, 1961. Edmondson personal files.

74 Memorandum, Menzel to 20 persons, December 18, 1961. Subject: Combined Meeting of Scientific Committee and ad hoc commiteee for Selecting Associate Director for Space. Edmondson personal files.
    Memorandum, Miller to Executive Committee, Scientific Committee, Ad Hoc Committee for Selecting an Associate Director-Space, January 4, 1962. Edmondson personal files.

75 Minutes, joint meeting of the Scientific Committee and the ad hoc Committee for Selecting an Associate Director for Space of AURA, Inc., January 14, 1962, Washington, DC.

76 Op. cit. Reference 66.

77 Minutes, Scientific Committee of AURA, Inc., January 15, 1962, Washington, D.C. Edmondson personal files.

78 Memorandum of phone call – Geoffrey Keller to Donald H. Menzel. Date aproximately January 10, 1962. Memo written by Menzel dated January 17, 1962. Edmondson personal files.

79 Minutes, joint meeting of the ad hoc Committee for Selecting an Associate Director for

Space, and the Scientific Committee of AURA, Inc., January 16, 1962, Washington, DC. Edmondson personal files.

80 Transcript of Clark interview with Edmondson, December 2, 1981.

81 Letter, Newell to Edmondson, February 20, 1981. Edmondson personal files.

82 Minutes, Executive Committee of AURA, Inc., January 16, 1962, Washington, D.C. Edmondson personal files.

83 Memorandum, Nidey to Mayall, January 22, 1962. Edmondson personal files.

84 Letter, Nidey to Edmondson, January 23, 1962. Edmondson personal files.

85 Minutes, written by J.M. Miller, Meeting held at National Science Foundation, Washington, DC, Friday, January 26, 1962 at 2:30 p.m. Edmondson personal files.

86 Notes taken by W.A. Hiltner at AURA–NSF Meeting on January 26, 1962. Edmondson personal files.

87 The author was at Western Illinois University in Macomb, Illinois on January 25–26 as American Astronomical Society Visiting Professor. Hiltner phoned him at his motel after the meeting at the National Science Foundation to describe what had taken place. Hiltner said that two members of the Nominating Committee (Hiltner and Carson) would like to nominate him for President if he would be willing to serve, and that they had already informed Menzel of their decision.

88 Letter, Mayall to J.W. Chamberlain, January 30, 1962. Edmondson personal files. This letter was a formal offer of employment, as approved by the ad hoc Committee.

Chapter 12

1 Op. cit. 11, Reference 33.
Note: This was one of the criticisms in a March 1960 audit of NSF–AURA by the GAO. Letter, Waterman to Joseph Campbell, April 28, 1961. Edmondson personal files.

2 Letter, Shane to Patey, September 14, 1959. Edmondson personal files.

3 Op. cit. 11, Reference 35.

4 Bell, an accountant, was Lee's assistant at the University of Michigan. He went to the University of Chicago in 1967 with Lee, when Harrell retired and Lee succeeded him as Vice-President for Business Affairs. Lee Took Harrell's place on the AURA Board, and Bell took Lee's place when Lee became the first full-time President of AURA in 1972.

5 Op. cit. 11, Reference 34.

6 Minutes, Annual Meeting of the Board of Directors of AURA, Inc., March 14, 1960, Tucson, Arizona. Edmondson personal files.
Note 1: Plan I provided for an Executive Vice-President who was not an astronomer and to whom the Observatory Director, a scientist, would report. The National Science Foundation objected to this, and so did several Board members.
Note 2: Meinel wrote his letter of resignation during the Executive Session of the Executive Committee.
Note 3: The AURA–NSF news release about Meinel's resignation was to have been delayed until after the dedication. Meinel chose to give it in advance to John Riddick of the *Tucson Daily Citizen*, who printed it on the day of the dedication.
Letter, Morgan Monroe to Embrey, March 17, 1960. Another, March 28, 1960. Edmondson personal files.

7 Letter, McMath to Mrs. Walter S. Adams, March 26, 1960. Robert R. McMath Collection, Michigan Historical Collections, Bentley Historical Library, University of Michigan.

8 Letter, McMath to L.A. Hyland (Hughes Aircraft), March 26, 1960. Robert R. McMath Collection, Michigan Historical Collections, Bentley Historical Library, University of Michigan.

9 Letter, Shane to Menzel, April 3, 1960. Edmondson personal files.
Letter, McMath to Shane, April 9, 1960. Robert R. McMath Collection, Michigan Historical Collections, Bentley Historical Library, University of Michigan.

10 Memorandum, Menzel to AURA Committee for the Selection of a Director of Kitt Peak National Observatory: Frank K. Edmondson, Arthur Code, George Herbig, William W. Morgan, Robert R. McMath (*ex officio*), C. Donald Shane (*ex officio*), March 28, 1960. Edmondson personal files.

11 Letter, Edmondson to Menzel, March 30, 1960. Edmondson personal files.

12 Op. cit. Reference 9.

13 Memorandum, Menzel to AURA Committee for the Selection of A Director of Kitt Peak National Observatory: F.K. Edmondson, A.D. Code, G.H. Herbig, W.W. Morgan, R.R. McMath, C.D. Shane, April 6, 1960. Edmondson personal files.
    Note: Edmondson's three choices reported in this memo are not included in the four names that were ranked in Reference 11.

14 Shane and McMath, Op. cit. Reference 9; Menzel, Op. cit. Reference 13.

15 Letter, Shane to McMath, April 15, 1960. Robert R. McMath Collection, Michigan Historical Collections, Bentley Historical Library, University of Michigan.

16 Longhand Letter, Mayall to McMath, April 26, 1960. Robert R. McMath Collection, Michigan Historical Collections, Bentley Historical Library, University of Michigan. This letter was written while Mayall was in Washington to attend the annual meeting of the National Academy of Sciences. He took time to visit NSF and talk to Keller and Mulders, and he met Luton.

17 Memorandum, Menzel to AURA Committee for the Selection of a Director of Kitt Peak National Observatory: F.K. Edmondson, A.D. Code, G.H. Herbig, W.W. Morgan, R.R. McMath, C.D. Shane, May 11, 1960. Edmondson personal files.

18 Minutes, Special Meeting, Board of Directors of AURA, Inc., June 6, 1960, Ann Arbor, Michigan. Edmondson personal files.

19 Letter, Shane to Waterman, August 1, 1960. Robert R. McMath Collection, Michigan Historical Collections, Bentley Historical Library, University of Michigan.

20 Observatory Reports: Kitt Peak National Observatory, C.D. Shane, Acting Director. *Astronomical Journal*, **65**, November, 1960, 525–528.

21 Transcript of Abt interview with Edmondson, March 27, 1979.

22 Minutes, Executive Committee of AURA, Inc., June 30, 1960, Tucson, Arizona. Edmondson personal files.

23 Minutes, Conference on the Chile Observatory, August 10, 1960, Tucson, Arizona. Edmondson personal files.

24 Minutes, Executive Committee of AURA, Inc., November 9, 1960, Kitt Peak, Arizona. Edmondson personal files.

25 Op. cit. 11, Reference 44.

26 Op. cit. 11, Reference 46.

27 Minutes, joint meeting of the Scientific Committee and the Space Subcommittee of AURA, Inc., March 11, 1961, Tucson, Arizona. Edmondson personal files.

28 Synopsis of Actions taken re AURA Space Program, March 1961 to March 5, 1962. Miller prepared this at Menzel's request, and sent a copy to Shane with a cover memorandum on March 6, 1962. Edmondson personal files.

29 Op. cit. 11, References 49 and 50. Miller also sent an unofficial copy to Edmondson.

30 Letter, Mayall to Reynolds, May 1961. Typed copy (without date) in Edmondson personal files. Mayall wrote: "If Miller's over-aggressiveness in the AURA is curbed, I suspect he may look around for another job. If he left, I think it would not be too easy to replace him. Both Dr. Shane and I, however, have candidates in mind if a successor is needed."

31 Letter, McMath to Harrell, September 5, 1961. Robert R. McMath Collection, Michigan Historical Collections, Bentley Historical Library, University of Michigan.

32 Op. cit. 11, Reference 54. (a) The author remembers Shane's personal attack on Miller, who was not present during the Executive Session. From this time on Shane was barely

on speaking terms with Miller and McMath. (b) Reynolds retired from the Board at this meeting, and served on the Reorganization Committee as an Administrative Consultant.

33 Letter, McMath to Harrell, August 3, 1961. Robert R. McMath Collection, Michigan Historical Collections, Bentley Historical Library, University of Michigan.

34 Meeting Schedule, AURA Board of Directors, Organization Review Committee, September 8 and 9, 1961. Edmondson personal files.

35 Letter, Menzel to McMath, September 29, 1961. Robert R. McMath Collection, Michigan Historical Collections, Bentley Historical Library, University of Michigan.

36 Minutes, Special Meeting, Board of Directors AURA, Inc., October 27, 1961, Tucson, Arizona. Edmondson personal files.

37 Reynolds was not able to attend, owing to a conflicting AUI meeting at Green Bank.

38 Letter, Menzel to Harrell (copies to Committee), November 1, 1961.
Memorandum, Peterson to Committee, November 3, 1961.
Memorandum, Edmondson to Committee, November 9, 1961. Edmondson personal files.

39 Letter, McMath to Harrell, September 5, 1961. Robert R. McMath Collection, Michigan Historical Collections, Bentley Historical Library, University of Michigan. McMath wrote: "Rumor has it that in the present situation there is a minimum of two telephone calls a day between San Jose and Tucson. This seems to me to be some sort of a remote control operation, and not necessary."

40 Memorandum, Reynolds to Committee, November 15, 1961. Edmondson personal files.

41 Letter, Menzel to Harrell, November 27, 1961. Edmondson personal files.

42 Letter, Harrell to Menzel, December 1, 1961. Edmondson personal files.

43 Letter, Keller to Dear Colleague, November 13, 1961. Edmondson personal files.

44 Op. cit. 11, Reference 45.

45 Letter, Mayall to McMath, April 5, 1961. Robert R. McMath Collections, Michigan Historical Collections, Bentley Historical Library, University of Michigan.

46 Letter, McMath to Mayall, June 27, 1961. Robert R. McMath Collections, Michigan Historical Collections, Bentley Historical Library, University of Michigan.

47 Letter, McMath to Mayall, August 3, 1961. Robert R. McMath Collections, Michigan Historical Collections, Bentley Historical Library, University of Michigan.

48 Op. cit. 11, Reference 67.

49 Telegram, Mohler to Edmondson, January 3, 1962. Edmondson personal files.

50 Letter, Menzel to Harrell, December 20, 1961. Edmondson personal files. Menzel was Chairman of the Nominating Committee; his nomination was proposed by the other members of the Committee (Hiltner and Carson).
Important background for this nomination is in: Letter, Menzel to Edmondson, April 12, 1961; Letter, Edmondson to Menzel, April 14, 1961; Letter, Menzel to Edmondson. April 29, 1961; Letter, Menzel to Shane, April 28, 1961; Letter, Shane to Menzel, May 5, 1961; Letter, Menzel to Shane, September 13, 1961; Letter, Elliott to Menzel, September 18, 1961; Memorandum, Menzel to Hiltner, Shane, McMath, Carson, Subject: Nominating Committee, September 14, 1961; Letter, Mohler to Menzel, September 19, 1961; Letter, Menzel to McMath, September 29, 1961; Letter, Miller to Menzel, November 22, 1961. Edmondson personal files.

51 Letter, Edmondson to Harrell, December 11, 1961. Edmondson personal files.

52 Op. cit. 11, References 85 and 86.

53 Op. cit. 11, Reference 87.

54 Record of Discussion at the Meeting of the Reorganization Committee, February 15, 1962. (The document reads February 25; this is a typo). Edmondson personal files.
Note: This was written by Miller, and is limited to his discussion with the Committee and Mayall to "clarify the role of the Associate Director-Administration under the proposed new By-laws."

55 Letter, Edmondson to Harrell, February 19, 1962. Edmondson personal files.

Note: This letter summarizes the results of the meeting: (a) Further consideration of changes in the by-laws. (b) The decision not to re-elect Jim Miller as AURA Secretary. (c) The decision not to propose Keith Pierce for the position of Deputy Director, and to say nothing about this position in the revised by-Laws. (d) The decision to support Mayall in his position as Director, and the decision to have the Committee meet in Tucson on February 24 to talk to Mayall.

56 Longhand notes about phone call, Edmondson to Scherer, February 19, 1962. Edmondson personal files.

57 Minutes, Board of Directors of AURA, Inc., March 12, 1962, Tucson, Arizona. Edmondson personal files.

58 Op. cit. Reference 50.

59 Letter, Mayall to Shane, February 7, 1962. Edmondson personal files.

60 Minutes, Scientific Committee of AURA, Inc., March 9 and 10, 1962, Tucson, Arizona. Edmondson personal files.

61 Letter, Mayall to Chamberlain, January 30, 1962. Edmondson personal files.

62 Letter, Chamberlain to Mayall, March 8, 1962. Edmondson personal files.

63 Op. cit. Reference 27.

64 Minutes, 150-inch Telescope Committee, September 23, 1961, Tucson, Arizona. Edmondson personal files. Staff members and others in attendance were: Abt, Baustian, Crawford, Livingston, Ludden, Pierce, Schulte, Trumbo, Waddell, Serurrier (Consultant) and Elliott (Assistant Secretary–AURA).

65 Minutes, Executive Committee of AURA, Inc., October 26, 1961, Tucson, Arizona. Edmondson personal files.

66 Op. cit. 11, Reference 77.

67 Op. cit. 11, Reference 82.

68 Op. cit. Reference 57.

69 KPNO Monthly Report, June 1961. Edmondson personal files.

70 KPNO Monthly Reports, October and December 1961. Edmondson personal files.

71 Notes on a Meeting of the Solar Telescope Review Committee, October 7, 1959, Tucson, Arizona. Edmondson personal files.

72 KPNO Monthly Report, February 1961. Edmondson personal files.

73 Operations Report, Month of November 1959. Edmondson personal files.

74 KPNO Monthly Report, May 1962. Edmondson personal files.

75 KPNO Monthly Report, August 1962. Edmondson personal files.

76 KPNO Monthly Reports, October 1962 and April 1963. Edmondson personal files.

77 KPNO Monthly Reports, November and December 1963.

78 KPNO Monthly Reports, April and May 1964. Edmondson personal files.

79 Astronomical Journal, 69, 1964, 527; SCIENCE, 145, September 18, 1964, 1285; et al.

80 KPNO Monthly Report, September 1964. Edmondson personal files. The author recalls that someone expressed the hope that the 150-inch telescope would be in operation before van Biesbroeck's 150th birthday.

81 KPNO Monthly Report, October–December 1964. Edmondson personal files.

82 Minutes, Executive Committee of AURA, Inc., June 12, 1962, Tucson, Arizona. Edmondson personal files.

83 Op. cit. 10, Reference 70.

84 Op. cit. Reference 18.

85 KPNO Monthly Report, June 1960. Edmondson personal files.

86 Op. cit. Reference 22.

87 KPNO Monthly Report, July 1960. Edmondson personal files.

88 Transcript of Hurdle interview with Edmondson, September 13, 1980.

89 KPNO Monthly Report, October 1960. Edmondson personal files.

90 KPNO Monthly Report, November 1960. Edmondson personal files.

91 KPNO Monthly Report, February 1961. Edmondson personal files.

92 KPNO Monthly Report, June 1961. Edmondson personal files.

93 KPNO Monthly Report, October 1961. Edmondson personal files.

94 KPNO Monthly Report, February 1962. Edmondson personal files.

95 Minutes, Executive Committee of AURA, Inc., June 12, 1962, Tucson, Arizona. Edmondson personal files.

96 Minutes, Executive Committee of AURA, Inc., September 22, 1962, Tucson, Arizona. Edmondson personal files.

97 KPNO Monthly Report, November 1962. Edmondson personal files. The author's correspondence files and memory have served as a supplement to the information in the Monthly Report.

98 Minutes, Scientific Committee of AURA, Inc., November 3, 1962, Tucson, Arizona. Edmondson personal files.
Note: There was a football game (Idaho vs. Arizona) on the evening of November 3 and President Harvill's office told Edmondson that the members of the Scientific Committee were invited to attend the game and sit in the President's box. Edmondson, no football fan himself, said "thanks" but he didn't think anyone would be interested. When the other Board members learned about this, there were demands for impeachment. Fortunately, the seats were still available, and the automobile pass included the privilege of parking in the stadium. Arizona lost the game. Letter, Edmondson to Harvill, November 7, 1962. Edmondson personal files.

Chapter 13

1 Memorandum summarizing results of May–June 1958 trip to the United States, F. Rutllant, March 1959. Kuiper Archives, Special Collections, University of Arizona Library. The author is indebted to Dr. Kathleen Sideli, Assistant Professor of Spanish, Indiana University, for an English translation of this document.

2 Letter, Kuiper to Shane, July 7, 1958. Kuiper Archives, Special Collections, University of Arizona Library.

3 Beginnings of the Cerro Tololo Inter-American Observatory, by C.D. Shane, October 9, 1978. Edmondson personal files. This is a xerox copy of a 19-page longhand document sent to Edmondson via Mayall.

4 Minutes, Executive Committee of AURA, Inc., March 1, 1959, Tucson Arizona. Edmondson personal files.

5 Letter, Kuiper to G.R. Miczaika, June 30, 1958. Kuiper Archives, Special Collections, University of Arizona Library.

6 Transcript of Hiltner interview with Edmondson, July 10, 1979.

7 Letter, Kuiper to Harrison, November 28, 1958. Kuiper Archives, Special Collections, University of Arizona Library.

8 Op. cit. 8, Reference 38.

9 Letter, Kuiper to Milton Greenberg, Director GRD, September 17, 1958. Kuiper Archives, Special Collections, University of Arizona Library.

10 Letter, Eaton to Kuiper, October 21, 1958. Kuiper Archives, Special Collections, University of Arizona Library.
Note: Earlier copies of (9) and (10) were obtained from the personal files of Gordon Wares at the time he gave an oral history on May 19, 1980.

11 Letter, Kuiper to Rutllant, November 6, 1958. Kuiper Archives, Special Collections, University of Arizona Library.

12 Letter, Rutllant to Kuiper, November 19, 1958. Kuiper Archives, Special Collections, University of Arizona Library.

13 Minutes, Department of Astronomy, Meeting No. 19, Friday, December 12, 1958 at 10:00 a.m. Kuiper Archives, Special Collections, University of Arizona Library.

14 *Big and Bright, A History of the McDonald Observatory*, by David S. Evans and J. Derral Mulholland, p. 133. University of Texas Press, Austin, 1986.
Warning: Owing to numerous substantial errors of fact, this book is not a reliable source of information about the chronology of events in 1959 and 1960 that led to Kuiper's departure.

15 Letter, Dean Whaley to President Logan Wilson, via Vice-President and Provost of the Main University H.H. Ransom, November 14, 1958 (Revised December 3, 1958). Kuiper Archives, Special Collections, University of Arizona Library.

16 Transcript of Miczaika interview with Edmondson, March 1, 1979.
Note: At the end of the interview Miczaika said: "But I want to recommend very strongly that you make an attempt to get hold of the original documentation, which I don't have, which should be at some Air Force Archive in Bedford, Massachusetts, which might be interesting and might shed light on other things." The author wrote to the Commanding Officer of the Air Force GRD, and received the following reply: "Although the official contract file has been long destroyed (done automatically six years after the file is retired), the contract monitor, Dr. Gordon Wares, now retired, retained his own file on the contract and still has it at his home in Newton, Massachusetts. I have talked to Dr. Wares about your request, and he would be happy to show you the contract file and to discuss the history of Cerro Tololo with you."
Letter, Ronald T. Podsiadlo to Edmondson, June 20, 1979. Edmondson personal files. The author visited Dr. Wares on May 17–19, 1980. Most of the time was spent selecting critical material from the contract file for retention by the author. Also, the detailed notes that Wares had written about phone calls and other conversations were given to the author. The taped interview on May 19 was an hour and sixteen minutes long.

17 Chile Observatory Project, General Report No. 1, G.P. Kuiper, March 1, 1960. Edmondson personal files.

18 Letter, Kuiper to Miczaika, November 28, 1958. From Gordon W. Wares personal files (now in Edmondson's personal files. Reference 16, Note).

19 Letter, Kuiper to Whaley, February 17, 1959. From Gordon W. Wares personal files.

20 Letter, Kuiper to Miczaika, February 18, 1959. From Gordon W. Wares personal files.

21 Letter, Miczaika to Kuiper, March 30, 1959. From Gordon W. Wares personal files.

22 Letter, Kuiper to Miczaika, September 14, 1959. From Gordon W. Wares personal files.

23 Memorandum, John M. McKeon, Contract Specialist, to Nannielou H. Dieter, Contract Monitor; Memorandum, Dieter to McKeon, December 3, 1959. From Gordon W. Wares personal files.
Dr. Dieter replaced Miczaika, who had taken a high level position with TRW in California.

24 Op. cit. Reference 17.

25 Op. cit. Reference 16.

26 *New York Times*, March 25, 1959; *Chicago Daily Tribune*, March 25, 1959; *El Mercurio*, Santiago, Chile, March 26, 1959. Kuiper Archives, Special Collections, University of Arizona Library.
The story in the *Chicago Tribune* emphasized the plans for the Joint Astronomy Department more than the plans for the Chile telescope.

27 Letter, Miczaika to Kuiper, March 30, 1959. Kuiper Archives, Special Collections, University of Arizona Library.

28 Letter, Kuiper to Whaley, November 27, 1958. From Gordon W. Wares personal files.
Kuiper had offered this position to Tom Gehrels. He was not interested and recommended Jurgen Stock, who had previous experience site testing in South Africa. See p. 107, *On the Glassy Sea*, by Tom Gehrels, American Institute of Physics, 1988.

29 Op. cit. Reference 13.

30 Letter, Stock to Kuiper, February 26, 1959. Kuiper Archives, Special Collections, University of Arizona Library.

31 Letter, Kuiper to Stock, March 3, 1959. Kuiper Archives, Special Collections, University of Arizona Library.

32 Letter, Kuiper to Rutllant, April 20, 1959. Kuiper Archives, Special Collections, University of Arizona Library.

33 Letters, Stock to Kuiper, April 25, 28, May 10, 12. Kuiper Archives, Special Collections, University of Arizona Library.
   These letters were the precursors of the "Stock Reports" that were first sent to Kuiper, and later to Shane and Edmondson, from February 1960 to February 1963.

34 Letter, Kuiper to Miczaika, May 28, 1959. Kuiper Archives, special Collections, University of Arizona Library.

35 Second Astrometric Conference, May 17–21, 1959. Cincinnati, Ohio. *Astronomical Journal*, 65, Number 4, May 1960.

   (a) Rutllant gave a paper, "The Observatory in Santiago", describing the history of the Observatory, the move to a new site in progress, the Chicago–Texas–Chile plans for a large telescope, and the beginning of the site survey by Stock.
   (b) A.A. Nemiro and M.S. Zverev gave a paper "A Plan of U.S.S.R. Participation in Astrometric Observations in the Southern Hemisphere." Their work in Chile will be described in Chapter 16.

36 Letters, Kuiper to Rutllant, June 26 and August 5, 1959. Kuiper Archives, Special Collections, University of Arizona Library.

37 Op. cit. Reference 17.

38 Diary Note, Seeger, September 11, 1958. NSF History Files.

39 Minutes, Meeting of Ad Hoc Committee on Astronomy in South America, October 19, 1959, Washington, DC. International Relations Records Series, The Collections of the Archives of the National Academy of Sciences.

40 Letter, Keller to list October 7, 1959. Edmondson personal files, from C.D. Shane.

41 Op. cit. Reference 3.

42 Report on the November 9, 1959 Meeting in Chicago on the 60-inch Chile Telescope Project, G.F.W. Mulders. Edmondson personal files, from C.D. Shane.

43 Shane had received "a note from Mulders dated November 3, 1959 in which he said that Morgan had told him the Yerkes staff was in open revolt because Kuiper had not kept them informed or asked their advice about the Chile project. Morgan thought Kuiper would not retain the directorship much longer." This was a Note appended to Reference 3.

44 The author was at the University of Chicago on January 4, 1960 to attend an AURA Committee meeting. During dinner Harrell told us that Kuiper had been relieved of all his administrative duties that very day, but he didn't say why. This information was found later in the longhand notes that Wares gave to the author (see Reference 16): "8 August '60. Kuiper's leaving Yerkes. Lunch at Indian Village, Lake Delevan, Hiltner, Morgan, Wares. 'Last straw: Kuiper went to Dean Whaley, U of Texas, and proposed their taking over McDonald from Yerkes and hiring Kuiper.' " This effort to find a way to retain administrative control of the Chile project backfired when word reached the University of Chicago administration.

45 The AURA group walked to the Quadrangle Club for dinner. Kuiper was waiting in the hallway and asked Meinel to have dinner with him. Meinel told us later that Kuiper had asked him to go the University of Arizona and find out if they would like to hire him. Meinel went to Dean David Patrick the next day, and the wheels began to turn. A letter from Dean Whaley to Kuiper, dated February 1, 1960, says that he heard of Kuiper's decision to move to the University of Arizona "last week." Kuiper Archives, Special Collections, University of Arizona Library.

46 Morgan gave the Dedication Address at the Dedication of the Kitt Peak National Observatory on March 15, 1960. The Planning Committee chose him on December 10, 1959. See 10, Reference 120.

47 Report on the January 5, 1960 Meeting in Chicago of the Policy Advisory Board – Chilean Observatory Project, Nannielou H. Dieter. Edmondson personal files.

48 Memorandum, Shane to Members of the AURA Scientific Committee and Bowen, Mayall, Meinel, Rule, January 21, 1960. Edmondson personal files. Sketches and specifications of the Rademakers design were enclosed. Shane Wrote: "As a member of the Advisory Committee on this telescope I agreed to submit the sketches and specifications to the Scientific Committee, Dr. Meinel, and certain AURA consultants for comment."

49 Memorandum, Warren C. Johnson to members of the PAB-COP, March 8, 1960. Edmondson personal files.

50 Op. cit. Reference 17.

51 Minutes, Scientific Committee of AURA, Inc., March 12, 1960, Tucson, Arizona, Edmondson personal files.

52 Letter, Shane to Warren C. Johnson, March 21, 1960. Edmondson personal files.

53 Report on the March 30, 1960 Meeting in Chicago of the Policy Advisory Board – Chilean Observatory Project, Nannielou H. Dieter. Kuiper Archives, Special Collection, University of Arizona Library.

54 Letter, Kuiper to Stock (in Chile), April 1, 1960. Kuiper Archives, Special Collections, University of Arizona Library.

55 Letter, Stock to Kuiper, April 7, 1960. Edmondson personal files, from Shane.

56 Letter, Kuiper to Shane, April 18, 1960. Edmondson persoanl files, from Shane.

57 Letter, Kuiper to Stock, May 1, 1960. Kuiper Archives, Special Collections, University of Arizona Library.

58 Letter, Stock to Kuiper May 9, 1960. Edmondson personal files, from Shane.

59 Minutes, Board of Directors of AURA, Inc., June 6, 1960, Ann Arbor, Michigan. Edmondson personal files.

60 Letter, Vice-President Warren C. Johnson to Board of Directors, AURA, Inc., June 17, 1960; Letter, Johnson to Shane, June 22, 1960. Edmondson personal files, from Shane. Dr. Dieter said it was not possible to transfer funds from one contract to another. She suggested that the University should subcontract to AURA for the duration of the present contract.

61 Letter, Johnson to Shane, June 27, 1960. Edmondson personal files, from Shane.

62 Minutes, Executive Committee of AURA, Inc., June 30, 1960, Tucson, Arizona. Edmondson personal files.

63 Minutes, Conference on the Chile Observatory, August 10, 1960, Tucson, Arizona. Edmondson personal files.

64 Op. cit. Reference 33.

65 Op. cit. 8, Reference 43. Congress turned down NSF funding for the Southern Hemisphere Astrograph two times, and NSF decided not try a third time. Letter, Keller to P.B. Pearson (Ford Foundation), October 5, 1959. NSF History Files. The Ford Foundation provided the necessary funds, $750,000, in 1960, and the Astrograph was installed at El Leoncito in Western Argentina, north of San Juan. Letter, J.M. McDaniel, Jr. (Ford Foundation) to A.W. Griswold (Yale University), June 29, 1960. Ford Foundation Archives.

66 Op. cit. Reference 35, item (b).

67 Letter and Proposal, Mayall to Mulders, October 17, 1960. Edmondson personal files.

68 Letter, Miller to Stock, November 1, 1960. Edmondson personal files, from Mayall.

69 Supplement to A Proposal to the National Science Foundation for Assistance in Developing a Southern Hemisphere Observatory in Chile, November 30, 1960. Edmondson personal files.

70 Letter, Luton to Miller, November 10, 1960. Edmondson personal files.
"It would be difficult for the Foundation to make basic research funds available in the fiscal year 1961 to initiate a facilities project without some prior clearance from the Appropriations Committee. ... As you know, we have requested on two different occasions funds for a Southern Hemisphere astrograph. In both instances, these funds have been denied. There

is sufficient similarity between the astrograph project and the Chilean project to cause the Committee to feel that we have not been sensitive to their wishes if they find that we have initiated such a project without prior clearance with them as a part of the budget process. In view of the above, there is no assurance that you will get any funds in the fiscal year 1961 from the National Science Foundation for the Chilean project."

71 Minutes, Executive Committee of AURA, Inc., November 9, 1960, Tucson, Arizona. Edmondson personal files.

72 Letter, Miller to R.W. Harrison (Chicago), December 13, 1960; Letter, Harrison to Miller, December 19, 1960. Edmondson personal files.

73 1960 Chile Trip, C.D. Shane and N.U. Mayall. Edmondson personal files.

74 *Eye on the Sky, Lick Observatory's First Century*, Donald E. Osterbrock, John R. Gustafson and W.J. Shiloh Unruh, University of California Press, 1988.

75 Op. cit. 11, Reference 44, Attachment II.

76 Monthly Report, April 1961. Edmondson personal files.

77 Stock Report No. 13. Edmondson personal files.

78 During the site survey all shipments were identified as gifts to the University of Chile, which avoided payment of duty. Sr. Enei, who was in charge of importations for the University of Chile, received and certified the site survey shipments.

79 Minutes, Executive Committee of AURA, Inc., June 14, 1961, Washington, DC. Edmondson personal files.

80 Monthly Report, April 1961 and May 1961. Edmondson personal files.

81 Minutes, Executive Committee of AURA, Inc., October 26, 1961, Tucson, Arizona. Edmondson personal files.

82 1961 Visit to Chile, C.D. Shane; Monthly Report, December 1961. Edmondson personal files.

83 Paul Kuiper, the teenage son of Gerard Kuiper, was employed as an observer on the site survey starting with his arrival in Chile on July 24, 1961 (Stock Report No. 17). A letter from Shane to Stock, July 16, 1961, says: "In talking to Nick [Mayall], I approved your hiring Paul but subject to the conditions that his transportation both ways be paid for by himself or parents and that he be treated exactly like any other employee." Shane expressed some reservations about doing this. Miller wrote to Stock on July 18, 1961: "Dr. Shane has instructed me to employ young Paul Kuiper and he should be leaving for Chile within the next 4 or 5 days." Paul proved to be a good hard-working observer. Unfortunately, he joined forces with two other employees, Bill and Sherry Mathias, in bad-mouthing Stock to his father in a series of letters. Gerard Kuiper sent duplicates to Shane and later to the author. The departure of the Mathias couple, and later the departure of Paul Kuiper, brought an end to the conflict.

84 Stock Report No. 20. Edmondson personal files.

Chapter 14

1 Minutes, Scientific Committee of AURA, Inc., March 10, 1962, Tucson, Arizona. Edmondson personal files.

2 Minutes, Board of Directors of AURA, Inc., March 12, 1962, Tucson, Arizona. Edmondson personal files.

3 Op. cit. 13, Reference 82. Note: Shane's full report was sent only to Mayall, who showed it to Edmondson in June. Sensitive material was omitted from the copy that was duplicated and distributed to the AURA Board after the June meeting of the Executive Committee. Letter, Mayall to Edmondson, July 19, 1962. Edmondson personal files.

4 Memorandum, Astronomy in Venezuela, C.D. Shane, May 4, 1962. Edmondson personal files.
   Also see *Sky and Telescope*, November 1961, p. 251.

5 Report of Visit to Chile – April 14–21, 1962, F.K. Edmondson, Attachment I, Minutes,

Executive Committee of AURA, Inc., June 12, 1962, Tucson, Arizona. Edmondson personal files.

Stock Report No. 21. Edmondson personal files.

6 Letter, Shane to Edmondson, June 26, 1962; Letter, Mayall to Edmondson, June 21, 1962. Edmondson personal files.

7 Letter, Edmondson to Shane, July 5, 1962. Edmondson personal files.

8 Memorandum, Edmondson to Harrell, Mayall, and Miller, July 5, 1962. Edmondson personal files.

9 Letter, Harrell to Edmondson, July 11, 1962. Edmondson personal files.

10 Letter, Edmondson to Miller, July 7, 1962. Edmondson personal files. Edmondson and Miller had discussed the need for this trip in a telephone call on July 5. The confirming letter was addressed to Miller as Secretary of AURA, to avoid the possibility that Mayall might veto sending the Associate Director for Administration to Chile. Miller began this trip during Edmondson's trip to France and returned on July 30.

11 Letter, Miller to Stock, July 7, 1962; Letter, Miller to Rutllant, July 7, 1962. Edmondson personal files.

12 Letters, Harrell to Edmondson, July 3 and August 3, 1962. Edmondson personal files.

13 Letter, Mulders to Edmondson, July 28, 1962. Edmondson personal files.

Note: Waterman had already complied with the legal requirements by asking for formal State Department approval of the Chile project.

Letter, Waterman to Rusk, November 6, 1961; Letter, Whitman to Waterman, January 27, 1962. Edmondson personal files.

14 Letter, Edmondson to Mulders, July 30, 1962. Edmondson personal files.

15 Minutes, AURA–NSF Meeting, Chile Project, 9:35 a.m.–12:30 p.m. August 13, 1962. Edmondson personal files.

   AURA: Edmondson, Harrell, A.G. Smith, Hiltner, Woodrow, Whaley, Clemence.
   KPNO: Mayall, Pierce, Miller, Stock.
   NSF: Robertson, Scherer, Mulders, Hoff, Friend, Greenwood.
   NAS: Lewis Slack, Assistant Executive Secretary, Division of Physical Sciences.

16 Report on Chile Trip, July 16–30, 1962. J.M. Miller, August 13, 1962. Edmondson personal files.

17 Minutes, AURA–NSF Meeting, NASA Relationships, 3:30–4:40 p.m. August 13, 1962. Edmondson personal files.

18 Minutes, Executive Committee of AURA, Inc., with several members of the Chile Committee in attendance, September 22, 1962, Tucson, Arizona. Edmondson personal files.

19 This is the author's recollection of what Menzel told him. Menzel did not hold a grudge about the AURA President nomination, and was very supportive of the author during his term as President.

20 Letter, Wiggins to Edmondson, October 5, 1962. Edmondson personal files.

21 Letter, Edmondson to Wiggins, October 16, 1962. Edmondson personal files.

22 Letter, Stock to Miller, October 23, 1962. Edmondson personal files. Lorenz was added on Miller's copy of this letter, following a phone call from Edmondson on October 29.

23 Op. cit. 12, Reference 92.

24 Memorandum to files, Chile Legislation, J.M. Miller, October 9, 1962. Edmondson personal files.

25 This is the way the author recalls it.

26 Minutes, Meeting of the Chile Site Survey Team, November 23, 1962. Santiago, Chile. Edmondson personal files.

27 *El Mercurio*, November 24, 1962, Santiago, Chile. Edmondson personal files. President Alessandri made good on his promise, and the law was passed by the Chamber of Deputies on January 9, 1963 [Letter, Stock to Miller, January 15, 1963] and by the Senate on January 31. The law as proposed by AURA was modified by the Congress to allow duty-free imports

related to cultural, scientific or teaching agreements between the University of Chile and other foreign organizations. The President signed the law on February 27, and it was promulgated in the DIARIO OFFICIAL on March 7. It should be noted that the law was passed by the Congress before the National Science Foundation signed the AURA contract (NSF-C300) on February 11 to provide $1,000,000 to start construction of the observatory. Edmondson personal files.

28 This was the 28th wedding anniversary of the author and his wife. Lee had told the Panagra agent about this when he left on an earlier flight, and a wedding cake decorated with bride and groom dolls was on the dessert cart on the evening flight.

29 Stock Report No. 27; Monthly Report, December 1962; Report on AURA Group Visit to Chile, by N.U. Mayall (based on records of NUM and JMM). Edmondson personal files.

30 Stock wrote: "At noon sharp 450 square kilometers of land became the property of AURA. The event was duly photographed and recorded by newspapers etc. We celebrated it with six bottles of warm beer, which we split between a dozen persons. Then we went to have lunch at the Club Social. We were practically the only guests there, and we had to wait for two hours before we were served. The first one to be served got a soup bowl full of potatoes. Then the rations became smaller and smaller, and the last one – unfortunately me – got only one small potato. Apart from that it was a fine lunch." Stock Report No. 30. Edmondson personal files.

31 Minutes, Executive Committee of AURA, Inc., December 1, 1962, Tucson, Arizona. Edmondson personal files.

32 Letter, Stock to Mayall, December 21, 1962. Edmondson personal files.

33 Stock Report No. 29. Edmondson personal files.

34 Letter, Stock to Mayall, December 25, 1962. Edmondson personal files.

35 Minutes, Executive Committee of AURA, Inc., February 9, 1963, Tucson, Arizona. Edmondson personal files.

36 Letter, Moore (AURA Treasurer) to Korp, October 7, 1963. Edmondson personal files. Note: Rolf Korp was hired by Stock effective May 13, 1963 to fill the position of Administrator that had been recommended by the site survey team (Reference 26). Letters, Stock to Mayall, May 9, 1963, and June 14, 1963; Letter, Korp to Miller, May 23, 1963. Edmondson personal files.

37 "The Construction of the Road to Cerro Tololo", by Jurgen Stock, December 23, 1963. Edmondson personal files.

38 Letters, Mayall to Kroecker, October 1, 1963; Mayall to Kroecker Contracting Company, October 14, 1963. Edmondson personal files.

39 Memorandum: Re: Conference 11-7-63-Kroecker dispute. Present: AURA (N.U. Mayall, Richard M. Bilby); Kroecker Construction Co. (Richard Kroecker, Elmer Courtland). Edmondson personal files.

40 Op. cit. Note, Reference 36.

41 Letter, Stock to Mayall, October 21, 1963. Edmondson personal files.

42 KPNO Monthly Report, September 1963. Edmondson personal files.

43 Letter, Edmondson to Stock, July 9, 1963. Edmondson personal files.

44 Letter, Edmondson to Gonzalez, August 30, 1963; Letter, Gonzalez to Edmondson, September 11, 1963. Edmondson personal files.

45 Letter, Gonzalez to Edmondson, October 21, 1963. Edmondson personal files. Letters from Stock to Edmondson (September 14 and October 17, 1963) provided more information about Rutllant's hospitalization ("two operations") and problems with the University.

46 Minutes, Executive Committee of AURA, Inc., November 22, 1963, Tucson, Arizona. Edmondson personal files.

47 KPNO and CTIO Monthly Report, CTIO – February 1964. This covered the period October 1963 through January 1964. Edmondson personal files. The road crew under

Korp's supervision had improved the road after the first access by a motor vehicle on September 10, and the road to the summit was open to all vehicular traffic by the beginning of December.

Press reports about the ceremony were in the La Serena newspaper, *El Dia*, on December 14, 15, and 16, and in the important Santiago newspaper, *El Mercurio*, on December 15. Edmondson personal files.

48 Report, Miller to Mayall, December 27, 1963. Edmondson personal files.

49 Op. cit. Reference 47.

50 Letter, Mayall to Stock, March 18, 1963. Edmondson personal files.

51 KPNO and CTIO Monthly Report, CTIO–June 1964. Edmondson personal files.

52 *Science*, March 29, 1963. Edmondson personal files.

53 Minutes, Executive Committee of AURA, Inc., September 20, 1963, Washington, DC. Edmondson personal files.

54 KPNO and CTIO Monthly Report, KPNO – October 1964. Edmondson personal files. The NSF Advisory Panel for Astronomy met on November 13–15, and the NSB met on November 19–21.

55 Minutes, Executive Committee of AURA, Inc., November 19, 1964, Tucson, Arizona. Edmondson personal files.

56 Letter, Father Theodore M. Hesburgh to Edmondson, May 12, 1978. Transcript, Hesburgh interview with Edmondson, November 1, 1979. Edmondson personal files.

57 Minutes, National Science Board, Open Session, November 19–21, 1964. NSB Archives. Copy sent to author by Vernice Anderson, Executive Secretary of the NSB.
Note: The AURA contract for $1,000,000, NSF-C300, was signed February 11, 1963 (Reference 27). Amendment No. 1 added $1,000,000 to the contract on April 2, 1964. Amendment No. 4 added $1,300,000 on March 31, 1965. Edmondson personal files.

58 Letter, Edmondson to Haworth, October 22, 1964. Edmondson personal files.

59 KPNO and CTIO Monthly Report, KPNO – December 1964. Edmondson personal files.

60 Report of Trip to Yale–Columbia Southern Observatory by AURA–NSF Group (November 30–December 3, 1964), by J. Stock. [with KPNO and CTIO Monthly Report, October–December 1964.] Edmondson personal files.

61 KPNO and CTIO Monthly Report, CTIO – November and December 1964. Edmondson personal files.

62 *El Mercurio*, December 12, 1964. Edmondson personal files.

63 Minutes, Board of Directors of AURA, Inc., March 23, 1965, Tucson, Arizona. Edmondson personal files.

64 Memorandum, Miller to Edmondson, Hiltner, Lee, McCuskey, Mohler, Slettebak, August 20, 1965. Edmondson personal files.

65 Minutes, Executive Committee of AURA, Inc., November 11, 1965, Tucson, Arizona. Edmondson personal files.

66 Minutes, Special Meeting of the Chile Subcommittee of AURA, Inc., December 21, 1965, Chicago, Illinois. Edmondson personal files.

67 Minutes, Chile Subcommittee of AURA, Inc., January 19k 1966, Tucson, Arizona. Edmondson personal files. Stock shortly thereafter accepted a research position on the staff of the University of Chile Observatory.

68 Minutes, Executive Committee of AURA, Inc., January 27, 1966, Tucson, Arizona. Edmondson personal files. Wildt wrote to Elvis J. Stahr, President of Indiana University, the day after this meeting to express the appreciation of AURA for the cooperation of the University in coping with this emergency. Edmondson personal files.

69 "Cerro Tololo Program – 1966", by F.K. Edmondson, Acting Director, March 1966. Edmondson personal files.

70 Letter, J.T. Wilson (NSF Deputy Director and Contracting Officer) to Miller, May 6, 1966. Edmondson personal files.

71 Information is from the author's pocket diary. The tension was high during the two days on Tololo, but never became explosive.

72 Monthly Report, January and February 1966. Edmondson personal files.

Note: Dungan became a member of the staff of Senator John F. Kennedy in 1956, and was one of the key organizers in the 1960 campaign. He became a special assistant to President Kennedy in 1962, and handled Latin-American affairs. He continued in this capacity in the Johnson administration until he was appointed Ambassador to Chile in the fall of 1964 (*New York Times*, and *Washington Post*, October 3, 1964). A letter from Mulders to Miller, October 27, 1964 describes Dungan's background, and says he will take up his new post on or about December 7. The letter concludes: "Dr. Edmondson, President of AURA, accompanied by Dr. Harold H. Lane of NSF, had an interview with Ambassador Dungan at the White House on Friday, October 23, which I had been very fortunate to arrange with the aid of Dr. Ralph Richardson of the State Department (Chile Desk)." Dungan was on the job in Chile at the time of the December 1964 AURA visit (References 58–62).

73 Op. cit. Reference 65.

Transcript, Hurdle interview with Edmondson, September 12, 1980. Edmondson personal files.

Other KPNO staff members who helped in Chile for short periods include: D.J. Ludden, senior engineer; Don Cassidy, field engineer and surveyor; D.V. Graham, drafting clerk; Roy Williams, crane operator; Robert Harrison, instrument maker; Charles Hobbs, senior technical associate; and Joe Wilson, supervisor of the KPNO instrument shop. Bi-Monthly Reports, 1967. Edmondson personal files.

74 Minutes, Executive Committee of AURA, Inc., June 16, 1967, Chicago, Illinois. Edmondson personal files. Additional information is in Bi-Monthly Reports, May–June and July–August 1967.

75 Bi-Monthly Reports, October–November 1967, Special Dedication Supplement, November 1967, V.M. Blanco. Edmondson personal files. The guests included John T. Wilson, Deputy Director of the National Science Foundation, and three members of Congress, George P. Miller, John Pettis and Olin Teague.

Chapter 15

1 Op. cit. 12, Reference 22.

2 Op. cit. 12, Reference 59.

3 Op. cit. 12, Reference 60.

4 Op. cit. 11, Reference 77 and Reference 82.

5 Minutes, Meeting with representatives of the Corning Glass Works, February 6, 1963. Edmondson personal files.

Corning Glass Works: Mr. E.L. Phillips, Optical Sales Department; Mr. Conrad Sternski, Manager, Pyrex Manufacturing Plant; Mr. George Mann, Manager, Fused Silica Manufacturing Plant.
AURA Inc.: Dr. F.K. Edmondson, President.
Mt. Wilson and Palomar Observatories: Dr. I.S. Bowen, Director.
KPNO: Dr. Mayall, Dr. Pierce, Mr. Miller, Mr. Baustian, Mr. Ludden, Mr. Novak, Miss Elliott.

6 Minutes, Executive Committee of AURA, Inc., February 9, 1963, Tucson, Arizona, Edmondson personal files.

7 Minutes, Scientific Committee of AURA, Inc., April 15, 1963, Tucson, Arizona. Edmondson personal files.

8 Minutes, Board of Directors of AURA, Inc., April 16, 1963, Tucson, Arizona. Edmondson personal files.

9 Minutes, Executive Committee of AURA, Inc., Executive Session, June 28, 1963, Tucson, Arizona. Edmondson personal files.

10 Letter, Mayall to Hoag, September 19, 1964; Letter, Hoag to Mayall, October 13, 1964; Memorandum, Mayall to AURA Scientific Committee, October 19, 1964. Edmondson personal files.

11 KPNO Monthly Report, September 1963. Edmondson personal files.

12 Minutes, Executive Committee of AURA, Inc., September 20, 1963. Tucson, Arizona. Edmondson personal files.

13 Minutes, Executive Committee of AURA, Inc., November 22, 1963, Tucson, Arizona. Edmondson personal files.

14 Letter, Mayall to Phillips, September 4, 1963. Edmondson personal files. The AURA visitors were Edmondson, Mayall, Miller, Crawford, and Baustian. The visit to Corning followed a KPNO–NRAO meeting at Greenbank on September 23–24, and a visit to the site of the ill-fated 600-foot Sugar Grove radio telescope on September 25.

15 Report on Meetings regarding 150-inch fused silica blank, by J.M. Miller. Minutes Executive Committee of AURA, Inc., November 22, 1963, Tucson, Arizona. Edmondson personal files.

16 Minutes, Executive Committee of AURA, Inc., January 31, 1964, Tucson, Arizona. Edmondson personal files.

17 Minutes, Board of Directors of AURA, Inc., March 10, 1964, Tucson, Arizona. Edmondson personal files.

18 Minutes, Third Meeting of 150-inch Telescope Advisory Committee, October 12, 1964, Tucson, Arizona. Edmondson personal files.

19 Minutes, Executive Committee of AURA, Inc., November 19, 1964, Tucson, Arizona. Edmondson personal files.

20 A framed copy of the first and last pages was presented to the author at the end of his term as President of AURA, and it hangs in a prominent place in his office.

21 Op. cit. 12, Reference 22.

22 Minutes, Board of Directors of AURA, Inc., March 10, 1964, Tucson, Arizona. Edmondson personal files.

23 Minutes, Board of Directors of AURA, Inc., March 23, 1965, Tucson, Arizona, Edmondson personal files.

24 Minutes, Scientific Committee of AURA, Inc., November 10, 1965, Tucson, Arizona. Edmondson personal files.

25 Diary Note: Status of UC-Australia 150-inch telescope, Mulders, August 12, 1966. NSF History Files.

26 Summary statement of actions taken by Trustees and Staff of the Carnegie Institution toward planning for Carnegie Southern Hemisphere Observatory (CARSO), undated. Carnegie Institution of Washington Archives.

27 "A Proposal for the Construction of a 200-inch Telescope in the Southern Hemisphere," Revised June 25, 1964; Letter, Edward A. Ackerman to Allan R. Sandage, July 6, 1964, Carnegie Institution of Washington Archives.

28 Docket Excerpt, Board of Trustees Meeting – March 14–15, 1966. Science and Engineering, Carnegie Institution of Washington, Carnegie Southern Observatory. Ford Foundation Archives.

ACTION: Tabled for further study and exploration of alternate means of support of astronomy.
COMMENT: The amount was more than the Foundation should devote to astronomy, and the proposed provision of access for astronomers other than those on Mount Wilson–Palomar staff was considered inadequate.

29 Letter, Joseph M. McDaniel, Jr. to Caryl P. Haskins, April 25, 1967. Carnegie Institution of Washington Archives.

30 Letter, Carl W. Borgmann to Wildt, January 23, 1967. Edmondson personal files.

31 Minutes, Executive Committee of AURA, Inc., January 26, 1967. Edmondson personal files.

32 Letter, Wildt to Borgmann, February 11, 1967. Edmondson personal files.

33 Letter, McDaniel to Wildt, April 19, 1967. Edmondson personal files.

34 Minutes, Scientific Committee of AURA, Inc., June 12, 1967, Fontana, Wisconsin. Edmondson personal files.

35 Minutes, Executive Committee of AURA, Inc., June 16, 1967, Chicago, Illinois. Edmondson personal files.

36 Bi-Monthly Report, July and August 1967. Edmondson personal files.

37 Minutes, Executive Committee of AURA, Inc., September 21, 1967, Tucson, Arizona. Edmondson personal files.

38 Bi-Monthly Report, September and October 1967. Edmondson personal files.

39 Minutes, Executive Committee of AURA, Inc., November 10, 1967, Santiago, Chile. Edmondson personal files.

40 Bi-Monthly Report, July and August 1969. Edmondson personal files.

41 Bi-Monthly Report, November and December 1969. Edmondson personal files.

42 Op. cit. Reference 38.

43 Bi-Monthly Reports, May and June, November and December 1967. Edmondson personal files.

44 Bi-Monthly Reports, January and February, March and April 1968. Edmondson personal files.

45 Transcript, Hurdle interview with Edmondson, September 12, 1980. Edmondson personal files.

46 Quarterly Report, July–September, 1970. Edmondson personal files.

47 Quarterly Report, October–December 1970. Edmondson personal files.

48 This information comes from the Monthly, Bi-Monthly, and Quarterly Reports. Edmondson personal files.

49 Bi-Monthly Report, September–October, 1968. Edmondson personal files.

50 Minutes, Executive Committee of AURA, Inc., November 16, 1971, Tucson, Arizona. Edmondson personal files.

51 Status Report on the 150-inch Telescope, David L. Crawford, in KPNO–CTIO Quarterly Bulletin, January–March 1972. Edmondson personal files.

52 Minutes, Executive Committee of AURA, Inc., September 17, 1974, Tucson, Arizona. Edmondson personal files.

53 Quarterly Bulletin, January–March 1973. Edmondson personal files.

54 Minutes, Executive Committee of AURA, Inc., December 3, 1974, Tucson, Arizona. Edmondson personal files.

Additional details of the two projects are given in the following publications: (1) KPNO Contributions Nos. 92 (The Kitt Peak 150-inch Telescope), 98 (Problems of Constructing Large Telescopes), and 294 (Optical Astronomy's Two New 150-inch Telescopes); (2) IAU Symposium No. 27 (The Construction of Large Telescopes); Proceedings of the Symposium on Support and Testing of Large Astronomical Mirrors, held by KPNO and the University of Arizona (Tucson, 1968); (3) *Sky and Telescope*, 38, No. 3, September 1969 ("Giant Mirror Blanks Poured for Chile and Australia"); and *Sky and Telescope*, 38, No. 5, November 1969 ("Photo Album of Kitt Peak's 150-inch Telescope Building").

Chapter 16

1 See Chapter 19.

2 Letter, Ackerman to Carl W. Borgmann, Director, Science and Engineering Program, Ford Foundation, February 28, 1964, CIW archival files.

3 Letters, Ackerman to Allan R. Sandage, July 6, 1964; Ackerman to Marshall Hornblower, July 6, 1964. CIW archival files.

4 Op. cit. 18, Reference 22.

5 Discussion Paper, Ground-based Astronomical Facilities, Carl W. Borgmann, December 10–11, 1964. Ford Foundation Archives.
Note: The $1 million grant to ESO had been announced in the *New York Times* on October 21.

6 Memorandum, Borgmann to S & E Files, December 23, 1964. Ford Foundation Archives.

7 Letter, Borgmann to Haskins, December 23, 1964. Ford Foundation Archives.

8 Memorandum, Discussion on January 11, 1965 with Caryl Haskins and Edward Ackerman on Carnegie Institution–Cal Tech plans for a 200-inch optical telescope in the Southern Hemisphere, Borgmann to S & E Files, January 11, 1965. Ford Foundation Archives.
Note: The inclusion of Cal Tech in the above title shows that it had not been made clear to Borgmann that Cal Tech would not have a role in CARSO. This led to some embarrassment when Borgmann visited Pasadena and "messed things up" during his conversation with Cal Tech's Provost Robert Bacher, who had not been told about the CARSO plans. Letter, Borgmann to Babcock, February 8, 1965. Ford Foundation Archives. A revision of several pages of the CARSO proposal was received by Borgmann on March 23, 1965. His longhand marginal note reads: "By hand from Ackerman. Has not been discussed with Bacher or other Cal-Tech authorities. Ackerman expects to do so next Monday."

9 Memorandum, Investigation of background of possible grant toward construction and operation of 200-inch optical telescope in the Southern Hemisphere by the Carnegie Institution, Borgmann to S & E Files, April 30, 1965. Ford Foundation Archives.

10 Memorandum, Possibility of Ford Foundation interest in Proposal by the Carnegie Institution of Washington for a 200-inch Optical Telescope in the Southern Hemisphere, Borgmann to Heald, April 30, 1965. Ford Foundation Archives.
Note: Borgmann also wrote: "Closer contact with the Mount-Wilson–Palomar staff has convinced me even further that this group is the world's best in optical astronomy." This extreme personal opinion governed his later actions.

11 Letter, Babcock to Borgmann, August 6, 1965. Ford Foundation Archives.
Memorandum, Borgmann to Clarence Faust, August 10, 1965. Ford Foundation Archives.

12 Carl W. Borgmann, Itinerary for Chile Trip, October 9–22, 1965. Ford Foundation Archives.

13 *El Mercurio*, October 16, 1965. "New Observatory in the North of Chile." Translation in Edmondson personal files.

14 Memorandum, Notes on CARSO–Chile trip October 11–15, 1965, Borgmann to S & E Files, October 27, 1965. Ford Foundation Archives.

15 Letter, Babcock to Borgmann, November 5, 1965. Ford Foundation Archives.

16 Memorandum, Status of Discussions with Carnegie Institution of Washington on the use of major telescopes by astronomers other than those on the staff of the Mount Wilson–Palomar and proposed Carnegie Southern Observatories. Borgmann to Faust, December 8, 1965. Ford Foundation Archives.

17 Memorandum for discussion with Dr. Edward A. Ackerman, November 29, 1965, C.W. Borgmann. Ford Foundation Archives.

18 Letter, Ackerman to Borgmann, December 3, 1965. Ford Foundation Archives.

19 DRAFT, Resolution of CIW Board of Trustees, March 9, 1966. CIW archival files.

20 Letter, Bundy to Edmondson, October 3, 1984. Edmondson personal files. "You are quite right in remarking that applicants for grants are very good at reading 'I will put it forward' as 'you will get it'."

21 Memorandum, Subject: Construction of 150" or larger optical telescope in Chile, Keller to Haworth, November 29, 1965. NSF History Files.
Note: This memorandum shows that Keller did not understand that CARSO was strictly a project of the Carnegie Institution of Washington.

22 Letter, Keller to Stratton, draft dated November 26, 1965. NSF History Files.

23 Op. cit. 14, References 67 and 68.

24 Memorandum, CARSO Negotiations, Borgmann to S & E Files, December 14, 1965. Ford Foundation Archives.

25 Memorandum, Carnegie Southern Observatory (CARSO), Borgmann to S & E Files, December 14, 1965. Ford Foundation Archives.

26 Memorandum, CARSO–AURA relationships, Borgmann to S & E Files, December 20, 1965. Ford Foundation Archives.

27 Letter, Babcock to Borgmann, December 23, 1965. Ford Foundation Archives.

28 Letter, Babcock to Borgmann, December 28, 1965. Ford Foundation Archives.

29 Minutes, Meeting of CARSO and AURA representatives, January 14, 1966, Tucson, Arizona. Edmondson personal files.
    Those in attendance were:

> *CARSO*
> Dr. E.A. Ackerman – Executive Officer, Carnegie Institution of Washington.
> Dr. H.W. Babcock – Director, Mt. Wilson and Palomar Observatories.
> Mr. James Boise – Bursar, Carnegie Institution of Washington.
> Mr. Bruce Rule – Mt. Wilson and Palomar Observatories.
> Mr. Fred Woodson – Administrative Assistant (Construction), Mt. Wilson and Palomar Observatories.
> *THE FORD FOUNDATION*
> Dr. Carl Borgmann – Director, Program in Science and Engineering.
> *NATIONAL SCIENCE FOUNDATION*
> Dr. Geoffrey Keller – Director, Division of Mathematical and Physical Sciences.
> Dr. G.F.W. Mulders – Head, Astronomy Section.
> *AURA*
> Dr. Rupert Wildt – President
> Mr. G.L. Lee – Vice President
> Mr. Minton Moore – Treasurer
> Dr. F.K. Edmondson – Chairman, Chile Subcommittee of the Scientific Committee.
> Dr. W. A. Hiltner – Chairman, Scientific Committee
> Dr. N.U. Mayall – Director, Kitt Peak National Observatory
> Mr. J.M. Miller – Associate Director – Administration, Kitt Peak National Observatory.
> Miss Julia Elliott – Assistant Secretary
> Dr. Wildt presiding.

30 Memorandum: Proposal for Carnegie Southern Observatory (CARSO) grant, Borgmann to W. McNeil Lowry, January 21, 1966. Ford Foundation Archives. Borgmann's account of the Tucson meeting included:

> The purposes of the meeting were to decide about how much and under what conditions land on Cerro Morado could be made available by AURA to CARSO and to explore possibilities of administrative, service, and planning collaboration between the two groups for the future. For several reasons the meeting completely bogged down on the first question – land. After three years of warm verbal assurances that CARSO could have any land needed, the AURA members and administrative officers started bringing obvious "red herrings" into the considerations, and no approach to a decision was made. The principal reason for this was that AURA had just learned from Keller of NSF that NSF budget decisions had been made which reduced AURA's desire for a 150- or 200-inch scope for Chile to a vague and distant future hope. Keller then asked AURA to consider how viable was their operation in Chile when the fact that a telescope larger than 60 inches was not likely for them in the foreseeable future. This

apparently cast AURA as the inconsequential little brother in Chile and dashed their hopes of being a more nearly equal partner.

31 Minutes, Chile Subcommittee of the AURA Scientific Committee, January 19, 1966, Tucson, Arizona. Edmondson personal files.

32 *Resolution adopted by the Chile Subcommittee, Jan. 19, 1966.* Edmondson personal files.

> WHEREAS, the development of major instrumentation in the Southern Hemisphere is vital to the advancement of astronomy,
> AND WHEREAS, AURA, with NSF support, is now constructing a facility primarily for U.S. and Latin American astronomers in Chile,
> AND WHEREAS, the Carnegie Institution of Washington has initiated a proposal to the Ford Foundation for a Carnegie Southern Observatory (CARSO),
> AND WHEREAS, it is anticipated that the long-range operating plans of both AURA and CARSO will require funding by the U.S. Government through the National Science Foundation,
> AND WHEREAS, the NSF may not be able to provide continuing separate support for two independent U.S. Southern Hemisphere observatories on a long-range basis,
> NOW, THEREFORE, the Chile Subcommittee of AURA resolves: That AURA and CARSO seek ways and means to cooperate in the establishment of a single major observatory in Chile.

33 Memorandum: Proposal for Carnegie Southern Observatory (CARSO) grant, Borgmann to Lowry, January 24, 1966. Ford Foundation Archives.

34 Longhand notes, Meeting with Ackerman, January 22, 1968. Edmondson personal files.

35 Memorandum, Conference call, R. Wildt, F.K. Edmondson, W.A. Hiltner, N.U. Mayall, January 23, 1966. Edmondson personal files.

36 Minutes, Meeting of CARSO and AURA representatives, Tucson, Arizona, January 26, 1966. Edmondson personal files.

37 Mulders had suggested this. Memo (longhand), Mulders to Wildt and Mayall, undated. Edmondson personal files.

38 Minutes, Executive Committee of AURA, Inc., January 27, 1966, Tucson, Arizona. Edmondson personal files.

39 Memorandum: CARSO proposal for telescope in the Southern Hemisphere, Borgmann to Lowry, January 27, 1966. Ford Foundation Archives.

40 Letters and enclosure, Ackerman to Edmondson and Miller in Chile, February 9, 1966. Edmondson personal files.

41 Draft, CARSO–AURA Agreement and Job Description of CAO, February 12, 1966. Carbon copy in Edmondson personal files.

42 Memorandum: Further information on the CARSO–AURA Negotiations, Borgmann to S & E Files, February 10, 1966. Ford Foundation Archives.

43 Memorandum, Elliott to Members of the Board of Directors, AURA, Inc., February 11, 1966. Edmondson personal files.

44 Longhand notes, Conference call about AURA and CIW, Mayall, Wildt, Lee, and Keller, February 8, 1966. Edmondson personal files.

45 Transcript of Borgmann interview with Edmondson, April 5, 1979. "In fact, when it was announced that Mr. Bundy was to be the President, I happened to have lunch with Walt Roberts. And we were talking about this and about the CARSO proposal, and he said: 'Well, I think you've lost it.' " Roberts recalled that when Bundy had been Dean at Harvard, he had "shut down some observatories that were operated by the Harvard department under Harlow Shapley."

46 Memorandum to Members of the National Science Board, Subject: Negotiations between AURA and the Carnegie Institution of Washington regarding the joint operation of an optical observatory in Chile, Leland J. Haworth, February 18, 1966. NSF History Files.

Note: Haworth's summary of the CARSO project, its relationship to the recommendations of the Whitford Report (Op. cit. 7, Reference 8), and the basic reason for the CARSO–AURA negotiations (Borgmann called it a "shotgun marriage." Op. cit. Reference 25) is couched in language that makes the role of NSF look better than the way AURA saw it at the time.

47 Memorandum about radio transmission from Chile, 9:00 a.m. Tucson time, Saturday, February 12, Elliott to Mayall, February 14, 1966. Edmondson personal files.

48 Memorandum about radio transmission from Chile, 11:30 a.m., Tuesday, February 15, Elliott to Mayall, February 15, 1966. Edmondson personal files.

49 Op. cit. 14, Reference 72.

50 Letter, Babcock to Haskins, February 18, 1966. CIW archival files. Babcock felt it was "only fair to state for the record that some of the terms of the Agreement represent concessions that I can accept only with serious misgivings." For example, he felt that "the common administrative arrangement may prove to be a source of serious friction." He said there were "fundamental differences between CIW and AURA as regards policies and philosophy of operation." He felt that CIW operated in a "relatively austere" way, while AURA's scale of operations was "relatively extravagant." The AURA "emphasis is more on expansion of physical facilities than on the science that is to be done." He was also concerned about the possibility of sharing Morado with AURA, and wrote: "I cannot avoid expressing the wish that we might have freedom to consider another site such as Campanita, not on the AURA property." This is the site that CARSO ultimately chose; its correct name was Las Campanas. Op. cit. 19, Reference 9.

51 Letter and enclosures, Ackerman to Mayall, February 21, 1966. Edmondson personal files.

52 Minutes, Special Meeting of the Board of Directors of AURA, Inc., March 1, 1966, Conference Room D-1, O'Hare Inn, Chicago, Illinois. Edmondson personal files.

Note: Most of the Board Members and Keller had an informal meeting after dinner on February 28, with no record taken. The author remembers that the discussion was exclusively about the CAO, with strong complaints about the intransigence of the CARSO representatives on this issue. He recalls vividly that Keller, Mayall and Wildt were looking at him during much of this discussion. After the meeting ended, Wildt and Mayall took the author to Wildt's room and asked him if he would be willing to take the position of CAO, if the Ford Foundation made the grant to CIW. The reply was to the effect that "when duty calls there can be only one answer."

53 Letter and enclosures, Elliott to Wildt, March 4, 1966. Edmondson personal files.

54 Letters and enclosures, Elliott to Ackerman and Borgmann, March 4, 1966. Edmondson personal files.

55 Letters, Wildt to Ackerman, March 7, 1966; Ackerman to Wildt, March 9, 1966. Edmondson personal files.

56 Letter and enclosures, Ackerman to Borgman, March 10, 1966. Ford Foundation Archives.

57 This sentence is copied from the background section of the Ford Foundation staff recommendation for a $5 million grant to AURA. Ford Foundation Archives. This point is also made in Stratton's interview with Edmondson, October 14, 1980.

58 Minutes, 1966 Annual Meeting of the Board of Directors of AURA, Inc., March 15, 1966, Tucson, Arizona. Edmondson personal files.

59 Telephone call, Wildt to Edmondson, March 21, 1966. Longhand note in Edmondson personal files. Ackerman phoned Wildt and told him the CARSO proposal had been tabled.

60 Memorandum, Miller to Members of the Board of Directors, AURA, Inc., March 25, 1966. Edmondson personal files.

61 Op. cit. Reference 45.

Transcript of Second Borgmann interview with Edmondson, November 10, 1983. "I suspect – and I firmly say I suspect – that if there had been no change of President, we'd have made the grant. But that's a suspicion."

Memorandum, Large Southern Hemisphere Telescope, Nancy G. Roman (NASA) to the File, June 14, 1966. NASA History Files (via J. Tatarewicz).

62 Letters, Goldberg to DuBridge, June 13, 1966; DuBridge to Goldberg, June 22, 1966. Edmondson personal files. Goldberg wrote: "If I understood you correctly, you felt that these negotiations were proceeding smoothly and successfully until AURA intervened as a competitor for the Ford funds. I have only just now become a member of the AURA Board of Directors and have not yet attended any meetings, but I was sufficiently concerned by your comments and also by similar remarks by one other member of the Mt. Wilson–Palomar staff to have made it a point to read all of the minutes of AURA meetings concerned with this matter and also to talk with various members of the organization.

It seems clear from what I have learned that AURA took no initiative whatsoever in intruding on the negotiations between CARSO and Ford and, far from expressing any opposition to the CARSO proposal, Drs. Wildt and Mayall have always made it plain that they supported the proposal and would welcome the presence of CARSO in Chile.

The AURA group entered the picture only because they were, in effect, ordered by the NSF to try to work out the best possible deal for the joint management with CARSO of the southern observatory project, since it is unlikely that the NSF will itself, provide support to AURA for a large southern telescope in the foreseeable future. Parenthetically, I have been told by Dr. Mulders within the last few days that the NSF sees no possibility of funding a large southern telescope in the next ten years. Thus, the arrangement that was finally consummated is a kind of shotgun marriage with which neither party is entirely happy, but both feel they can live with it. There is no doubt, however, that the negotiations were carried out in good faith and in friendly fashion."

DuBridge replied: "It was never our thought that AURA officially, or as an organization, intervened in the CARSO proposal to The Ford Foundation. We were informed – I thought reliably – that one or more individual members connected with AURA institutions did raise this question with some member of the Ford staff. Whether true or not, this report did stimulate immediate conversation between Carnegie and AURA officials and I hope satisfactory arrangements developed therefrom. In any case, all competition between the Carnegie Institution of Washington and AURA has, I think, been settled, and we hope that all astronomers in the country – and indeed throughout the world – will participate in the observing program when observatory facilities are developed in the Southern Hemisphere. . . . Thanks again very much for your letter, and please accept my apologies if I attributed to AURA something which may have been only an individual's unofficial comments – or which, indeed, might have been a false report which reached us."

A comment made 10 years later (December 10, 1976) repeats the charge: "A last minute ploy by the competition resulted in the diversion of a $5 million grant from the Ford Foundation; instead of coming to the Carnegie Institution, which had done all the ground work and provided the justification, the grant went to the Associated [sic] Universities for Research in Astronomy." Edmondson personal files (source and names withheld by Edmondson).

63 Transcript of Bundy interview with Edmondson, December 3, 1981.

64 Library of Congress, Leland J. Haworth Papers, Box 29, Telephone Log 1966 Jan.–June. Note: These papers were found inadvertently by J. Merton England, NSF Historian, while looking for unrelated materials. He phoned the author, who was in Tucson, and mailed a copy of his notes on October 14, 1982. The author visited the Library of Congress on November 15 and copied additional information from the Haworth papers.

65 Letter, Whitford to Babcock, April 15, 1966. CIW archival files.

About a week before the March meeting of the Ford trustees, President McGeorge Bundy called to ask my opinion of the CARSO proposal. After I had summarized the statements previously made to Dr. Borgmann, he raised the question of monopoly. Is it good policy, he asked, for one group to control the two largest telescopes in the world, one in each hemisphere? I replied that I was not personally worried about this danger, particularly since the proposal brought forward by the National Science Foundation had now led to a pact for joint operation with AURA, giving that organization freedom to use one-third of the observing time in any way that it may feel is wise. The Observatory Associates proposed by the Carnegie Institution had already provided another safeguard, and would draw in quite a number of astronomers of proven ability with good ideas on southern hemisphere research, in addition to the talented group that would automatically have access as staff members. Thus the use of the telescope would be diffused over a considerable spectrum of the eligible clientele and would not be restricted to members of a closed society. . . . From President Bundy's question to me, I suspect that this is the aspect of the total problem that is giving the Ford Foundation concern. It is a proper concern, because the continuing record of scientific achievement, and indeed the future scientific health of the country depend not only on the facilities but also on the development of new generations of top-quality scientists to use them effectively. Questions of organization, motivation, and incentives enter into such a decision.

Also, see Reference 71, in which Bundy tells Stratton he had phoned some of those who had written letters to Babcock.

66 Memorandum, Keller to Haworth, March 25, 1966. NSF History Files. This memo provided a list of 14 names. The author has not been able to find any record that such a committee was ever appointed.

67 Letter, Babcock to Haskins, April 21, 1966. CIW archival files.

68 Letter, Goldberg to Babcock, April 15, 1966. CIW archival files.

69 Letter, Goldberg to Bundy, May 5, 1966. Goldberg personal files.

70 Letter, Bundy to Goldberg, May 25, 1966. Goldberg personal files.

71 Letter, Bundy to Stratton, May 26, 1966. Ford Foundation Archives, courtesy of Dr. Stratton.

72 Memorandum: Thoughts on Observatories, Borgmann to Bundy, June 14, 1966. Ford Foundation archives.

73 Memorandum: Discussions with Carnegie Institution, Borgmann to S & E Files, July 1, 1966. Ford Foundation Archives.
Note: Bundy's views about the Carnegie–Cal Tech "monopoly" were widely shared in the astronomical community.

74 Memorandum: Notes on Recent Negotiations About a Southern Hemisphere Observatory, Item 1, October 3, 1966. CIW archival files.

75 Memorandum: AURA as possible builder and operator of a 200" Southern Hemisphere telescope, Borgmann to Bundy, September 12, 1966. Ford Foundation Archives.

76 Memorandum: Further Activities – Southern Hemisphere Telescope, Borgmann to S & E Files, September 28, 1966. Ford Foundation Archives. This memorandum also says: "I called Ackerman on Tuesday, September 27th and told him informally of our internal decision (partial grant to AURA, if any) and that I had had preliminary conversations with Haworth. I suggested that a formal denial letter would follow soon. He immediately asked for a few days, suggesting that they might have an acceptable alternative. I agreed, although expressing doubts that one existed." [Note: Ackerman's summary of this conversation is in Reference 71, Item 6.]

77 Diary Note: Status of UC-Australia 150-inch telescope, Mulders, August 12, 1966. NSF History Files. Longhand note written by Haworth says: "Robertson: What do we know from NASA about this? There should certainly be a full exchange of information and views about such things. LJH 8/29/66."

78 Letter, Newell to Haworth, October 12, 1966. NSF History Files.

79 Op. cit. Reference 74, Items 2, 4 and 7.
Memorandum: Comments on Sir Richard Woolley's Proposal for Cooperation on Southern Observatory, Ackerman to Haskins, September 2, 1966. CIW archival files.
Memorandum: Further consideration on Southern Hemisphere Telescope, Borgmann to Bundy (copy for Stratton), October 14, 1966. Ford Foundation Archives, courtesy of Dr. Stratton.
Letter, Borgmann to Haworth, November 3, 1966. NSF History Files.
80 Minutes, National Science Board, NSB-66-239, November 17–18, 1966. NSB Archives.
81 Letters, Haworth to Schultze, November 28, 1966; Haworth to Hornig, November 28, 1966. NSF History Files.
82 Administratively Confidential: Some Considerations Regarding a Large-aperture Telescope at Cerro Tololo, November 28, 1966. NSF History Files.
83 Discussion Paper, and Summary of Action, Ford Foundation Board of Trustees, December 8–9, 1966. Ford Foundation Archives.
84 Transcript of Hugh Loweth interview with Edmondson, December 4, 1981.
85 Transcript of Thomas O. Jones interview with Edmondson, April 23, 1979.
86 Transcript of Thomas O. Jones interview with Edmondson, October 20, 1978.
87 Transcript of Harold H. Lane interview with Edmondson, April 27, 1979.
88 Op. cit. Reference 64.
Haworth phoned Mayall at 1:45 p.m. on December 22. Mayall was driving to California for Christmas, and his secretary got word to him in Pasadena on December 23. Mayall called Haworth at 11:10 a.m. Haworth was on another call, and Mayall's call was completed at 11:18 a.m.
89 Op. cit. Reference 86.
90 Minutes, Executive Committee of AURA, Inc., June 27, 1966, Princeton, New Jersey. Edmondson personal files.
91 Minutes, Scientific Committee of AURA, Inc., August 30–31, 1966, Tucson, Arizona. Edmondson personal files.
92 Letter, Mayall to Mohler, September 13, 1966. Wildt Papers, Yale University Library.
93 Letter, Mohler to Mayall, September 19, 1966. Wildt Papers, Yale University Library.
94 Op. cit. Reference 76.
95 Memorandum: Telescope for the Southern Hemisphere – Conference with Prof. Rupert Wildt, Borgmann to Files, January 17, 1967. NSF History Files (Haworth's copy).
96 Memorandum: Telescope for the Southern Hemisphere – Telephone call to Dr. Ackerman, Borgmann to Files, January 17, 1967. NSF History Files (Haworth's copy).
97 Memorandum: Telescope for Southern Hemisphere – Telephone call from Dr. Ackerman, Borgmann to Files, January 17, 1967. NSF History Files (Haworth's copy).
98 Op. cit. Reference 64.
99 Letter, Borgmann to Wildt, January 23, 1967. Edmondson personal files. The complete text of the letter was:

> Dear Professor Wildt:
>   I am writing to confirm our recent conversation of January 12, 1967, and to indicate possible interest of the Ford Foundation in sharing with the National Science Foundation, and perhaps with others, the provision of funds for a large optical telescope to be used in the Southern Hemisphere. In consequence, we would be interested in receiving a proposal from AURA for the construction of such a telescope.
>   As I explained, there are two situations that should be considered. One of these depends on the ability of Carnegie Institute, Washington (CIW) to raise sufficient funds to cover the additional costs of a 200" telescope over those of a 150" telescope and to work out with AURA a mutually satisfactory modus operandi. In the case that CIW cannot raise such funds (and I hope to have an answer to the

question by January 26th), the Foundation would consider a proposal to fund in part the costs of a 150" telescope (probably a duplicate of the one under construction for Kitt Peak). In either case, the Ford Foundation would limit its participation to $5.0 million.

I shall look forward to learning of the interest of AURA and to considering its proposal.

100 Minutes, Chile Subcommittee of the Scientific Committee of AURA, Inc., January 24, 1967, Tucson, Arizona. Edmondson personal files.

101 Memorandum: Southern Hemisphere Telescope – Phone call from Dr. Ackerman, Borgmann to ASTEC Files, January 25, 1967. Ford Foundation Archives.

102 Memorandum: Southern Hemisphere Telescope – Telephone Conversation with Dr. Robertson, Borgmann to Files, January 26, 1967. Ford Foundation Archives.

103 Memorandum: Southern Hemisphere Telescope – Telephone call to Rupert Wildt, Borgmann to Files, January 26, 1967. Ford Foundation Archives.

104 Memorandum: Southern Hemisphere Telescope – Phone call from Dr. Robertson [January 24], Borgmann to ASTEC Files, January 26, 1967. Ford Foundation Archives.

105 Minutes, Executive Committee of AURA, Inc., January 26, 1967, Tucson, Arizona. Edmondson personal files.

106 Letter, Wildt to Borgmann, February 11, 1967. Edmondson personal files.

107 Letter, Borgmann to Wildt, March 2, 1967. Ford Foundation Archives.

108 Minutes, Annual Meeting of the Board of Directors of AURA, Inc., March 14, 1967, Tucson, Arizona. Edmondson personal files.

109 Letter, Grant Number 67-165, McDaniel (Secretary of the Ford Foundation) to Wildt, April 19, 1967. Ford Foundation Archives.

110 Op. cit. Reference 64.

111 Letter, McDaniel to Haskins, April 25, 1967. CIW archival files.

112 *El Mercurio*, Santiago, Chile; *Los Angeles Times*; *Wall Street Journal*; and many others, April 14, 1967. Edmondson personal files.

113 Memorandum, Harold A. Klein to Borgmann, June 2, 1967. Ford Foundation Archives. Roland Paine, NSF Public Information Officer, had called Klein to ask if the Ford Foundation had any plans to announce its grant for the telescope in Chile. Paine wanted to issue a fact sheet about the telescope project, on NSF press release letterhead, and wanted to coordinate this with the Ford Foundation.

114 AURA was not obligated to accept the grant in Ford Motor Company stock because it was not paid in a lump sum. See References 18 and 19.

115 Letter, Mayall to Howard R. Dressner, Secretary – Ford Foundation, September 25, 1967. Edmondson personal files.

116 Letter, Dressner to Wildt, December 5, 1967. Edmondson personal files.

117 Memorandum: Final appraisal of Grant #67-165 to the Association of Universities for Research in Astronomy, John J. Scanlon to File, March 13, 1975. Ford Foundation Archives.

118 Letter, Haworth to Edmondson, February 8, 1979. Edmondson personal files. Note: Haworth was seriously ill and having chemotherapy when he started to write this letter, and the handwriting shows that it took considerable effort. A postscript added by his wife said he was back in the hospital, and that "he wanted me to get this letter off to you." It was postmarked March 1, and received in Tucson on March 5, the day of his death.

Chapter 17

1 Letter, Francis P. Scott (US Naval Obs.) to Edmondson, September 12, 1962. Edmondson personal files.

2 Soviet Bloc Research, Geophysics-Astronomy and Space, No. 48, 1962. Typed copy in Edmondson personal files.

3 *Transactions of the International Astronomical Union*, **IV**, Fourth General Assembly, Cambridge, Massachusetts, September 2–9, 1932, pp. 225, 263, 282. Cambridge University Press.

4 *Transactions of the International Astronomical Union*, **X**, Tenth General Assembly, Moscow, 12–20 August 1958, pp. 74, 116. Cambridge University Press.
   A committee consisting of Dieckvoss, Scott, Stoy, Zverev, Wood and Brouwer (Chairman) was appointed by Commission 8 to prepare detailed plans for a project in the southern hemisphere.

5 Op. cit. 13, Reference 35. Nineteen participants were from the United States. There were three from Argentina, two from the USSR, two from Germany, two from Australia, and one each from Chile, France, the Netherlands, South Africa, and Venezuela. Note: Zverev, Nemiro, and Harley Wood came to Bloomington for the weekend after the conference, chiefly because of their interest in the Indiana University asteroid program. Social entertainment included a dinner party in the Edmondson residence. This level of personal contact was helpful later in Chile. We were not strangers to Zverev and Nemiro, nor they to us.

6 Report on the First Inter-American Conference on Astronomy at La Plata and Cordoba, October 30 to November 3, 1959. Op. cit. Reference 5. Note: Publication of the two reports in the same issue of the *Astronomical Journal* was arranged by the editor, Dirk Brouwer, who participated in both conferences. Twelve participants were from Argentina, eight from the United States, and one each from Brazil, Canada, Chile, Ecuador, and Uruguay. The directors of observatories in Mexico, Venezuela and Columbia were not able to attend.

7 Op. cit. 13, Reference 63.

8 "Reports on Astronomy," p. 24, *Transactions of the International Astronomical Union*, **XI A**, Eleventh General Assembly, Berkeley, 1961. Academic Press.
   Note: The name of Commission 8 was changed to "Positional Astronomy" at this meeting.

9 Agenda and Draft Reports, p. 44, International Astronomical Union, Twelfth General Assembly, Hamburg, 25 August – 3 September 1964. Academic Press.

10 Minutes, Scientific Committee of AURA, Inc., November, 18, 1964, Tucson, Arizona. Edmondson personal files.

11 On the Expedition of the Pulkovo Astronomers to Chile, by M.S. Zverev. Publications of the Pulkovo Observatory, No. 185, 1970. Edmondson personal files.

12 Letter, Polojentsev to Edmondson, August 31, 1991. Edmondson personal files.
   Note: This letter and the enclosures were a response to a letter that the author wrote to Zverev asking for information about the Pulkovo program. It was accompanied by a longhand note from Zverev saying he was not well and had been in the hospital when the author's letter arrived. He died at the age of 88 on November 17, 1991.

13 Op. cit. 14, Reference 62.

14 *Transactions of the International Astronomical Union*, **XIII** (Prague, 1967), **XIV** (Brighton, 1970), **XV** (Sydney, 1973), **XVI** (Grenoble, 1976), **XVII** (Montreal, 1979), **XVIII** (Patras, 1982), **XIX** (Dehli, 1985), **XX** (Baltimore, 1988), and **XXI** (Buenos Aires, 1991). D. Reidel Publishing Company.

15 "The $\Delta\alpha_\delta$ System of the FK4 Catalogue in the Southern Hemisphere as Derived from Observations in Chile," by C. Anguita, M.S. Zverev, G. Carrasco, P. Loyola, A.A. Naumova, A.A. Nemiro, D.D. Polojentsev, R. Taibo and V.N. Shishkina. Information Bulletin for the Southern Hemisphere, Nr. 22, April 1973. Edmondson personal files.

16 Memorandum, Polojentsev to Edmondson, August 29, 1991. Edmondson personal files.

17 Transcript, Stock interview with Edmondson, August 18, 1979. Edmondson personal files.

18 Op. cit. Reference 1.

Chapter 18

1 *ESO's Early History*, by Adriaan Blaauw. European Southern Observatory,

Karl-Schwarzschild-Str. 2, D-8046 Garching bei München, Germany. Copyright 1991, ISBN 3-923524-40-4.

The individual chapters were first published in ESO's journal *The Messenger*, No. 54, December 1988 to No. 64, June 1991.

2 *des hommes, des télescopes, des étoiles*, by Charles Fehrenbach. Editions du Centre National de la Recherche Scientifique, 15 quai Antatole France, 75700 Paris. Copyright 1990, ISBN 2-222-04459-6.

3 This document is reproduced on pp. 2 and 3 of Blaauw's book, and on p. 2 of *The Messenger*, No. 54. The author is indebted to his wife for French to English translations of this and other ESO documents.

4 Op. cit. 5, Reference 9.

5 The author is indebted to Dr. Blaauw and Mrs. Katgert (Leiden Observatory) for copies of these letters from the Oort archives:
Letter, Oort to Struve, February 11, 1954.
Letter, Struve to Oort, February 16, 1954.
Letter, Struve to Oort, March 19, 1954.
Letter, Oort to Struve, March 30, 1954.

6 Op. cit. 4, Reference 55.

7 Letter, Oort to Shepard Stone (Ford Foundation), August 2, 1956. Ford Foundation Archives.

8 Letter, McDaniel to Oort, August 14, 1956. Ford Foundation Archives.

9 Op. cit. Reference 5.

10 Transcript of Carl W. Borgmann interview with Edmondson, April 5, 1979.
Abstract, Contributions of the Ford Foundation to Astronomy in the Southern Hemisphere, F.K. Edmondson, Publications of the Astronomical Society of the Pacific, Vol. 101, p. 434, April 1989.
Four large grants to support astronomical programs, all in the southern hemisphere, were made during the period 1959 to early 1967: ESO, $1,000,000 promised in 1959 and paid in 1964; Yale–Columbia astrograph in Argentina, $750,000 in 1960; CSIRO for Australian Solar Radio Telescope (Radioheliograph), $550,000 in 1962 and $80,000 in 1966; and AURA for half the cost of the CTIO 150-inch telescope, $5,000,000 in 1967.
The Ford Foundation was restructured in March 1967 by Heald's successor, McGeorge Bundy, and the Program in Science and Engineering was terminated.

11 Letter, Lindblad to Adam Skapski (Ford Foundation consultant), March 1, 1958; Ford Foundation memoranda: William McPeak to Bill Nims, April 16, 1958; Stanley T. Gordon to Waldemar Nielsen, April 29, 1958; Letter, McDaniel to Lindblad, May 2, 1958. Ford Foundation Archives.
Gordon drafted the letter to Lindblad. His cover memo to Nielsen said: "Even if this project had some such attraction for us, however, I think we would do well to send the attached negative letter."

12 Op. cit. Reference 1, pp. 9–10. Woolley attended the April 1956 meeting with Redman, and Redman alone attended the April 1957 meeting. On July 27, 1960 Woolley informed Oort that "the British National Committee prefers participation in a Commonwealth telescope to participation in a European telescope." Woolley and Olin Eggen attended the January and June 1961 meetings, and A. Hunter attended the November 1961 meeting. This was the last time a British astronomer attended an ESO Committee meeting. Eggen, who was Chief Assistant at the Royal Greenwich Observatory at that time, shared Woolley's opposition to British participation in ESO.
Letter, Eggen to Shane, January 5(?), 1961. Copy given to the author by Shane. Also, transcript of Eggen interview with Edmondson, May 10, 1987.

13 Letter, Oort to Henry Heald, July 18, 1958. Ford Foundation Archives. This letter included Oort's reason for coming to the Ford Foundation: "I am very much afraid that, unless

some outside help could be obtained for the initial expense, it will prove impossible to realize the project on a satisfactory international basis. It was for this reason that in August 1956 I made an informal approach to the Ford Foundation. At that time, I was informed that the Foundation could not consider our request because it did not have a program in the natural sciences. It has now been brought to my attention that there has recently been a change in the general policy of your Foundation, which would permit it to consider also projects which, like ours, lie partly in the domain of science and partly in that of international relations."

14 Letter, Heald to Oort, July 22, 1958. Ford Foundation Archives.

15 Letter, Borgmann to Pearson, August 1, 1958. Ford Foundation Archives.

16 Memorandum, Pearson to Borgmann, August 2, 1958. Ford Foundation Archives.

17 Report (Confidential), Skapski to Borgmann, August 28, 1958. Ford Foundation Archives.

18 Letters, Oort to Heald, September 8, 1958; Heald to Oort, September 12, 1958, Ford Foundation Archives.

19 Letter, Borgmann to Lindblad, September 26, 1958. Ford Foundation Archives.

20 Op. cit. Reference 1, p. 14.

21 Letter, Oort to Heald, November 4, 1958. Ford Foundation Archives.

22 Letter, Berger to Stone, November 29, 1958. Ford Foundation Archives.

23 Letter, Borgmann to Oort, January 5, 1959. Ford Foundation Archives.

24 Letter, Oort to Borgmann, January 18, 1959. Ford Foundation Archives.

25 Letter, Borgmann to Oort, February 5, 1959. Ford Foundation Archives.
   The Nationalist Party, representing the fast-growing Afrikaner population, took control of the government in 1948 and began to expand and codify the laws separating the races ("apartheid"). Coloured voters were removed to a separate voters' roll in 1956, two years after the historic meeting in Leiden and three years before Borgmann wrote this letter. The Prime Minister from 1956 to 1965, Hendrik Verwoerd, created nine tribal "homelands." Blacks residing in the townships of the Republic had political rights only in their appropriate homeland. Responding to strong criticism from the British Prime Minister, Harold Macmillan, the National Government created the Republic of South Africa on May 31, 1961, and withdrew from the British Commonwealth.

26 Letter, Oort to Borgmann, February 17, 1959. Ford Foundation Archives.

27 Memorandum, Pearson to Science and Engineering Files, March 31, 1959. Ford Foundation Archives.

28 Letter, Borgmann to Oort, April 30, 1959. Ford Foundation Archives.

29 Docket Excerpt, Meeting of Board of Trustees, September 25, 1959. Ford Foundation Archives.

30 Memorandum, Borgmann to Records Center, October 5, 1959; Letter, Borgmann to Oort, October 2, 1959. Ford Foundation Archives. Oort had sent Shane a photocopy of Borgmann's letter, and the author has a copy of this.

31 Letter, Maurice Bayen (for the Director General of Higher Education) to Monsieur le Directeur de la Fondation Ford, June 22, 1960. Ford Foundation Archives. The author read a paper "The Ford Foundation and the European Southern Observatory" at the meeting of Commission 41 (History of Astronomy) during the IAU meeting in Baltimore, August 1988. A summary was printed in a "box" in The Messenger, No. 54, and as Annex 2 in Blaauw's book (Reference 1, p. 255).

32 Op. cit. Reference 1, p. 18.

33 Letter, Oort to Borgmann, February 10, 1964. Ford Foundation Archives.

34 Letter, Norman W. McLeod (Ford Foundation) to Stanley Millard (Morgan Guaranty Trust company), September 15, 1964. Ford Foundation Archives.

35 Memorandum, Pearson to Records Center, Sir Solly Zuckerman's comments on the European Southern Observatory Project, March 24, 1960. Ford Foundation Archives.

36 Memorandum, Oort to the members of the ESO Committee, April 22, 1960. Ford Foundation Archives.

37 Letter, Oort to Shane, April 28, 1960. The author has a copy from Shane's files. By the time this letter was written, Kuiper had accepted a position at the University of Arizona and the future of the Chile Observatory Project was under discussion.

38 Letter, Shane to Oort, May 6, 1960. The author has a copy from Shane's files. This letter was written a month and a half after Meinel had been removed as Director of KPNO; Shane signed the letter "Acting Director."

39 Letter, Shane to Oort, June 3, 1960. The author has a copy from Shane's files.

40 Letter, Oort to Shane, June 9, 1960. The author has a copy from Shane's file.

41 Letter, Blaauw to Stock, June 15, 1962. KPNO-Mayall Archives, Tucson.
Letter, Blaauw to Mayall, June 15, 1962. KPNO-Mayall Archives, Tucson.
Pearson, who was unaware of ESO's plans to check the observing conditions in Chile, talked to two key people in the State Department before writing: "It seems to me that the worsening of racial difficulties in South Africa warrants the Foundation raising with the ESO group the question of whether they have fully considered all of the potential complications of placing the Observatory in that country." Memorandum to Records Center, July 11, 1962. Ford Foundation Archives.

42 Letter, Blaauw to Edmondson, October 15, 1962. Edmondson personal files.
Letter, Edmondson to Blaauw, October 25, 1962. Edmondson personal files.
Letter, Blaauw to Edmondson, October 31, 1962. Edmondson personal files.

43 Op. cit. Reference 1, pp. 48–49.

44 The visit of ESO observers to Chile, Report No. 1, by A.B. Muller, Edmondson personal files (received March 18, 1963).

45 Op. cit. Reference 1, pp. 49–52.

46 Memorandum, Oort to ESO Committee, August 9, 1962. Ford Foundation Archives.

47 Letter, Edmondson to Heckmann, March 8, 1963. Edmondson personal files.
Letter, Heckmann to Edmondson, March 25, 1963. Edmondson personal files.
Letter, Edmondson to Heckmann, April 3, 1963. Edmondson personal files.
Letter, Heckmann to Edmondson, April 9, 1963. Edmondson personal files.
Letter, Edmondson to Heckmann, April 26, 1963. Edmondson personal files.
Letter, Heckmann to Edmondson, May 7, 1963. Edmondson personal files.
Letter, Heckmann to Edmondson, May 24, 1963. Edmondson personal files.

48 Letter and enclosure, Heckmann to Edmondson, June 15, 1963. Edmondson personal files. This was the first letter addressed "Dear Frank" and signed "Otto." Oort, Heckmann, and Edmondson had slept on straw mattresses on the dirt floor of a "mud hut" at Los Placeres on June 8. Before retiring, Edmondson suggested to Heckmann that they start using first names. After a brief initial shock, Heckmann smiled and cheerfully agreed. The first letter addressed to "Dear Nick" was written in October.

49 Letter, Edmondson to Shane, June 20, 1963. Edmondson personal files. A French translation of part of this letter is on pp. 380–381 of Fehrenbach's book (Reference 2).

50 Letter, Oort to Edmondson, November 17, 1963. Edmondson personal files.

51 Minutes, Executive Committee of AURA, Inc., June 28, 1963, Tucson, Arizona. Edmondson personal files.

52 Letter, Edmondson to Haworth, July 8, 1963. Edmondson personal files.

53 Letter, Haworth to Edmondson, August 13, 1963. Edmondson personal files.

54 Letter, Edmondson to Oort, August 20, 1963; Letter, Oort to Edmondson, August 23, 1963. Edmondson personal files.

55 Minutes, Executive Committee of AURA, Inc., September 20, 1963. Washington, DC. Edmondson personal files.

56 Letter, Mayall to Heckmann, September 20, 1963. Edmondson personal files.

57 Letters, H.W. Marck (ESO) to Mayall and Edmondson, September 23, 1963. Edmondson personal files.

58 Letter, Heckmann to Mayall, September 30, 1963. Edmondson personal files.

59 Minutes, Meeting regarding relationships between AURA and ESO, October 11–12, 1963, Tucson, Arizona. Edmondson personal files.

60 Letter, Wildt to Edmondson, October 16, 1963. Edmondson personal files.

61 Letter, Heckmann to Edmondson, October 20, 1963. Edmondson personal files.

62 Letter, Edmondson to Heckmann, October 23, 1963. Edmondson personal files.

63 Op. cit. Reference 50.

64 Minutes, Executive Committee of AURA, Inc., November 22, 1963, Tucson, Arizona. Edmondson personal files.

65 *El Mercurio*, November 7, 1963; Letter, Stock to Mayall, November 13, 1963. Edmondson personal files.

The next to the last paragraph of Oort's November 17 letter (Op. cit. Reference 50) reads:

> You will have heard that Heckmann has recently been in Chile. The purpose of his visit was to find out whether the Chilean Government would agree to the conditions which would be necessary for the establishment of an international observatory.
>
> During the visit it appeared that things could be arranged much more quickly and completely than anybody had anticipated, so that a full text for an agreement was drawn up and even signed by the Chilean Minister of Foreign Affairs. The document contains, of course, the evident proviso that it will only come into force after approval by the official ESO Council, and after ratification by the Chilean Government. The agreement, which appears entirely satisfactory, is very similar to the convention between Chile and the UNO Economic Commission for Latin America.

This is also discussed in Blaauw's book on pp. 54 and 56–57, and in *The Messenger*, No. 54, pp. 24–26. Fehrenbach's discussion on pp. 383 and 387–390 of his book provides some interesting details that are not in Blaauw's account.

Edmondson did not object to the signing ("Apparently events moved more rapidly than you had expected.") but he did complain that Stock had not been invited to attend: "Because we are friends, I feel that I can say to you that we are quite puzzled by your failure to invite an AURA representative (Dr. Stock) to attend the signing of this document, or at least the social event that followed. One of my students received a newspaper clipping from his parents, and this article said that Ambassador Cole was invited. This is very nice, but it is not equivalent to inviting an AURA representative." Heckmann replied: "Concerning the signing of this document and the following social event I have to say: It was not ESO who invited but the Chilean Government. They had not even invited any scientist of their own to the signing itself, and *they* had composed the list of the participants of the following lunch. So please believe me that on our side nothing of negligence was in the non-inviting of Dr. Stock."

Letters, Edmondson to Heckmann, November 27, 1964; Heckmann to Edmondson, December 4, 1963. Edmondson personal files.

66 "The members of the Scientific Committee assembled in Tucson on November 22, 1963, reaffirming the statement of objectives in the AURA Articles of Incorporation, recognize their responsibility to assist universities and other institutions and individuals in their development of astronomical research and related activities. Such assistance cannot but be rendered in accordance with established policies based upon the experience of AURA in the formation of the Kitt Peak National Observatory. AURA's Board of Directors administers the Kitt Peak National Observatory and the Cerro Tololo Inter-American Observatory for the benefit of all astronomers and astronomical institutions. In particular, no member university or tenant of AURA property is granted special privileges or benefits. The discovery and initial development of the El Totoral site calls for an extension of this policy

to the long-range development of AURA's property in Chile. The sale of part of this property would set a precedent making it difficult to refuse further requests by other interested groups. This could lead to fragmentation and would violate the trust vested in the Board of Directors. The members of the Scientific Committee recommend that a policy statement of this kind be considered by the Board of Directors."

67 Letters, Edmondson to Oort, November 26, 1963; Mayall to Oort, November 27, 1963. Edmondson personal files.

68 Letter, Oort to Edmondson, December 9, 1963.

69 Op. cit. Reference 65. (Letter, Heckmann to Edmondson, December 4, 1963.)

70 Minutes of the joint meeting of representatives of AURA and ESO, with two representatives of the Carnegie Institution and one representative of Denmark attending as observers, held at l'Observatoire de Paris, January 21 and 22, 1963.
(a) The ESO Minutes were taken by J.M. Ramberg, Assistant Director of ESO. (b) The AURA Minutes were taken by Julia Elliott, Assistant Secretary of AURA. Edmondson personal files.
The 12 ESO representatives came from all of the ESO member countries except Belgium. Edmondson, Harrell, Wildt and Mayall represented AURA. Babcock and Ackerman represented CARSO.

71 Minutes, Executive Committee of AURA, Inc., January 31, 1964, Washington, DC. Edmondson personal files.

72 Cablegram and letter, Edmondson to Oort, February 1, 1964. Edmondson personal files.

73 Letter, Oort to Edmondson, February 10, 1964. Edmondson personal files.

74 Op. cit. Reference 1, p. 59.

75 Memorandum, Mayall's "Radio Conversation with Dr. Heckmann in La Serena," March 24, 1969. Edmondson personal files.

76 Memorandum of discussion between ESO and AURA representatives on Cerro Tololo on April 12, 1964. Those present were Heckmann, Fehrenbach, Muller, Mayall, Hiltner, Schaeffer and Stock. Stock's longhand copy was revised by Counsel Puga on April 16. Edmondson personal files.

77 Letters, Heckmann to Mayall, May 5, 1964; Puga to Mayall, May 12, 1964; Mayall to Heckmann, May 18, 1964. Edmondson personal files.

78 Op. cit. Reference 2, pp. 386–387.

79 Op. cit. Reference 1, p. 85.

80 Letter, Oort to Edmondson, June 12, 1964. Edmondson personal files.

81 Letters, Mayall to Oort, June 20, 1964; Edmondson to Oort, July 7, 1964. Edmondson personal files.

82 Transcript of Oort interview with Edmondson, January 22, 1981.

83 The information in this paragraph and the two that follow comes from Blaauw's book (Reference 1).

Note: After Heckmann retired as Director General of ESO in 1970 he wrote a book: Sterne, Kosmos, Weltmodelle, Verlag Piper & Co., München-Zürich, 1976. Chapter XII, pp. 258–338, is entitled: Die Europäische Südsternwarte, European Southern Observatory (ESO), and is divided into 10 sections. The $1,000,000 grant from the Ford Foundation is discussed in Section 2, pp. 265–266. Section 5 has accounts of the June 1963 AURA–ESO meeting in Chile, the October 1963 discussions in Tucson, and the signing of the ESO–Chile Convention on November 6, 1965 followed by the reaction of the provisional ESO Council. Section 6 includes a description of the selection of La Silla by ESO, and mentions the CARSO choice of Las Campanas. Heckmann died on May 13, 1983 at the age of 82.

Chapter 19

1 Letter and Report, Pickering to the President of the Carnegie Institution, September 19, 1902. CIW archival files.

2 Letter, Hale to J.C. Merriam, July 21, 1925. CIW archival files.

3 Memorandum on a Sixty-inch Telescope in the Southern Hemisphere, W.S. Adams, September 20, 1926. CIW archival files.

4 Two letters, Adams to Merriam, December 16, 1929, CIW archival files.
Letter, Merriam to Adams, January 31, 1930. CIW archival files.
Telegram, Adams to Merriam, May 19, 1931. CIW archival files.
Letter, Merriam to Adams, May 19, 1931. CIW archival files.

5 Memorandum, Project for a large southern telescope, Henry Norris Russell, March 7, 1931. CIW archival files.
Note: Russell proposed a 150-inch telescope.
Memorandum of telephone conversation with Dr. Hale, J.C. Merriam, April 20, 1932. CIW archival files.
Note: Hale thought the Mount Wilson 60-inch should not be moved; he supported a 120-inch for the southern hemisphere.

6 Letter and enclosures, Adams to Merriam, September 17, 1934. The proposal was signed by the Advisory Committee on Large Telescope for the Southern Hemisphere: W.S. Adams, F.H. Seares, Fred E. Wright (chairman), F.G. Pease.
Letter, Merriam to Adams, October 3, 1934. CIW archival files.

7 Letter, Merriam to F.P. Keppel, January 14, 1936. CIW archival files.

8 Transcript of Mayall interview with Edmondson, March 17, 1979.

9 Carnegie Evening 1986: "An Eight-Meter Telescope for the 21st Century," by George W. Preston, III. Carnegie Institution of Washington, May 8, 1986.
Note: This booklet was prepared as a program souvenir, to be handed out to guests attending the lecture. It contains an unsigned historical essay written by Ray Bowers, CIW Editor, entitled: That Special Mountain. It describes the site survey and the selection of Las Campanas.

10 Letter, Bowen to Caryl P. Haskins, December 17, 1962. CIW archival files.

11 Letter, O.C. Wilson to Tuve, March 15, 1963. CIW archival files. This appears to be the first suggestion of a 200-inch for the southern hemisphere.

12 Letter, Babcock to Mayall, May 1, 1963. Edmondson personal files.
Note: Babcock had been appointed Director of the Mount Wilson and Palomar Observatories, effective July 1, 1964, and Associate Director, effective May 10, 1963.
Letter, Haskins to Babcock, June 3, 1963. CIW archival files.

13 Letter, Babcock to Rutllant, May 21, 1963. CIW archival files.

14 Letter, Babcock to Haskins, April 23, 1963. CIW archival files.

15 Letters, Mayall to Babcock, May 22, 1963; Babcock to Mayall, May 27, 1963. Edmondson personal files.

16 Letter, Babcock to Haskins, May 27, 1963. CIW archival files.

17 Letter, Mayall to Babcock, June 26, 1963. Edmondson personal files.

18 Minutes, Executive Committee of AURA, Inc., September 20, 1963, Tucson, Arizona. Edmondson personal files.

19 Letter and enclosure, Babcock to Haskins, October 22, 1963. CIW archival files.

20 Letter, Mayall to Babcock, October 24, 1963. Edmondson personal files.

21 Letter, Babcock to Mayall, October 28, 1963. Edmondson personal files.

22 A Proposal for the Construction of a 200-inch Telescope in the Southern Hemisphere, Carnegie Institution of Washington, undated. Edmondson personal files.
Note: Ackerman gave this copy to the author at the November 22, 1963 Executive Committee Meeting.

23 Letter, Mayall to Edmondson, October 31, 1963. Edmondson personal files.

24 Minutes, Executive Committee of AURA, Inc., November 22, 1963, Tucson, Arizona. Edmondson personal files.

25 Op. cit. 18, Reference 70.

26 Op. cit. 18, References 68 and 69.

27 Op. cit. 18, Reference 80.

28 H.W. Babcock, Report on Trip to Chile, Nov. 4 – Dec. 21, 1963. Edmondson personal files.

29 Op. cit. 18, Reference 65.

30 Op. cit. 14, Reference 47.

31 Letter, Babcock to Mayall, February 3, 1964. Edmondson personal files.

32 Memorandum for the Record, Field Trip Between February 8–19, 1964, to Santiago, La Serena, Tololo, Pachon, Hurtado, and Ovalle, Chile, for Examination of Possible Observatory Property, by Edward A. Ackerman. Edmondson personal files.

33 Letter, Ackerman to Mayall, March 17, 1964. Edmondson personal files.

34 Op. cit. Reference 9.

35 Op. cit. Reference 32. Ackerman's report said that Korp and the AURA staff were most helpful in "making arrangements for our field trip and other needs. Indeed, they were much more in evidence than our own representative, James Hanson. Had we depended on Mr. Hanson's assistance alone, we probably should have been much longer in the field than we were."

36 Letter, Stock to Bowen, April 7, 1964. Edmondson personal files.

37 Letters, Babcock to Mayall, April 22, 1964; Babcock to Stock May 11, 1964. Edmondson personal files.

38 Letter, Babcock to Mayall, June 1, 1967. Edmondson personal files.

39 Letters, Mayall to Babcock, July 23, 1964; Mayall to Ackerman, July 24, 1964; Mayall to Haskins, December 31, 1964. Edmondson personal files.

40 Minutes, Board of Directors of AURA, Inc., March 10, 1964, Tucson, Arizona. Edmondson personal files.

41 Letters, Babcock to Stock, May 11, 1964; Stock to Edmondson, May 19, 1964; Edmondson to Mayall, June 2, 1964; Mayall to Stock, June 11, 1964. Edmondson personal files.

42 Letter, Babcock to Mayall, April 11, 1967. Edmondson personal files.

43 Letter, Mayall to Babcock, April 14, 1967. Edmondson personal files.

44 Letter, Miller to Bilby, April 6, 1967. Edmondson personal files.

45 Letter, Bilby to Miller, April 12, 1967. Edmondson personal files.

46 Letter, Babcock to Ackerman, May 15, 1967. CIW archival files.

47 Memorandum, Nancy G. Roman to Associate Administrator for Space Science and Applications, June 22, 1967. NASA History Files, via J.N. Tatarewicz.

48 Transcript, Mohler interview with Edmondson, July 11, 1979. Mohler learned about this from W.K. Pierpont, who was on the Kresge Foundation Board. Pierpont was Vice-President and Chief Financial Officer of the University of Michigan, and was a member of the AURA Board from 1967 to 1977.

49 Letter, Babcock to Donald A. MacRae, December 20, 1967. CIW archival files.

50 Op. cit. Reference 9.

51 Minutes of Meeting on CARSO proposal, March 18, 1968, O'Hare Inn, Chicago, Illinois. Edmondson personal files.

52 Letter, Babcock to Mayall, April 3, 1968. Edmondson personal files.

53 Letter, Mayall to Babcock, April 12, 1968. Edmondson personal files.

54 Op. cit. 16, Reference 71, Item 3.

55 Op. cit. Reference 9.

56 Letter, Irwin to Babcock, September 10, 1966. CIW archival files.

57 Babcock had expressed an interest in Campanita in a letter to Haskins dated February 18, 1966. Op. cit. 16, Reference 47.

58 Letter, Irwin to Mayall, August 26, 1967. Edmondson personal files.

59 Op. cit. Reference 38.
    Letter, Babcock to Mayall, September 27, 1967. Edmondson personal files.

60 Letter and attachment, Miller to Bilby, June 5, 1968. Edmondson personal files.
61 Minutes, Executive Committee of AURA, Inc., June 20, 1968, Tucson, Arizona. Edmondson personal files.
62 Letter, Miller to Babcock, June 21, 1968. Edmondson personal files.
63 Letter, Babcock to Miller, July 2, 1968. Edmondson personal files.
64 Letter, Hiltner to Babcock, July 10, 1968. Edmondson personal files.
65 Minutes, Annual meeting of the Board of Directors of AURA, Inc., March 9, 1968, Tucson, Arizona. Edmondson personal files.
66 Letter, Mayall to Babcock, September 17, 1968. Edmondson personal files.
67 Letter, Babcock to Mayall, September 19, 1968. Edmondson personal files.
68 Letter, Babcock to Hiltner, October 3, 1968. Edmondson personal files.
69 Letter, Hiltner to Babcock, October 16, 1968. Edmondson personal files.
70 *Arizona Daily Star*, October 27, 1968. Edmondson personal files.
71 *El Mercurio*, November 20, 1968. Edmondson personal files.
72 Minutes, Executive Committee of AURA, Inc., November 21, 1968, Cerro Tololo, Chile. Edmondson personal files.
73 Letters, Mayall to Haskins, January 28, 1964; Haskins to Mayall, February 4, 1964. Edmondson personal files.
74 Op. cit. Reference 9, pp. 17–18.
75 Minutes, Organization Committee of AURA, Inc., December 13, 1968, O'Hare Inn, Chicago, Illinois. Edmondson personal files.
76 Letter, Babcock to Hiltner, December 4, 1968. Edmondson personal files.
77 Minutes, 1969 Annual Meeting of the Board of Directors of AURA, Inc., January 23, 1969, Tucson, Arizona. Edmondson personal files.
78 Letter, Buck to Blanco, April 29, 1969. Edmondson personal files.
79 Op. cit. Reference 9, pp. 19–20.
80 *Los Angeles Times*, July 1, 1980. Edmondson personal files.
81 News Release, Carnegie Institution of Washington, June 29, 1984. Copy supplied by Ray Bowers.
82 The President's Report, Year Book 1983, Carnegie Institution of Washington, 1983–84. Copy supplied by Ray Bowers.
83 Joint News Release, CIW and Mount Wilson Institute, October 27, 1987. Copy supplied by Ray Bowers.
84 CIW newsletter, Carnegie Institution of Washington, March 1987.
85 *SPECTRA*, The Newsletter of the Carnegie Institution of Washington, November 1991.
86 *SPECTRA*, The Newsletter of the Carnegie Institution of Washington, March 1992.
87 *SPECTRA*, The Newsletter of the Carnegie Institution of Washington, June 1993.

Chapter 20

1 Op. cit. 12, Ref 52.
2 Op. cit. 12, Reference 57.
3 Op. cit. 12, Reference 77. D.M. Hunten was the senior level appointee. J.C. Brandt, A.L. Broadfoot, Lloyd Wallace, and R.D. Wolstencroft were the non-tenure appointments.
4 Minutes, AURA–NSF Meeting, August 13, 1962, Washington, DC. Edmondson personal files.
5 Op. cit. 11, Reference 15.
6 KPNO Monthly Report, April 1963, pp. 5–7. Edmondson personal files.
7 KPNO Monthly Report, July 1963, pp. 4–6. Edmondson personal files.
8 KPNO Monthly Report, October 1963, pp. 3–4. Edmondson personal files.
9 KPNO Monthly Report, April 1964, pp. 3–5. Edmondson personal files.
10 KPNO Monthly Report, June 1964, pp. 7–9. Edmondson personal files.
11 KPNO Monthly Report, October, November, December 1964, pp. 11–15 and table.

Edmondson personal files. The table summarizes the six flights: instruments, scientific results, and comments about problems with each flight.

12 Minutes, Space Astronomy Subcommittee of the Scientific Committee of AURA, Inc., October 31, 1968, Tucson, Arizona. Edmondson personal files.

13 Minutes, Joint Meeting of the Scientific and Organization Committees of AURA, Inc., January 22, 1969, Tucson, Arizona. Edmondson personal files.

14 Minutes, 1969 Annual Meeting of the Board of Directors of AURA, Inc., January 23, 1969, Tucson, Arizona. Edmondson personal files.

15 Memorandum, Mayall to AURA Board of Directors, May 12, 1970. Edmondson personal files. Mayall felt that his successor should appoint Chamberlain's replacement, and the Scientific Committee agreed.

Minutes, Scientific Committee of AURA, Inc., June 24, 1970, Madison, Wisconsin. Edmondson personal files.

16 Minutes, Executive Committee of AURA, Inc., March 16, 1971, Tucson, Arizona. Edmondson personal files.

17 Letter, Whitford to List, June 1, 1970. Edmondson personal files.

18 Memorandum, Whitford to AURA Board of Directors, December 14, 1970. Edmondson personal files.

19 Op. cit. Reference 16.

20 KPNO and CTIO Quarterly Report, April–May–June 1971, pp. 4–6. Edmondson personal files.

21 Minutes, Executive Committee of AURA, Inc., September 9, 1971, Tucson, Arizona. Edmondson personal files.

22 Minutes, 1971 Annual Meeting of the Board of Directors of AURA, Inc., January 21, 1971, Tucson, Arizona. Edmondson personal files. It is the author's recollection that James A. Van Allen proposed naming the 158-inch telescope for Mayall at this meeting or at the meeting of the Scientific Committee on January 19, but this is not recorded in the open Minutes of either meeting. Van Allen responded to an inquiry from the author in a letter dated 11 November 1993: "I clearly recall having been the one who proposed that the 150-inch telescope at Kitt Peak be named the Nicholas U. Mayall telescope. I was a special admirer of Nick and thought that this was a well deserved recognition of his services as director upon the occasion of his retirement. My further recollection is that my proposal was greeted somewhat grudgingly by Leo Goldberg but was nonetheless adopted by the Executive Committee. Later, I took special pleasure in my suggestion and thought that it might have been my most durable contribution to the work of the AURA Board." Van Allen's three-year term as a director-at-large ended at this meeting.

23 Staff Memorandum, Leland J. Haworth, June 19, 1969. Edmondson personal files.

24 The Daddario Committee and its revision of the charter of the National Science Foundation is discussed in Chapter 12 of *A Minor Miracle*, by Milton Lomask (Op. cit. 2, Reference 2). Congressman Daddario was a friend of the National Science Foundation, and his reorganization plan was proposed with the best of intentions. Haworth thought it would be helpful, but the NSF Comptroller, Aaron Rosenthal, had a different view. [Transcript of Rosenthal interview with Edmondson, June 26, 1979.] He had been Comptroller of the Veterans Administration and NASA before coming to NSF. The VA had a $7 billion budget and NASA's budget was $2–3 billion. NSF, with a budget of $300 million, attracted him because of what it did. "One of the reasons I left NASA and came to the Foundation was the fact that at that time this was the only government organization that had the name 'foundation.' . . . NSF was in the business of giving away money. . . . And this was because it was felt that this was an organization that should support science, not do science but support science." The Daddario reorganization changed the NSF budget process from "continuing authorization" to "annual re-authorization." Rosenthal's experience in NASA and the VA gave him experience with both systems and he felt the change for NSF was a

mistake because it took away the Director's flexibility to reprogram funds to meet unexpected needs.

25 KPNO Phone Memorandum, Fleischer to Mayall, October 28, 1969. Edmondson personal files.

26 Transcript of Jones interview with Edmondson, October 20, 1978.

27 Daniel Hunt, Jr. was the Head of the Office of National Centers and Facilities Operations, under National and International Programs, and Gerald Anderson worked for Hunt as Project Officer for KPNO and CTIO.

   (a) Minutes, 1970 Annual Meeting of the Board of Directors of AURA, Inc., January 22, 1970, Tucson, Arizona. Edmondson personal files. Fleischer and Anderson, representing Jones, attended this meeting.

   (b) Minutes, Planetary Sciences Subcommittee of the Scientific Committee of AURA, Inc., October 21, 1970, Tucson, Arizona. Edmondson personal files. Lane and Anderson attended this meeting.

   (c) Minutes, Executive Committee of AURA, Inc., November 19, 1970, Cerro Tololo, Chile. Edmondson personal files. Fleischer and Hunt, representing Owen, attended this meeting.

28 Minutes, Executive Committee of AURA, Inc., June 24, 1971, Tucson, Arizona. Edmondson personal files.

29 Minutes, Organization Committee of AURA, Inc., July 22, 1971, Chicago, Illinois. Edmondson personal files.

30 Minutes, Executive Committee of AURA, Inc., November 16, 1971, Tucson, Arizona. Edmondson personal files.

31 Minutes, 1972 Annual Meeting of the Board of Directors of AURA, Inc., January 19, 1972, Tucson, Arizona. Edmondson personal files.

32 Minutes, Executive Committee of AURA, Inc., March 7, 1972, Tucson, Arizona. Edmondson personal files.

33 Op. cit. 8, Reference 7.

34 Minutes, Executive Committee of AURA, Inc., June 20, 1972, Tucson, Arizona. Edmondson personal files. The Minutes simply state that Wildt "wanted the Board Members to know that Dr. Goldberg had informed the other officers in great detail and that the officers completely concur in the evaluation." It is the author's recollection that Miller wrote a highly critical letter to an NSF official, and that Goldberg learned about the letter when the recipient read it to him on the telephone.

35 Miller's many successful defenses in contract disputes, the purchase of El Totoral in Chile, finding a less expensive source than Corning for the KPNO 150-inch fused quartz mirror blank (a saving of $500,000), and building several telescopes within budget were described in earlier chapters, and should not be forgotten. Aaron Rosenthal commented: "I dare say Jim Miller had more to do with getting those things built and in place than any other single individual, and I think that's a credit he never really got. . . . But when it came time to do the nitty gritty of letting the contracts and getting the work done and seeing the money was there on time when it was needed, Jim was a tremendous advocate in coming to Washington, fighting the battles for increased funds. . . . I don't mean to get off on a tangent of Jim Miller, but in any case I thought that ought to be in the record." Op. cit. Reference 24.

36 Minutes, 1973 Annual Meeting of the Board of Directors of AURA, Inc., January 17, 1973, Tucson, Arizona. Edmondson personal files.

37 Letter, Lee to Members of the Board of Directors of AURA, September 6, 1973.
    Minutes, Executive Committee of AURA, Inc., September 27, 1973, Tucson, Arizona. Edmondson personal files.

38 Minutes, Executive Committee of AURA, Inc., March 14, 1973, Tucson, Arizona. Edmondson personal files.

39 Letter, R.B. Leighton to Wildt, February 10, 1972. The Report of the AURA Visiting
    Committee, Meeting of November 9–10, 1971 was enclosed. Edmondson personal files.

40 Memorandum, Lorenz to Members of the Organization Committee of AURA, Inc., May
    3, 1973. Edmondson personal files.

41 Minutes, Executive Committee of AURA, Inc., June 18, 1973, Tucson, Arizona. Edmond-
    son personal files. In attendance, with their new titles, were: Arthur A. Hoag, Director,
    Stellar Program and Lloyd V. Wallace, Director, Planetary Program. A. Keith Pierce,
    Director, Solar Program was absent. Also in attendance were: Beverly T. Lynds, Assistant
    to the Director; Harry Albers, Director of Administrative Services, who had been appointed
    in 1972 after Miller left; and L.K. Randall, Director of the new Office of Engineering and
    Technical Services, which replaced the Research Support Division with Crawford as Associ-
    ate Director.
    Note: The Planetary Program and the positions of Program Director for Solar and for
    Stellar Programs were eliminated at the 1975 Annual Meeting. Minutes, 1975 Annual Meet-
    ing of the Board of Directors of AURA, Inc., February 26, 1975, Tucson, Arizona.
    Edmondson personal files.

42 Minutes, 1974 Annual Meeting of the Board of Directors of AURA, Inc., February 27,
    1974, Tucson, Arizona. Edmondson personal files.

43 Cf. References 31 and 37.

44 Minutes, Executive Committee of AURA, Inc., September 28, 1973, Tucson, Arizona.
    Edmondson personal files.

45 Letter, Owen to Lee, March 1, 1973. Edmondson personal files.

46 Minutes Scientific Committee Meeting, May 3–4, 1973, Tucson, Arizona. Edmondson per-
    sonal files.

47 Op. cit. Reference 39.

48 Minutes, Executive Committee of AURA, Inc., September 27, 1973, Tucson, Arizona.
    Edmondson personal files.

49 KPNO and CTIO Quarterly Bulletin, January–February–March 1973, p. 5. Edmondson
    personal files.

50 The printed program for the Dedication called it "The new 4-meter (or 158-inch) tele-
    scope." The change from English to Metric units for the KPNO and CTIO telescopes took
    place gradually in 1972. The Mayall Telescope was called a 150-inch telescope in the March
    7, 1972 Executive Committee Minutes and the May 25, 1972 Scientific Committee Minutes.
    150-inch and 4-meter were both used in the June 20, 1972 Executive Committee Minutes,
    and only 4-meter was used in the September 28, 1972 Executive Committee Minutes.

51 KPNO and CTIO Quarterly Bulletin, April–May–June 1973, p. 8. Edmondson personal
    files.

52 The author clearly remembers Wildt's concern about the impact of Goldberg's IAU presi-
    dency on his service as KPNO Director. Wildt was a close personal friend of Heckmann,
    and this gave him first hand knowledge about the amount of time Heckmann had to spend
    on IAU affairs when he was President in 1967–70.

53 Op. cit. Reference 42.
    Note: Greenstein's outspoken hostility to AURA and KPNO was well known and his nomi-
    nation to be Chairman of the AURA Board surprised many, including the author. To his
    credit, it should be said that he did not allow these prejudices to affect his performance of
    his duties as Chairman.

54 Minutes, Executive Committee of AURA, Inc., June 27, 1974, Tucson, Arizona. Edmond-
    son personal files.

55 Minutes, Executive Committee of AURA, Inc., September 17, 1974, Tucson, Arizona.
    Edmondson personal files.

56 Minutes, Executive Committee of AURA, Inc., December 3, 1974, Tucson, Arizona.
    Edmondson personal files.

57 Minutes, Executive Committee of AURA, Inc., June 4, 1975, Tucson, Arizona.
    Letters, Lee to Hunt, June 19, 1975; Lee to Hughes, July 30, 1975. Edmondson personal
    files.
58 Letters, Goldberg to Lee, January 28, 1976; Lee to Goldberg, January 30, 1976; Goldberg
    to Lee, January 31, 1976. Edmondson personal files.
59 Letter, Greenstein to Stever, February 23, 1976. Edmondson personal files. This letter was
    written following the exchange of letters between Goldberg and Lee.
60 Phone call, Lee to Edmondson, 10:55–11:28 a.m., March 10, 1976. Longhand notes in
    Edmondson personal files. The amount was $500 more than the AURA Executive Com-
    mittee had recommended for July 1, 1974 and $3,000 less than the amount recommended
    for July 1, 1975. Hughes told Lee he would put this in writing after the NSB meeting.
61 Letter, Greenstein to Edmondson, April 5, 1976. Edmondson personal files. The letter also
    mentioned Senator Proxmire's demand that the Director of NSF should send to Congress
    periodically a public document giving the names and salaries of all professors or scientists
    employed even part-time on grants or contracts whose total stipend (salary, benefits,
    retirement) exceeded $50,000 per year. The NSF concession on Goldberg's salary was going
    to put NSF in a delicate position.
62 Letter, Greenstein to Edmondson, Hiltner, Pierpont, and Wildt, January 22, 1975.
    Edmondson personal files.
63 Memorandum, Greenstein to Edmondson, Hiltner, Lee, and Pierpont, February 27, 1975.
    Edmondson personal files. Lee was not invited to attend this meeting, but he was informed
    about it in advance, and received a copy of this memorandum summarizing the discussion.
64 The Executive Committee met on June 4 and November 19, but not in September.
    Greenstein raised the question of Goldberg's retirement at the November 19 meeting. The
    final consensus was that the Chairman should bring this to the attention of the Board in
    January, proposing at that time the names of persons to constitute a search committee.
    Minutes, Executive Committee of AURA, Inc., November 19, 1975, Tucson, Arizona.
    Edmondson personal files.
65 Minutes, 1976 Annual Meeting of the Board of Directors of AURA, Inc., Appendix A,
    January 12, 1976, Santiago, Chile. Edmondson personal files. Appendix A is on file in the
    AURA Corporate Office as part of the official copy of the Minutes, but was not distributed
    with the Minutes.
66 Letter, Greenstein to Goldberg, March 3, 1976. Edmondson personal files.
67 Phone call, Lee to Edmondson, 1:08–1:31 p.m., March 12, 1976. Longhand notes in
    Edmondson personal files.
68 Letter, Margaret Burbidge to Greenstein, March 16, 1976. Edmondson personal files.
69 Memorandum, Lee to Members of the AURA Board of Directors, April 1, 1976. Edmond-
    son personal files.
70 Op. cit. Reference 61.
71 Minutes, Special Meeting of the Board of Directors of AURA, Inc., May 3, 1976, Tucson,
    Arizona. Edmondson personal files.
72 Minutes, Special Meeting of the Board of Directors of AURA, Inc, December 3, 1976,
    Tucson, Arizona. Edmondson personal files.
73 Minutes Executive Committee of AURA, Inc., December 3, 1976, Tucson, Arizona.
    Edmondson personal files.
74 Minutes, Scientific Committee of AURA, Inc., June 3, 1975, Tucson, Arizona.
    Minutes, Executive Committee of AURA, Inc., June 4, 1975, Tucson, Arizona. Edmondson
    personal files.
75 Minutes, Executive Committee of AURA, Inc., November 19, 1975, Tucson, Arizona.
    Edmondson personal files.
76 Minutes, Executive Committee of AURA, Inc., September 28, 1976, Tucson, Arizona.
    Edmondson personal files.

77 Minutes, 1977 Annual Meeting of the Board of Directors of AURA, Inc., March 9, 1977, Tucson, Arizona. Edmondson personal files.

78 Minutes, Special Meeting of the Board of Directors of AURA, Inc., September 16–17, 1977, Tucson, Arizona. Edmondson personal files.

79 Op. cit. Reference 71.

80 Minutes, 1978 Annual Meeting of the Board of Directors of AURA, Inc., March 8–9, 1978, Tucson, Arizona. Edmondson personal files.

81 Minutes, Executive Committee of AURA, Inc., June 11, 1977, Atlanta, Georgia. Edmondson personal files.

82 Op. cit. Reference 78.

83 Op. cit. Reference 80.

84 Op. cit. Reference 75, Report of the Director.

85 Minutes, 1976 Annual Meeting of the Board of Directors of AURA, Inc., January 14, 1976, Cerro Tololo, Chile. Edmondson personal files.

86 Op. cit. Reference 71.

87 Op. cit. Reference 85, Attachment C.

88 Op. cit. Reference 76. The Walker Committee is mentioned in Attachment 2, the Minutes of the September 28, 1976 Meeting of the Scientific Committee.

89 Op. cit. Reference 73.

90 Op. cit. Reference 73, Attachments 1 and 2.

91 Op. cit. Reference 77.

92 Op. cit. Reference 78.

93 Op. cit. Reference 80.

94 Minutes, Executive Committee of AURA, Inc., December 4, 1978, Tucson, Arizona. Edmondson personal files.

95 Op. cit. 14, Reference 27 and 57. NSF thought two contracts would protect KPNO from possible future Congressional hostility to funding an overseas operation.

96 Op. cit. Reference 72. The Report of the Visiting Committee, dated November 4–6, 1976, is Attachment 1.

97 Op. cit. Reference 78.

98 Minutes of the Executive Committee of AURA, Inc., March 6, 1979, Tucson, Arizona. Edmondson personal files.

(a) Attachment 7 is a six page letter from Teem to W.E. Howard III giving detailed information about the AURA management of SPO.

(b) Attachment 12 is a summary of Burbidge's reorganization of KPNO administrative and support services.

99 Minutes, 1979 Annual Meeting of the Board of Directors of AURA, Inc., March 7–8, 1979, Tucson, Arizona. Edmondson personal files. The report of the joint meeting is Attachment 8.

100 Minutes, 1980 Annual Meeting of the Board of Directors of AURA, Inc., March 26–27, 1980, Tucson, Arizona. Edmondson personal files. The word "Budgets" is what differentiated COOB from the Organization Committee. The Committee was set up to work with NSF on the serious budget problems that were developing, especially those for CTIO resulting from economic conditions in Chile.

101 Minutes, Executive Committee of AURA, Inc., July 14, 1980, Tucson, Arizona. Edmondson personal files.

102 Minutes, Board of Directors of AURA, Inc., September 10–11, 1980, Tucson, Arizona. Edmondson personal files.

103 Minutes, Executive Committee of AURA, Inc., December 15, 1980, Baltimore, Maryland. Edmondson personal files. Bell resigned from the AURA Board on October 31, due to his leaving the University of Chicago to take a position in San Francisco. The position of Vice-Chairman was vacant until the 1981 Annual Meeting.

104 Op. cit. Reference 102.

105 Minutes, 1981 Annual Meeting of the Board of Directors of AURA, Inc., April 1–2, 1981, Tucson, Arizona. Edmondson personal files. Harlan Smith was re-elected Chairman and Reuben Lorenz was elected Vice-Chairman at this meeting.

106 Minutes, Executive Committee of AURA, Inc., June 19, 1981, Tucson, Arizona. Edmondson personal files. This position had been recommended by COOB and the Organization Committee, and was approved at the 1981 Annual Meeting. A revised job description was approved at this meeting, and the title was changed to Corporate Staff Scientist.

107 Minutes, Executive Committee of AURA, Inc., September 11, 1981, Tucson, Arizona. Edmondson personal files.

108 Minutes, 1982 Annual Meeting of the Board of Directors of AURA, Inc., April 1–2, 1982, Tucson, Arizona. Edmondson personal files.

109 Minutes, Executive Committee of AURA, Inc., October 1–3, 1982, La Serena, Chile. Edmondson personal files.

110 Minutes, Special Meeting of the Board of Directors of AURA, Inc., November 3–4, 1982, Chicago, Illinois. Edmondson personal files.

111 Minutes, Executive Committee of AURA, Inc., February 15, 1983, Tucson, Arizona. Edmondson personal files.

112 Minutes, Executive Committee Teleconference, January 19, 1983. Edmondson personal files.

113 Minutes, Executive Committee of AURA, Inc., April 15, 1983, Chicago, Illinois. Edmondson personal files.

114 Letter, Knapp to Teem, May 24, 1983. Edmondson personal files.

115 FOR RELEASE: June 10, 1983. AURA, Inc. Announces Appointment of Dr. John T. Jefferies as Director for the National Optical Astronomy Observatories.

116 Minutes, Executive Committee Teleconference, May 10, 1983. Edmondson personal files.

117 Minutes, Executive Committee of AURA, Inc., June 16, 1983, Baltimore, Maryland. Edmondson personal files.

118 Minutes, Executive Committee of AURA, Inc., October 3, 1983, Chicago, Illinois. Edmondson personal files.

119 Minutes, Executive Committee of AURA, Inc., December 12, 1983, Tucson, Arizona. Edmondson personal files.

120 Minutes, 1984 Annual Meeting of the Board of Directors of AURA, Inc., March 21–23, 1984, Tucson, Arizona. Edmondson personal files.

121 Minutes, Executive Committee Teleconference, April 11, 1984. Edmondson personal files.

122 Minutes, Annual Meeting of the Board of Directors of AURA, Inc., March 31–April 2, 1982, Tucson, Arizona. Edmondson personal files.

123 Op. cit. Reference 42.

124 Minutes, Scientific Committee of AURA, Inc., November 18, 1975, Tucson, Arizona. Edmondson personal files. Three months before this meeting, C.R. O'Dell (Project Scientist for the Space Telescope) and G. Pieper (Goddard Director of Earth and Planetary Sciences) made a presentation to the Council of the American Astronomical Society:

> O'Dell discussed the LST operations as he saw it, emphasizing the need for a completely independent scientific institute to manage the operations. O'Dell stressed that this would be analogous to national laboratories, giving the great flexibility needed in hiring a good staff. He envisioned a staff of twelve scientists and forty to sixty technical people and that this institute would be a focal point for the operations. . . . Pieper discussed the manner in which his Center is looking at the LST and pointed out that they had changed their way of thinking and now agreed that an institute for the management of the scientific operations of LST seemed a good route to go. His point was that the institute should be located near the

Goddard Space Flight Center for economic reasons. He envisioned a staff similar to that discussed by O'Dell.

The Council discussed the NASA–LST briefing, and passed the following Resolution:

> The Council of the American Astronomical Society recognizes that the Large Space Telescope will initiate a new era in observational astronomy, in its unprecedented capabilities for research both in foreseeable and in hitherto unforeseen directions, as well as in its long life and potentiality for updating and refurbishment. The Council recommends that the LST be operated by an independent Space Astronomy Institute, be managed by a consortium of universities forming a corporation under contract with NASA.

Those in attendance included AAS President Robert P. Kraft, Vice-President Victor M. Blanco, President-elect E. Margaret Burbidge, Outgoing Treasurer Frank K. Edmondson, Incoming Treasurer William E. Howard III, and Goetz K. Oertel "of NSF who attended as an agency observer." Note: Oertel left NSF a few weeks after this meeting. Howard resigned as AAS Treasurer when he became the first Director of the Division of Astronomical Sciences in 1977.

Minutes, Council of the American Astronomical Society, 146th Meeting, San Diego State University, San Diego, California, August 17, 1975. From files of the AAS Secretary, Roger A. Bell (May 1994).

125 Op. cit. Reference 75.

126 Op. cit. Reference 85.

127 Minutes, Scientific Committee of AURA, Inc., May 2, 1976, Tucson, Arizona. Edmondson personal files.

128 Minutes, Executive Committee of AURA, Inc., May 3, 1976, Tucson, Arizona. Edmondson personal files.

129 Op. cit. Reference 76.

130 Op. cit. Reference 77. Attachment 1, Report of the Scientific Committee.

131 Op. cit. Reference 78.

132 Op. cit. Reference 99.

133 Minutes, Executive Committee of AURA, Inc., June 5, 1979, Tucson, Arizona. Edmondson personal files.

134 Op. cit. Reference 100.

135 Transcript of Welch interview with Edmondson, May 16, 1990.

136 Op. cit. Reference 101.

137 *AURA, The First Twenty-five Years, 1957–1982*, p. 38. The Association of Universities for Research in Astronomy, Inc., Tucson, Arizona, February, 1983.

138 Minutes, Executive Committee of AURA, Inc., June 19, 1981, Tucson, Arizona. Edmondson personal files.

139 Minutes, Executive Committee Teleconference, May 28, 1981. Edmondson personal files.

140 Minutes, Executive Committee of AURA, Inc., September 11–12, 1981, Baltimore, Maryland. Edmondson personal files.

141 Op. cit. Reference 108.

142 Op. cit. Reference 110.

Chapter 21

1 Op. cit. 8, Reference 8.

2 Op. cit. 9, Reference 10.

3 Minutes, Annual Meeting of the Board of Directors of the Association of Universities for Research in Astronomy, Inc., April 1–2, 1982, Tucson, Arizona. Edmondson personal files.

Edmondson also had been the Chairman of the Planning Committee that planned the March 15, 1960 Dedication of KPNO. Op. cit. 10, Reference 97.

4 Longhand notes, Edmondson, April 2, 1982. Edmondson personal files.

5 *AURA, The First Twenty-five Years, 1957–1982.* The Association of Universities for Research in Astronomy, Inc., Tucson, Arizona, February, 1983.

6 The basic list included:

    (a) AURA: The current Board of Directors, all living members of the original Board, the original officers, former Presidents and Board Chairmen, and selected early Board members. Early site survey and KPNO employees, current senior KPNO and CTIO scientific staff, former and current KPNO and CTIO Directors. Current and former Chairmen of the AURA Visiting Committee. Presidents of AURA member universities – no alternates.

    (b) NSF: Selected members of the 1957–58 National Science Board and all current NSB members. The Director of the National Science Foundation, selected former NSF staff and current NSF senior staff.

    (c) Government: The Arizona congressional delegation. Members of Congress and their staff members who were directly involved in committees that handled appropriating funds for science. The Directors of the Office of Management and Budget and the Congressional Budget Office. The President's Science Advisor. The Administrator of NASA and NASA senior staff.

    (d) Miscellaneous: Officers of the Papago Tribe. Presidents of AAAS, AAS, ASP, AIP. The U.S. Vice President of the International Astronomical Union. Selected other individuals.

7 From the table seating list in Edmondson's 25th Anniversary Committee files.

8 This is the way the author remembers the press conference.

9 Professor Spicer died on April 5, 1983.

10 This was reported to the NSB at their February 17–18, 1983 meeting. Page 4 of the NSB Minutes was sent to the author.

11 Minutes, 1983 Annual Meeting of the Board of Directors of the Association of Universities for Research in Astronomy, Inc., February 17, 1983, Tucson, Arizona. Edmondson personal files.

12 Minutes, Executive Committee of the Association of Universities for Research in Astronomy, Inc., February 17, 1983, Tucson, Arizona. Edmondson personal files.

13 President's Annual Report to AURA Board of Directors, John M. Teem, February 28, 1984. Edmondson personal files.

14 Report to the Board of Directors, Goetz K. Oertel, President, March 19, 1987. Edmondson personal files. The move was scheduled for March 28, 1987. A memo from Fults to the author, dated 4/3/87 says: "We are moved in, just barely. All the phones don't work, and nothing is in the right place."

15 Minutes, Organization Committee of AURA, Inc., July 22, 1971, Chicago, Illinois. Edmondson personal files.

16 Minutes, Executive Committee of the Association of Universities for Research in Astronomy, Inc., July 14, 1980, Tucson, Arizona. Edmondson personal files.

17 Minutes, 1981 Annual Meeting of the Board of Directors of the Association of Universities for Research in Astronomy, Inc., April 1–2, 1981, Tucson, Arizona. Edmondson personal files.

18 Minutes, Executive Committee of the Association of Universities for Research in Astronomy, Inc., June 19, 1981, Tucson, Arizona. Edmondson personal files.

19 Organization Committee Report, Meeting of December 3–4, 1981, Baltimore, Maryland. Edmondson personal files.

20 Minutes, 1968 Annual Meeting of the Board of Directors of AURA, Inc., March 12, 1968, Tucson, Arizona. Edmondson personal files.

21 Letter, Harvill to Hiltner (the new AURA President), April 22, 1968. Edmondson personal files.

22 Minutes, Organization Committee of AURA, Inc., December 13, 1968, Chicago, Illinois. Edmondson personal files.

23 Minutes, 1969 Annual Meeting of the Board of Directors of AURA, Inc., January 23, 1969, Tucson, Arizona. Edmondson personal files.

24 Minutes, 1970 Annual Meeting of the Board of Directors of AURA, Inc., January 22, 1970, Tucson, Arizona. Edmondson personal files.

25 Minutes, Executive Committee of AURA, Inc., November 19, 1970, Cerro Tololo, Chile. Edmondson personal files.
    Note: Fleischer was appointed NSF Program Director for Astronomy in 1968, when Mulders retired.

26 Minutes, 1971 Annual Meeting of the Board of Directors of AURA, Inc., January 21, 1971, Tucson, Arizona. Edmondson personal files.

27 Minutes, Special Meeting of the Board of Directors of AURA, Inc., March 16, 1971, Tucson, Arizona. Edmondson personal files.

28 Minutes, Executive Committee of AURA, Inc., March 16, 1971, Tucson, Arizona. Edmondson personal files.

29 Minutes, 1972 Annual Meeting of the Board of Directors of AURA, Inc., January 19, 1972, Tucson, Arizona. Edmondson personal files.
    Note: The AURA Board also voted at this meeting to establish a Corporate Office with a full-time paid President, as reported in Chapter 20.

30 Minutes, Executive Committee of AURA, Inc., June 20, 1972, Tucson, Arizona. Edmondson personal files.

31 Minutes of the Annual Meeting, Board of Directors, Association of Universities for Research in Astronomy, Inc., January 17, 1973. Edmondson personal files.

32 Minutes of the Annual Meeting, Board of Directors, Association of Universities for Research in Astronomy, Inc., March 9, 1977, Tucson, Arizona. Edmondson personal files.
    Note: A committee to search for a scientist President of AURA was appointed at this meeting.

33 Minutes of the Annual Meeting, Board of Directors, Association of Universities for Research in Astronomy, Inc., March 8–9, 1978, Tucson, Arizona. Edmondson personal files.
    Note: John M. Teem had replaced G.L. Lee, Jr. as President of AURA on October 1, 1977.

34 Action Items, Board of Directors, Association of Universities for Research in Astronomy, Inc., March 26–27, 1980, Tucson, Arizona. Edmondson personal files.

35 Minutes of the Annual Meeting, Board of Directors, Association of Universities for Research in Astronomy, Inc., April 1–2, 1981, Tucson, Arizona. Edmondson personal files.

36 Minutes of the Annual Meeting, Board of Directors, Association of Universities for Research in Astronomy, Inc., March 31–April 2, 1982, Tucson, Arizona. Edmondson personal files.

37 Minutes of the Annual Meeting, Board of Directors, Association of Universities for Research in Astronomy, Inc., March 22–23, 1984, Tucson, Arizona. Edmondson personal files.

38 Thirty-five Board Members attended the 1984 Annual Meeting, and 33 attended in 1985; 30 of these attended both meetings. The 1986 Board had only 17 of these as members, plus the newly elected directors-at-large.

39 Memorandum, Jefferies to AURA Board of Directors, March 13, 1985. Edmondson personal files.

40 Minutes, Executive Committee of AURA, Inc., March 27, 1985, Tucson, Arizona. Edmondson personal files.

41 Minutes, 1985 Annual Meeting of the Board of Directors of AURA, Inc., March 28–29, 1985, Tucson, Arizona. Edmondson personal files. The quotation is how the author remembers what Conti said.

42 The author recalls that Conti's generous description of Howard's leak to the NSO Users Committee was not shared by a number of Board Members, who called it "intentional sabotage."

43 Op. cit. Reference 39.

44 Minutes, Executive Committee of AURA, Inc., June 27–28, 1985, Baltimore, Maryland. Edmondson personal files.

45 Minutes, Special Meeting of the Executive Committee of AURA, Inc., August 14, 1985, Dallas, Texas. Edmondson personal files.

46 Letter, Senator Pete Domenici and Representative Joe Skeen to Roland W. Schmitt, September 18, 1985. Edmondson personal files.

47 Letter, Teem to Domenici, September 26, 1985. Edmondson personal files.

48 Letter, Domenici to Schmitt, November 27, 1985. Edmondson personal files.

49 Letter, Bautz to Teem, November 15, 1985. Edmondson personal files.

50 Parker to Bautz, December 12, 1985. Edmondson personal files.

51 Letter, Teem to Edmondson, September 16, 1985. Edmondson personal files. The second paragraph of this two-page letter reads:

> My current term of appointment has three years remaining, and I had intended to serve out that term as President, or at least two more years. However, I have recently concluded that it is in both AURA's and my own best interests that I make the transition to an emeritus status at the end of FY 1986. If there is to be a change in AURA's presidency within the next three years, several upcoming events make it most timely that this be accomplished within the coming year. These are discussed below.

The upcoming events were:

> First, the NSF has informed us that it will be conducting a major review of NOAO and AURA's management of it toward the end of FY 1987 or early in FY 1988, in connection with the possible renewal of AURA's contract.
> Second, during late FY 1987 or early FY 1988, NASA will be deciding whether or not to exercise its initial option for renewing AURA's contract for STScI management.
> Third, by August, 1987, AURA must relocate its Corporate Office from the Joseph Henry Building, since the NAS is moving and ending our sublease.

The first item and the timing of this decision make it clear that the NOAO–SPO fiasco was weighing heavily on Teem's mind.

52 Minutes, 1986 Annual Meeting, Board of Directors, Association of Universities for Research in Astronomy, Inc., April 11–12, 1986, Washington, DC. Edmondson personal files.

53 Based on notes by the author, who was invited to listen to the telephone meeting. Edmondson personal files.

54 Jefferies resigned as Director of NOAO in 1987, and was succeeded by Sidney Wolff, the Director of KPNO he had appointed. Minutes, Executive Committee of AURA, Inc., December 11–12, 1986, Tucson, Arizona; Minutes, Executive Committee Teleconference, August 4, 1987. Edmondson personal files.

Appendix I
Taped interviews

Name	Date	Length
Helmut A. Abt Site survey; KPNO Staff Astronomer	1979, Mar. 27	(1:38:41)
Vernice Anderson Executive Secretary, National Science Board	1980, Dec. 12	(28:10)
Claudio Anguita University of Chile, AURA Director-at-large	1979, Mar. 9	(18:24)
Horace W. Babcock Mt. Wilson and Palomar Observatories, CARSO	1979, Jan. 31	(27:04)
Richard M. Bilby Second AURA Counsel in Tucson	1979, Mar. 22	(29:37)
Victor M. Blanco Second Director, CTIO	1979, Sept. 13	(53:55)
Carl W. Borgmann #1 Carl W. Borgmann #2 Director, Science & Engineering Program, Ford Foundation	1979, Apr. 5 1983, Nov. 10	(41:56) (22:10 +01:30) (1:04:06 + 01:30)
William E. Brunk NASA Headquarters	1982, Nov. 18	(46:20)
McGeorge Bundy President, Ford Foundation	1981, Dec. 3	(24:34)
Herbert E. Carter National Science Board	1979, Sept. 10	(48:57)

Appendix I *cont.*

Name	Date	Length
Walker L. Cisler	1980, July 31	(25:36)
President, Detroit Edison, and close friend of Robert McMath		
John F. Clark	1981, Dec. 2	(1:21:03)
J.F. Clark and Mrs. Clark	1981, Dec. 2	(0:29:17)
NASA Headquarters; Director, Goddard Space Flight Center		(1:50:20)
Thomas E. Cooney	1981, Jan. 23	(44:43)
Ford Foundation staff		
David L. Crawford	1979, Mar. 21	(1:06:28)
KPNO staff (Project Manager for 150-inch telescopes)		
John H. Denton	1984, Dec. 5	(1:16:44)
Counsel for Papago Tribe, University of Arizona Professor		
Richard L. Doane	1979, Mar. 20	(59:40)
KPNO Mountain Superintendent		
Olin J. Eggen	1987, May 10	(18:04)
CTIO staff		
Julia I. Elliott	1983, Feb. 16	(1:01:50)
AURA Assistant Secretary		
Lee Anna Embrey	1979, Apr. 25	(44:56)
NSF Deputy Public Information Officer		
J.A. Franklin	1979, Nov. 6	(1:09:17)
Member of Original AURA Board (Indiana University)		
Henry L. Giclas #1	1978, Aug. 18	(18:00)
Henry L. Giclas #2	1990, May 15	(19:34)
Lowell Observatory		(37:34)
Leo Goldberg #1	1978, Sept. 6	(40:55)
Leo Goldberg #2	1979, Feb. 13	(10:52)
University of Michigan; Harvard; AURA Board; Third Director, KPNO		(51:47)
J.C. Golson	1979, Mar. 15	(28:20)
Night Assistant, KPNO		
Laurence M. Gould	1980, Sept. 8	(32:36)
National Science Board, and Ford Foundation Board		

Jesse L. Greenstein   1980, July 21   (1:39:17)
AURA Board (California Institute of Technology)

Norman Hackerman   1981, May 29   (46:40)
AURA Board Consultant (University of Texas, and Rice University);
National Science Board

Philip Handler   1980, Dec. 12   (58:38)
National Science Board

Richard A. Harvill   1978, Aug. 30   (35:52 + 1:28)
President, University of Arizona

Emil W. Haury   1979, Mar. 16   (19:10)
University of Arizona, Professor of Anthropology

Paul Herget   1979, Aug. 7   (57:50)
NSF Advisory Panel for Astronomy, (University of Cincinnati)

Rev. Theodore M. Hesburgh   1979, Nov. 1   (1:00:22)
National Science Board; President, University of Notre Dame

Roger W. Heyns   1981, July 27   (59:15)
National Science Board

Chester J. Higman   1979, July 25   (59:42)
Business Manager, Papago Tribe (salary paid by American Friends Service
Committee)

W.A. Hiltner   1979, July 10   (48:35 + 1:40)
AURA Board (University of Chicago and University of Michigan)

Arthur A. Hoag   1980, Jan. 29   (24:44)
KPNO, Associate Director, Stellar Division

William J. Hoff   1978, July 6   (17:48)
NSF General Counsel

Helen S. Hogg   1979, Aug. 14   (49:13)
Second NSF Program Director for Astronomy

Stuart R. Hurdle   1980, Sept. 1   (49:56)
KPNO Mountain Superintendent. In charge of major construction in
Chile

John B. Irwin   1978, Aug. 22   (45:22)
Proposed a cooperative Photoelectric Observatory (Indiana University)

Robert W. Johnston   1980, June 24   (56:35)
Special Assistant to NSF Director Leland J. Haworth

Thomas O. Jones #1   1978, Oct. 20   (30:27)

Appendix I *cont.*

Name	Date	Length
Thomas O. Jones #2 NSF (various top level positions)	1979, Apr. 23	<u>(12:01)</u> (42:28)
Philip C. Keenan Member of original AURA Board (Ohio State)	1983, Feb. 14	(22:25)
Geoffrey Keller Fourth NSF Program Director for Astronomy	1978, Nov. 4	(1:03:41)
Paul E. Klopsteg Associate Director, NSF	1979, Feb. 28	(27:20)
Francis Knuckey McFadden Peak information-site survey	1991, Apr. 8	(24:20)
Arlo U. Landolt Louisiana State University (first guest observer on Kitt Peak)	1979, Jan. 14	(20:37)
Harold H. Lane NSF Astronomy Program	1979, Apr. 27	(20:58)
Gilbert L. Lee, Jr. #1 Gilbert L. Lee, Jr. #2 Member of original AURA Board (University of Michigan); First full-time President of AURA	1978, June 26 1983, Oct. 30	(0:32:24) <u>(1:31:22)</u> (2:03:46)
Per Olof Lindblad ESO–Ford Foundation information	1988, Aug. 8	(06:42)
Albert P. Linnell AURA Board – Director-at-large (Michigan State University)	1979, Aug. 22	(59:42 + 1:46)
Hugh F. Loweth Bureau of the Budget	1981, Dec. 4	(15:57)
Nicholas U. Mayall Second Director, KPNO	1979, Mar. 17	(2:14:07)
Paul A. McKalip Senior Editor, Tucson Citizen	1982, Sept. 15	(27:12)
Aden B. Meinel US site survey; First Director of KPNO	1979, Feb. 5	(1:29:40)
G.R. Miczaika Air Force Scientist – GRD Cambridge	1979, Mar. 1	(35:02)

Orren C. Mohler   1979, July 11   (2:11:24)
AURA Board (University of Michigan)

Hugo Moreno   1979, Aug. 16   (1:06:23)
University of Chile, AURA Director-at-large

William W. Morgan   1980, May 28   (38:10)
NSF Advisory Panel for Astronomy (University of Chicago); KPNO
Dedication speaker

Gerard F.W. Mulders   1979, Apr. 24   (56:51)
National Science Foundation, Associate Program Director for Astronomy
and Program Director for Astronomy

Homer E. Newell   1981, May 29   (21:38)
NASA Headquarters

Edward C. Olson   1978, Nov. 24   (13:26)
University of Illinois; Summer assistant on site survey while a graduate
student at Indiana University

Jan H. Oort   1981, Jan. 22   (1:04:07)
Director, Leiden Observatory; First President of ESO

Juan Pascoe   1985, Mar. 23   (46:32)
Advisor to Papago Tribe

Ralph Patey   1978, Aug. 31   (1:22:38)
Business Manager, KPNO

A.G. Davis Philip   1979, Aug. 20   (13:57)
Guest observer, KPNO

A. Keith Pierce   1979, Mar. 21   (37:48)
KPNO, Associate Director, Solar Division

Floyd Roberson   1982, Sept. 15   (25:26)
First employee hired after selection of Kitt Peak

Randal M. Robertson   1979, Apr. 24   (43:01)
NSF Assistant Director, MPE

Nancy G. Roman   1982, Nov. 17   (1:01:07)
NASA Headquarters

Aaron Rosenthal   1979, June 26   (1:55:29)
Comptroller, NSF

Raymond J. Seeger   1979, Apr. 27   (1:08:22 + 3:38)
NSF Acting Assistant Director, MPE

C.D. Shane #1   1978, Aug. 5   (34:12 + 3:07)

Appendix I *cont.*

Name	Date	Length
C.D. Shane #2	1980, July 24	<u>(25:20)</u>
Member of original AURA Board (University of California)		(59:32 + 3:07)
T.K. Shoenhair	1979, Mar. 23	(30:05)
First AURA counsel in Tucson		
Alex G. Smith	1979, Aug. 20	(1:02:43)
AURA Board – Director-at-large (University of Florida)		
William Hawes Smith	1980, Mar. 22	(30:58)
Tucson business man		
Edward H. Spicer #1	1979, Jan. 25	(20:24)
Edward H. Spicer #2	Feb. 8	<u>(28:59)</u>
University of Arizona, Professor of Anthropology		(49:23)
Lyman Spitzer, Jr.	1983, Feb. 16	(28:05)
AURA Board (Princeton)		
Jurgen Stock	1979, Aug. 18	(1:22:18)
Chile site survey; First Director CTIO		
Julius A. Stratton	1980, Oct. 14	(31:37)
National Science Board; Chairman of the Board, Ford Foundation		
(Massachusetts Institute of Technology)		
C.E. Sunderlin	1980, Dec. 13	(1:01:29)
NSF Deputy Director		
Harold J. Thompson	1979, Mar. 30	(54:45)
KPNO staff		
Peter van de Kamp	1978, July 9	(43:31)
First NSF Program Director for Astronomy		
Gordon W. Wares	1980, May 19	(1:16:00)
Air Force Scientist – GRD Cambridge		
David F. Welch	1990, May 16	(43:29)
AURA Corporate Office		
John T. Wilson	1980, Nov. 12	(1:20:08)
NSF Assistant Director Biological and Medical Sciences; Deputy Director		

NOTE: The AURA Board Dinner in Tucson on April 1, 1981 featured an illustrated talk by Aden B. Meinel about the first ascent to the summit of Kitt Peak 25 years earlier. This talk was taped, and also some brief remarks by Helmut A. Abt. (22:38)
Plans were being made to visit Leland J. Haworth at his home and tape an interview in April 1979. His last letter was received in Tucson on March 5, the day of his death.

Appendix II
Senior officers of AURA

Former Presidents & Board Chairmen			Former Vice-Presidents & Vice-Chairmen		
Robert R. McMath	President	1957–59	Frank K. Edmondson	Vice-President	1957–61
	Chairman	1959–62			
C.D. Shane	President	1959–62	Nicholas U. Mayall	Vice-President	1961–62
Frank K. Edmondson	President	1962–65	William B. Harrell	Vice-President	1962–65
Rupert Wildt	President	1965–68	G.L. Lee, Jr.	Vice-President	1965–68
W.A. Hiltner	President	1968–71	J.A. Franklin	Vice-President	1968–71
Rupert Wildt	President	1971–72	Reuben H. Lorenz	Vice-President	1971–72
	Chairman	1972–74		Vice-Chairman	1972–74
Gilbert L. Lee, Jr.	President	1972–77			
Jesse L. Greenstein	Chairman	1974–77	Wilbur K. Pierpont	Vice-Chairman	1974–77
Arthur D. Code	Chairman	1977–80	Richard A. Rossi	Vice-Chairman	1977–80
John M. Teem	President	1977–86			
Harlan J. Smith	Chairman	1980–83	Harold E. Bell	Vice-Chairman	1980–81
			Reuben H. Lorenz	Vice-Chairman	1981–84
Peter S. Conti	Chairman	1983–86			
			Richard A. Rossi	Vice-Chairman	1984–87
Robert W. Noyes	Chairman	1986–89			
			David W. Morrisroe	Vice-Chairman	1987–88
Goetz K. Oertel	President	1986–	Harry Albers	Vice-Chairman	1988–91
Robert M. MacQueen	Chairman	1989–92			
			Richard Margison	Vice-Chairman	1991–94
Maarten Schmidt	Chairman	1992–95			
			Richard Zdanis	Vice-Chairman	1994–
Bruce Margon	Chairman	1995–			

Index

AAS (American Astronomical Society), appointed special committee to study relations of astronomers with the NSF, 7; McMath began two-year term as President in 1952, 17; suggestion that AAS might operate new telescope, 20; McMath was President when asked to chair NAO Panel, 27; memorandum from Helen Hogg invited AAS members to submit proposals to NSF, 42; Meinel investigated access to Junipero Serra Peak during August 1956 AAS meeting in Berkeley, 49; Whitford gave talk about work of NAO Panel at November 1955 AAS meeting in Troy, New York, 60; reorganization of KPNO, 126; AAS Council heard C.R. O'Dell and G. Pieper (NASA) discuss Space Telescope management. Passed resolution recommending management by a consortium of universities under contract with NASA, 342N124

Abdala, Jose, 149

Abt, Helmut A., xiv; participated in site survey, 37–8, 40–5; returned to position at Yerkes Observatory, 47; visited Junipero Serra Peak, 47; 49, 138

Ackerman, Edward A., approached Ford Foundation, February 28, 1964, 178; AURA–CARSO negotiations, 180–1, 183–90, 193; attended AURA Executive Committee meeting with Bowen, November 22, 1963, 217; attended at AURA–ESO meeting in Paris with Babcock, January 21–2, 1964, 217; trip to Chile, February 1964, 218; discussion of lease versus purchase of Morado and CARSO long-term plan for Morado, Chicago, March 18, 1968, 219–20

Adams, Walter S., 55, 215

Adams, Mrs. Walter S., 125

Advisory Panel for National Astronomical Observatory (see NAO Panel)

Advisory Panel for Radio Astronomy (see Radio Astronomy Panel)

AIP (American Institute of Physics), Center for the History of Physics, xi

Albers, Harry, 231

Alessandri, President Jorge, 153–4, 201

Allende, President Salvadore, 201

American Astronomical Society (see AAS)

American Friends Service Committee xii, xiii

American Institute of Physics (see AIP)

Anderson, Gerald, 230

Anderson, Vernice, xii, 103–4

Anguita, Claudio, 163, 167, 202

Archival sources, xiv

Arizona, Indiana, Ohio State Proposal, submitted to NSF May 15, 1952, 14; discussed by Advisory Panel for Astronomy, August 1, 1952, 16; declined September 11, 1952, 16

Associated Universities, Inc. (see AUI)